Stability of Drugs and Dosage Forms

Stability of Drugs and Dosage Forms

Sumie Yoshioka
National Institute of Health Sciences
Tokyo, Japan

and

Valentino J. Stella
The University of Kansas
Lawrence, Kansas

Springer Science+Business Media, LLC

Library of Congress Cataloging-in-Publication Data

Yoshioka, Sumie, Ph.D., and Stella, Valentino J., Ph.D.
Stability of drugs and dosage forms/authored by Sumie Yoshioka and Valentino J. Stella.
p. cm.
Includes bibliographical references and index.

DOI 10.1007/978-0-306-46829-2
1. Drug stability. 2. Drugs—Dosage forms. I. Stella, Valentino J. II. Title.

RS424 .Y674 2000
615.'18—dc21

00-035697

Originally published by Kluwer Academic / Plenum Publishers, New York in 2000
MyCopy version of the original edition 2000

http://www.wkap.nl/

10 9 8 7 6 5 4 3 2 1

A C.I.P. record for this book is available from the Library of Congress

Preface

A thorough knowledge of the chemical and physical stability of drugs and dosage forms is critical in the development and evaluation of pharmaceuticals. Although a very large number of studies on the subject of stability have appeared in the primary pharmaceutical literature, books on the subject have not been comprehensive. Therefore, researchers and students have had to rely on individual papers for an in-depth analysis of the subject. The objective of this book is to bring together and analyze, in a systematic fashion, some examples relating to the stability of drugs from the work of others as well as from studies that have been performed in the laboratories of the authors.

This text is organized by first presenting the major mechanisms contributing to chemical instability, followed by discussion of factors affecting degradation based on kinetic theory and of ways in which problematic pharmaceutical products might be stabilized. Predictions of stability are covered, from basic theory to practical solutions.

Unlike earlier books in which chemical stability profiles and the effects of factors such as pH and other catalytic contributions on single substances were described extensively, the present text attempts to take a more global approach. Specifically, analysis of the factors affecting drug stability based on basic kinetic theory allowed for a sound theoretical treatment of available information. Furthermore, the bases for physical degradation kinetics, which have generally been treated empirically, are also covered, and sound bases for the observations are also presented and discussed.

The chemical and physical stability of protein and peptide drugs is considered in a separate chapter of this book. Although some newer texts have comprehensively addressed the difficult subject of protein stability, it was felt that no drug stability text would be complete without this subject.

Drug products are complex mixtures of drug and excipients, and, as such, their chemical and physical stability kinetics are complex. The chemical and physical stability of these complex dosage forms, starting with preformulation studies and continuing through to studies of the final products, including the role of packaging, are discussed. Information on the stability of novel drug delivery systems such as biodegradable microspheres is also included where possible. Issues of quality assurance, the estimation of shelf life, and the relevant regulatory requirements are described. The most recent information on International Harmonised Guidelines for stability testing is also provided, along with a brief discussion on conflicts that exist between the requirements of different countries.

As stated earlier, this book attempts to present a reasonably systematic and comprehensive approach to the subject of chemical and physical drug stability. Efforts have also been made to provide a fairly comprehensive listing of references that could be used by the reader to access the primary literature. Our understanding of the chemical and physical stability of drugs in solid dosage forms is still quite incomplete, and good comprehensive studies on the stability of proteins are only now providing the type of information from which the prediction of physical and chemical stability of proteins might be possible. Additionally, we do not know what the stability problems of the drugs of the future, especially the products of genomic research, will be. Therefore, no book on the subject of drug stability should be considered complete. Opinions and interpretations of any scientific study also differ. The emphases presented here represent the biases of the authors, who welcome constructive comments and criticisms on any of the work presented in this text.

Finally, the authors would like to thank their colleagues for their contributions to the studies presented in this book. They would like to especially thank Drs. Aso and Izutsu, who contributed significantly to a number of the studies presented from Dr. Yoshioka's laboratory. Contributors from Professor Stella's laboratory include various students, technicians (especially Ms. Waugh), and postdoctoral and visiting scientists. The authors would also like to thank Ms. Kawai and Ms. Nakamura of Nankoudo Publishers, who helped with the original Japanese version of this book.

Sumie Yoshioka
Valentino J. Stella

Contents

Chapter 1

Introduction

In this book, we define "pharmaceuticals" as drug substances having pharmacological effects and the dosage forms containing these drug substances, which are intended for therapeutic use. Drug substances used at the present time range from small-molecular-weight chemicals to polymers such as proteins. In the future, products derived from genomic research will have to be included. Some drug substances are susceptible to chemical degradation under various conditions owing to their fragility of their molecular structure. Other drug substances undergo physical degradation changes rather than chemical degradation, leading to various changes in their physical state.

Chemical degradation and physical degradation of drug substances may change their pharmacological effects, resulting in altered efficacy therapeutic as well as toxicological consequences. Because pharmaceuticals are used therapeutically based on their efficacy and safety, they should be stable and maintain their quality until the time of usage or until their expiration date. The quality should be maintained under the various conditions that pharmaceuticals encounter, during production, storage in warehouses, transportation, and storage in hospital and community pharmacies, as well as in the home. Therefore, understanding the factors that alter the stability of pharmaceuticals and identifying ways to guarantee their stability are critical.

Since the early 1950s, many studies on the stability of pharmaceuticals, degradation pathways, rates of reaction, and the means of stabilizing drugs have been well documented in the primary literature. Ongoing studies, especially those with complex new drugs such as proteins, are continuously adding to our knowledge base. New assay methodologies are being developed, and new ways of treating stability data are also evolving. This is especially the case with the newer complex drugs and dosage forms.

This book examines the stability of pharmaceuticals. In Chapters 2 and 3, the chemical and the physical stability of drug substances are described. In each of these chapters, degradation pathways, probable mechanisms, factors affecting stability, methods of stabilization, and prediction methodologies are discussed. Chapter 4 describes the physical and chemical stability of drugs in complex, heterogeneous dosage forms containing both the drug and excipients. Here, both preformulation stability studies and those performed on the final dosage form, including the effect that packaging might have on the stability, are evaluated. Chapter 5 covers the rapidly changing area of protein and peptide pharmaceuticals. The physical and chemical stability of these newer biotechnology products is often difficult to characterize. Therefore, a range of methodologies is needed. The final chapter concerns regulatory requirements for stability testing.

Chapter 2

Chemical Stability of Drug Substances

The most easily understood and most studied form of drug instability is the loss of drug through a chemical reaction resulting in a reduction of potency. Loss of potency is a well-recognized cause of poor product quality.

In this chapter, the quantitation of chemical drug loss is discussed and analyzed. However, loss of drug potency *per se* by various pathways is only one of many possible reasons for quantitating drug loss. Identification of the product(s) formed provides a better understanding of the mechanism(s) of these chemical reactions as well as other valuable information. Other reasons for quantitating drug loss include the following.

1. The drug may degrade to a toxic substance. Therefore, it is important to determine not only how much drug is lost with time but also what are its degradants. In some cases, the degradants may be of known toxicity. For example, the drug pralidoxime degrades via two parallel, pH-sensitive pathways. Under basic pH conditions, the toxic product cyanide is formed (Scheme 1).[1] For other drugs, the toxicity of degradants is initially unknown. For example, a degradant of tetracycline is epianhydrotetracycline, known to cause Fanconi syndrome (Scheme 2).[2,3]

Sometimes, reactive intermediates are formed that are known or suspected to be toxic. For example, penicillins rearrange under acidic pH conditions to penicillenic acids, which are suspected to contribute to the allergenicity of penicillins (Scheme 3).[4] Gosselin *et al.*

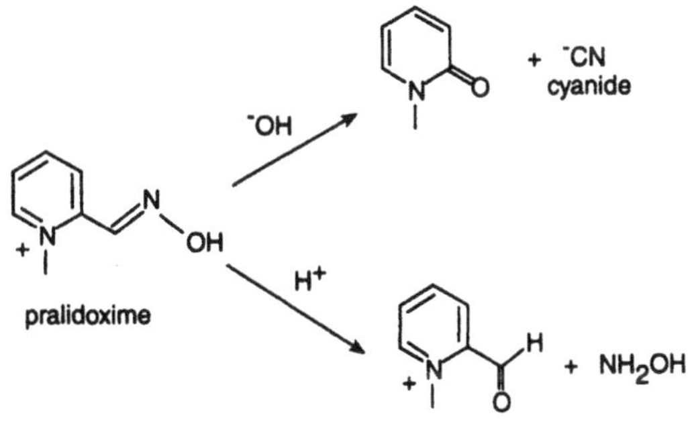

Scheme 1. Parallel degradation pathways for pralidoxime leading to cyanide formation under basic pH conditions. (Reproduced from Ref. 1 with permission.)

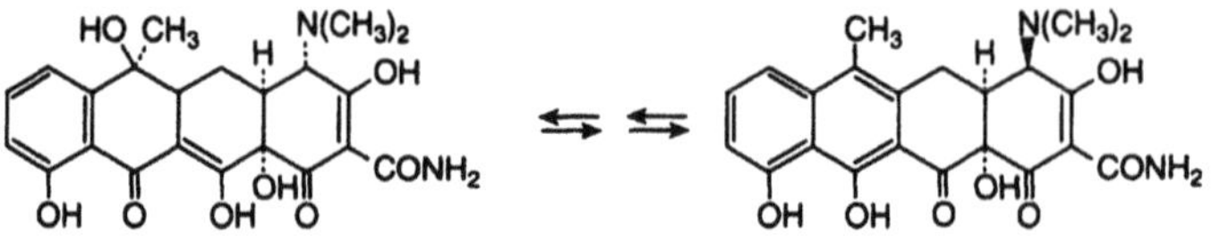

tetracycline epianhydrotetracycline

Scheme 2. Dehydration and epimerization of tetracycline, leading to formation of epianhydrotetracycline, known to be associated with Fanconi syndrome. (Reproduced from Refs. 2 and 3 with permission.)

proposed a protecting group for phosphates that produces episulfide, a sulfur analog of ethylene oxide of unknown toxicity.[5]

2. Degradation of the drug may make the product esthetically unacceptable. Products are presumed to be adulterated if significant changes in, for instance, color or odor have occurred with time. For example, epinephirine is oxidized to adrenochrome (Scheme 4), a highly colored red material. Any epinephrine-containing product that develops a significant pink tinge is usually considered adulterated.

Recently, one of the authors was asked to comment on the acceptability of a drug substance that degraded to volatile, odor-producing, sulfur-containing degradant. Even minor degradation of the drug produced an unacceptable odor. This was of specific concern because one intended route of drug administration was via a nasal spray.

3. Even though a drug may be stabilized in its intended formulation, the formulator must show that the drug is also stable under the pH conditions found in the gastrointestinal tract, if the drug is intended for oral use. Most drug substances are fairly stable at the neutral pH values found in the small intestine (disregarding enzymatic degradation) but can be unstable at pH values found in the stomach. Examples of drugs that are very acid-labile are various penicillins,[4,6] erythromycin and some of its analogs,[7] and the 2′,3′-dideoxypurine nucleoside anti-AIDS drugs.[8] Knowledge of the stability of a drug in the pH range of 1–2 at 37°C is important in the design of potentially acid-labile drugs and their dosage forms.

2.1. Pathways of Chemical Degradation

Drug substances used as pharmaceuticals have diverse molecular structures and are, therefore, susceptible to many and variable degradation pathways. Possible degradation pathways include hydrolysis, dehydration, isomerization and racemization, elimination, oxidation, photodegradation, and complex interactions with excipients and other drugs. It would be very useful if we could predict the chemical instability of a drug based on its molecular structure. This would help both in the design of stability studies and, at the earliest

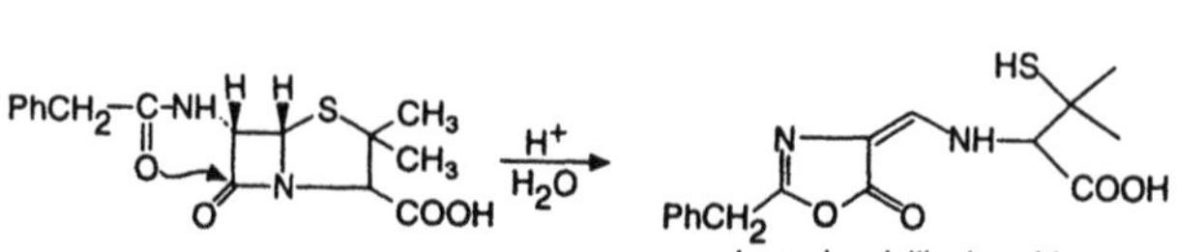

benzylpenicillin benzylpenicillenic acid

Scheme 3. Representative example of the rearrangement of penicillins to their penicillenic acids under acidic pH conditions. (Reproduced from Ref. 4 with permission.)

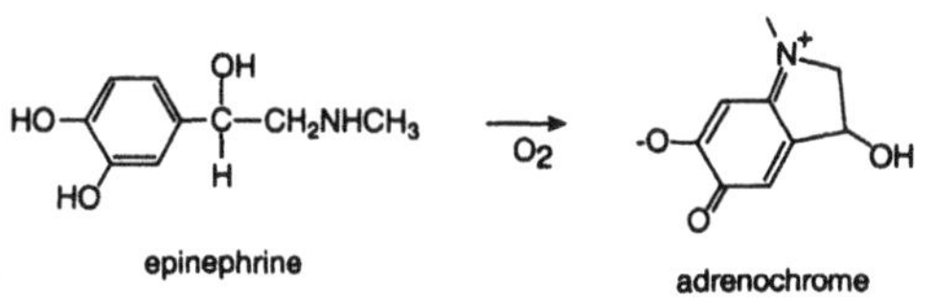

epinephrine $\xrightarrow{O_2}$ adrenochrome

Scheme 4. Oxidation of epinephrine to the highly colored adrenochrome.

stages of drug development, in identifying ways in which problematic drugs could be formulated to minimize chemical degradation. The immense chemical and pharmaceutical literature is probably underutilized as a source of such information. Expert systems are also being developed for predicting stability.

Below, the major degradation pathways in relation to molecular structure are discussed and examples provided.

2.1.1. Hydrolysis

For most parenteral products, the drug comes into contact with water and, even in solid dosage forms, moisture is often present, albeit in low amounts. Accordingly, hydrolysis is one of the most common reactions seen with pharmaceuticals. Many researchers have reported extensively on the hydrolysis of drug substances. In the 1950s, elegant studies, especially considering the lack of high-throughput analytical techniques, concerning the hydrolysis of procaine,[9,10] aspirin,[11,12] chloramphenicol,[13–15] atropine,[16–18] and methylphenidate[19] were reported. Hydrolysis is often the main degradation pathway for drug substances having ester and amide functional groups within their structure.

2.1.1.1. Esters

Many drug substances contain an ester bond. Traditional esters are those formed between a carboxylic acid and various alcohols. Other esters, however, include those formed between carbamic, sulfonic, and sulfamic acids and various alcohols. These ester compounds are primarily hydrolyzed through nucleophilic attack of hydroxide ion or water at the ester, as shown in Scheme 5 for the case of a carboxylic acid ester.

The degradation rate depends on the substituents R_1 and R_2, in that electron-withdrawing groups enhance hydrolysis whereas electron-donating groups inhibit hydrolysis. As shown in Table 1, substituted benzoates having an electron-withdrawing group, such as a nitro group, in the *para* position of the phenyl ring (R_1) exhibit higher decomposition rates than the unsubstituted benzoate. On the other hand, the decomposition rate decreases with increasing electron-donating effect of the alkyl group (in the alcohol portion of the ester (R_2)) (e.g., it decreases in the order methyl > ethyl > *n*-propyl). Replacing a hydrogen atom

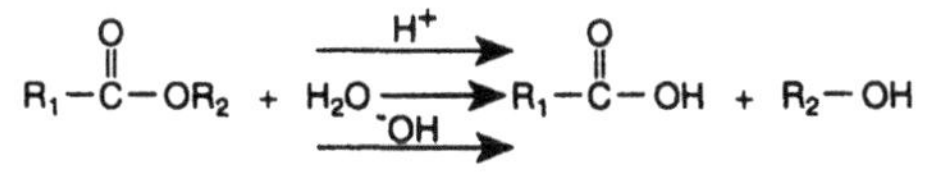

$$R_1-\overset{O}{\overset{\|}{C}}-OR_2 + H_2O \xrightarrow[{}^{-}OH]{H^+} R_1-\overset{O}{\overset{\|}{C}}-OH + R_2-OH$$

Scheme 5. Hydrolysis of a carboxylic acid ester.

Table 1. Second-Order Rate Constants for the Hydrolysis of Various Benzoic Acid Esters through Nucleophilic Attack of Hydroxide Ion, in Accordance with Scheme 5 ($R_1 = R'$–C$_6$H$_4$–)

R′	R_2	Second-order rate constant K_{OH}^{-} ($\times 10^{-4}\ M^{-1}\ s^{-1}$)[a]
H	CH_3	6.08
H	C_2H_5	1.98
H	*n*-C_3H_7	1.67
H	*iso*-C_3H_7	0.319
H	Phenyl	33.6
H	C_2H_4Cl	12.4
CH_3	CH_3	2.65
F	CH_3	12.1
Cl	CH_3	19.1
Cl	C_2H_5	6.51
Cl	*n*-C_3H_7	5.11
Cl	*iso*-C_3H_7	1.21
Cl	Phenyl	103
NO_2	CH_3	276
NO_2	C_2H_5	98.8
NO_2	*n*-C_3H_7	76.0
NO_2	*iso*-C_3H_7	19.6
NO_2	Phenyl	1140

[a]In 50% acetonitrile–0.02*M* phosphate buffer solution; 25°C.

with an electron-withdrawing halogen such as chlorine, e.g., $-C_2H_5$ versus $-C_2H_4Cl$, also increases the rate of decomposition.[20]

Another way of viewing this reaction is by considering leaving-group ability. The mechanism of ester hydrolysis can be considered an addition/elimination reaction, the leaving group being R_2OH. The rate of the elimination step will be determined in part by the ability of the leaving alcohol to sustain the buildup of negative charge on the oxygen atom. This will also be reflected in the pK_a of the alcohol. For example, hydrolysis of phenyl benzoate is much faster than that of ethyl benzoate (Table 1) because the pK_a values of ethanol and phenol are 18 and 10, respectively.

Steric factors also play a role. Bulky groups on either R_1 or R_2 decrease the decomposition rate. For example, when an iso-propyl group is substituted for an *n*-propyl group on R_2, the decomposition is five times slower (Table 1).

Attack of hydroxide ion on an ester bond is also affected by the presence of neighboring charges. For example, the hydrolysis rates of all ester bonds within poly(butylene tartrate) are not equal; the ester bonds close to the negatively charged, terminal carboxylate group are less reactive toward hydroxide-ion attack than are the ester groups removed from the negatively charged carboxylate group.[21]

2.1.1.1.a. Carboxylic Acid Esters of Pharmaceutical Relevance. Representative examples of carboxylic acid esters that are susceptible to hydrolysis are shown in Fig. 1. These include ethylparaben,[22] benzocaine,[10,23,24] procaine,[9,10] oxathiin carboxanilide (NSC-

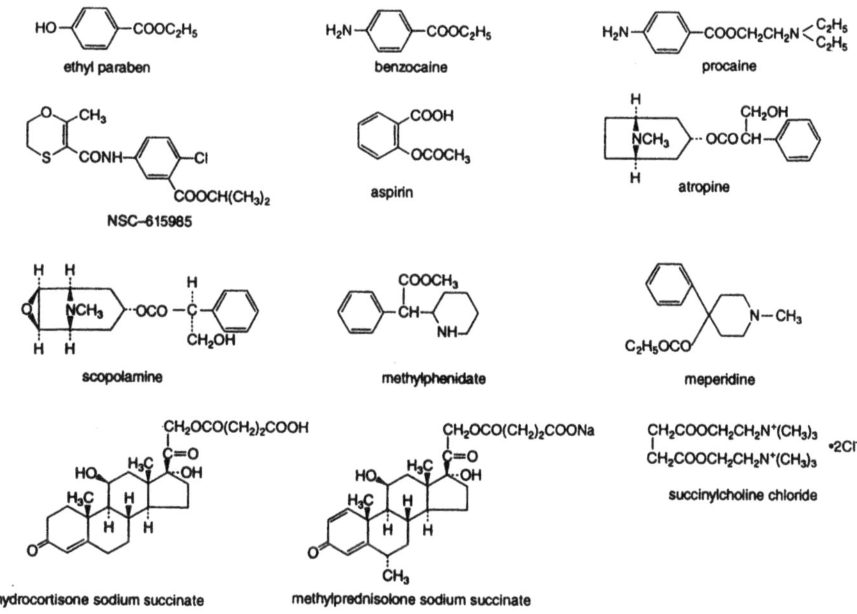

Figure 1. Representative examples of carboxylic acid esters of pharmaceutical interest, susceptible to hydrolysis.

615985),[25] aspirin,[11,12] atropine,[16–18,26] scopolamine,[27] methylphenidate,[19] meperidine,[28] steroid esters such as hydrocortisone sodium succinate[29,30] and methylprednisolone sodium succinate,[31] and succinylcholine chloride.[32,33] Cocaine has two ester bonds that hydrolyze to produce benzoylecgonine or ecgonine methyl ester, as shown in Scheme 6.[34,35] It is said to undergo parallel pathways of degradation. Shown in all future reaction schemes are the primary reaction pathways. As such, these are not meant to be complete; that is, some compounds undergo other competing reactions.

Based on the structures of these various esters, it can be readily seen that having information on the reactivity of one ester should provide valuable insight into that of a second

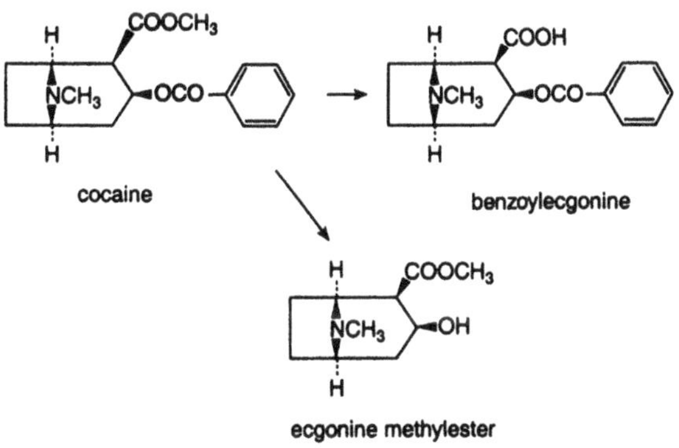

cocaine → benzoylecgonine

cocaine → ecgonine methylester

Scheme 6. Parallel hydrolysis pathways for cocaine.

ester. For example, ethylparaben and benzocaine are very similar in structure; both have a *para* electron-donating group and both are ethyl esters. Therefore, information about the reactivity of one of them could be the basis for predicting the stability of the other. Similarly, ester group hydrolysis in atropine should be similar in rate and pH dependency to that in scopolamine. Is it not reasonable to expect the hydrolysis of methylprednisolone sodium succinate to be similar to that of hydrocortisone sodium succinate? Therefore, if one is presented with a new drug substance containing a hydrolyzable ester moiety, it should be possible, using appropriate literature examples of similar drugs, to make a good estimate of the sensitivity of the ester group to hydrolysis.

Lactones, or cyclic esters, also undergo hydrolysis. As shown in Fig. 2, pilocarpine,[36–38] dalvastatin,[39] warfarin,[40,41] and camptothecin[42] exhibit ring opening due to hydrolysis. Note that, unlike linear esters, lactones often exist in dynamic equilibrium with their carboxylic acid/carboxylate forms.

Apparent rate constants for the hydrolysis of various carboxylic acid esters are shown in Table 2 for the comparison of their reactivities. As these values were obtained under different conditions of temperature, pH, ionic strength, and buffer species, they are for rough comparison only. Nevertheless, they do point out the role that structure plays in the relative reactivity of the ester bond.

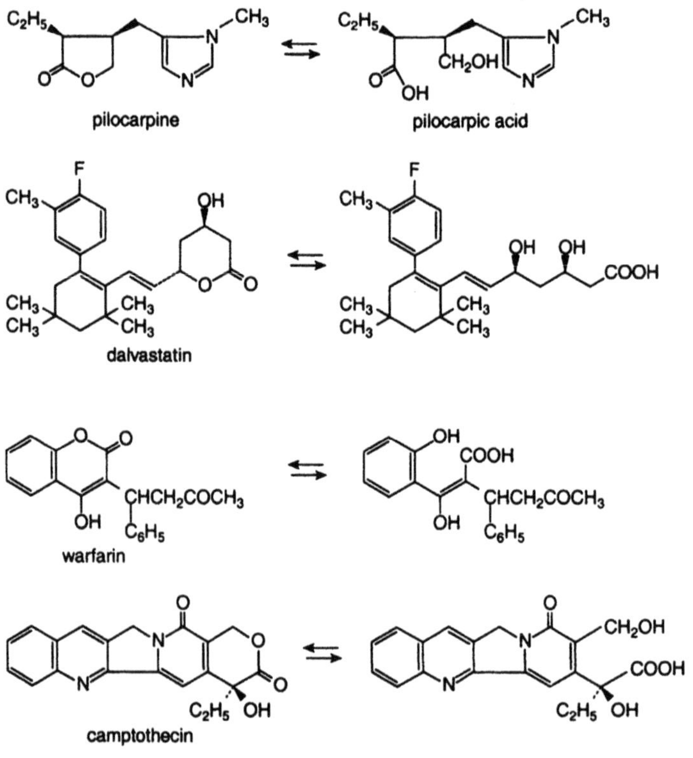

Figure 2. Representative lactones of pharmaceutical interest susceptible to hydrolysis. Note that, unlike esters, lactones often exist in dynamic equilibrium (pH dependent) with their carboxylate forms.

Table 2. Apparent Rate Constants for the Hydrolysis of Various Carboxylic Acid Esters

	k (s^{-1})	pH	Reference
Camptothecin	6.0×10^{-5} (25°C)	7.13	42
Aspirin	3.7×10^{-6} (25°C)	6.90	12
Methylprednisolone sodium succinate	2.5×10^{-7} (25°C)	7.30	31
Oxathiin carboxanilide	1.8×10^{-7} (25°C)	6.92	25
Benzocaine	5.7×10^{-8} (25°C)	9.2	4
Ethylparaben	4.2×10^{-8} (25°C)	9.16	22
Cocaine	4.97×10^{-6} (30°C)	7.25	34
Succinylcholine	5.0×10^{-5} (40°C)	8.00	32
Procaine	6×10^{-6} (40°C)[a]	8	9
Pilocarpine	1.7×10^{-6} (40°C)[a]	8	36
Atropine	1.8×10^{-7} (40°C)	7.01	17
Methylphenidate	3.2×10^{-6} (50°C)	6.07	19
Hydrocortisone sodium succinate	9.0×10^{-6} (65.2°C)	7.0	29
	1×10^{-7} (25°C)[b]		29
Meperidine	1.8×10^{-7} (89.7°C)	6.192	28

[a]Value of k estimated from plots in the reference.
[b]Value of k estimated using the reported value of the activation energy (E_a).

2.1.1.1.b. Other Esters. Carbamic acid esters such as chlorphenesin carbamate[43] and carmethizole,[44] shown in Scheme 7, are known to undergo hydrolysis in strongly acidic and neutral-to-alkaline solutions, respectively. The two carbamate ester groups in carmethizole undergo hydrolysis at significantly different rates owing in large part to completely different mechanisms.[44] The first carbamate group is cleaved by more of an elimination reaction via carbonium formation whereas the second carbamate linkage appears to hydrolyze via a normal hydrolysis mechanism.

Cyclodisone,[45] a sulfonic acid ester, and sulfamic acid 1,7-heptanediyl ester (NSC-329680),[46] a sulfamic acid ester, have been reported to hydrolyze in the neutral-to-alkaline pH range (Scheme 8). Both hydrolyze via carbon–oxygen bond cleavage rather than sulfur–oxygen bond cleavage.[45,46]

chlorophenesin carbamate

carmethizole

Scheme 7. Representative carbamic acid esters of pharmaceutical relevance susceptible to hydrolysis.

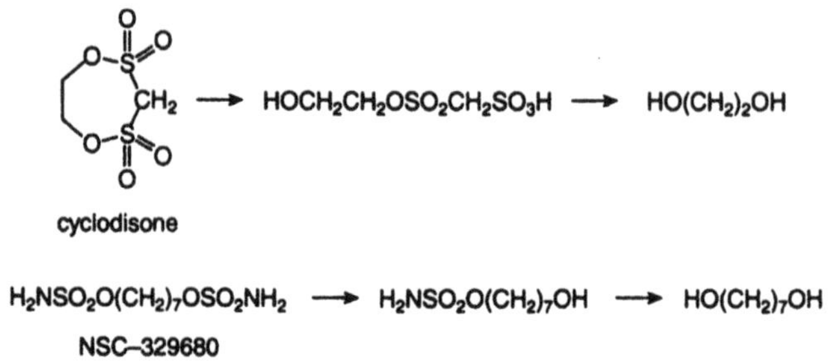

cyclodisone → $HOCH_2CH_2OSO_2CH_2SO_3H$ → $HO(CH_2)_2OH$

$H_2NSO_2O(CH_2)_7OSO_2NH_2$ → $H_2NSO_2O(CH_2)_7OH$ → $HO(CH_2)_7OH$

NSC-329680

Scheme 8. Representative sulfonic esters and sulfamic esters susceptible to hydrolysis.

Phosphoric acid esters such as hydrocortisone disodium phosphate[47,48] and echothiophate iodide[49] are known to hydrolyze (Scheme 9). Although nitric esters such as nitroglycerin[50] and nicorandil[51] undergo hydrolysis, nitroglycerin is relatively stable (Scheme 9). Phosphatidylcholine and phosphatidylethanolamine in intravenous lipid emulsion and aqueous liposome dispersions have been reported to hydrolyze in the neutral pH range.[52,53]

2.1.1.2. *Amides*

Amide bonds are commonly found in drug molecules. Amide bonds are less susceptible to hydrolysis than ester bonds because the carbonyl carbon of the amide bond is less electrophilic (the carbon-to-nitrogen bond has considerable double bond character) and the leaving group, an amine, is a poorer leaving group (Scheme 10). Figure 3 shows the structure

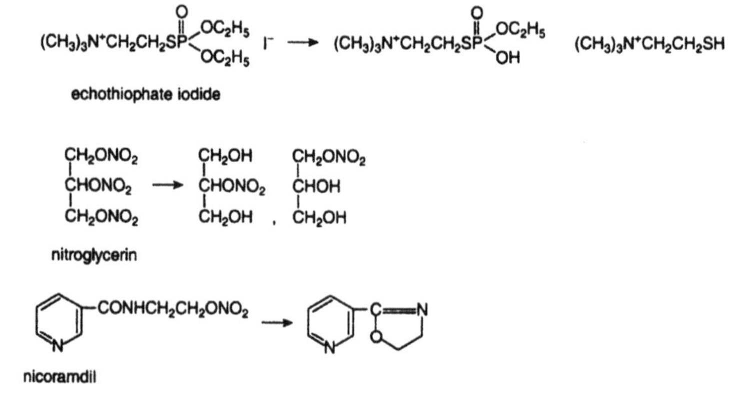

hydrocortisone disodium phosphate

echothiophate iodide

$(CH_3)_3N^+CH_2CH_2SP(O)(OC_2H_5)_2\ I^-$ → $(CH_3)_3N^+CH_2CH_2SP(O)(OC_2H_5)OH$ $(CH_3)_3N^+CH_2CH_2SH$

nitroglycerin

CH_2ONO_2–$CHONO_2$–CH_2ONO_2 → CH_2OH–$CHONO_2$–CH_2OH , CH_2ONO_2–$CHOH$–CH_2OH

nicoramdil

$-CONHCH_2CH_2ONO_2$ →

Scheme 9. Other esters of pharmaceutical relevance susceptible to hydrolysis.

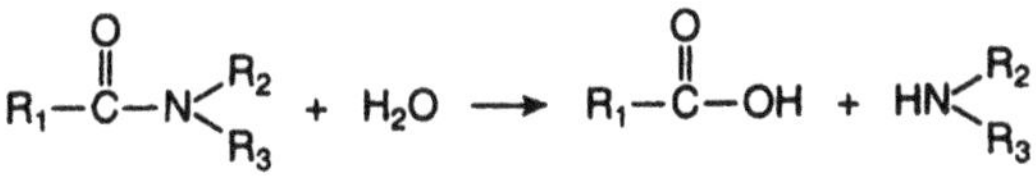

$$R_1-\overset{O}{\overset{\|}{C}}-N\begin{matrix}R_2\\R_3\end{matrix} + H_2O \longrightarrow R_1-\overset{O}{\overset{\|}{C}}-OH + HN\begin{matrix}R_2\\R_3\end{matrix}$$

Scheme 10. Hydrolysis of amides.

of acetaminophen,[54] chloramphenicol,[13–15] lincomycin,[55] indomethacin,[56–59] and sulfacetamide,[60] all of which are known to produce an amine and an acid through hydrolysis of their amide bonds; moricizine, a derivative of phenothiazine, which undergoes hydrolysis of its amide bonds followed by oxidation[61]; and HI-6, a bis(pyridimium)aldoxime having an amide bond, which exhibits fast hydrolysis in concentrated aqueous solutions owing to the acidifying effect of a strongly acidic oxime group.[62]

β-Lactam antibiotics such as penicillins and cephalosporins, which are cyclic amides or lactams, undergo rapid ring opening due to hydrolysis. Ring opening of the β-lactam group has been reported for penams, such as, benzylpenicillin,[63,64] ampicillin,[65] amoxicillin,[66] carbenicillin,[67] phenethicillin,[68] and methicillin[69] (Scheme 11), and for cephems, such as cephalothin,[70] cefadroxil,[71,72] cephradine,[70] and cefotaxime[73–75] (Scheme 12). These drug substances have both a lactam and an amide bond in their molecular structure, the former being considerably more susceptible to hydrolysis. Cephalothin and cefotaxime are also acetoxy esters, and opening of their lactam ring competes with hydrolysis of the ester bond. Decomposition products produced by hydrolysis of penam and cephem β-lactams are still reactive and undergo various side reactions. For example, condensation products were formed upon hydrolysis of cefaclor,[76] and dimeric products were detected upon hydrolysis of loracarbef,[77] as shown in Scheme 13, as well as of ampicillin.[78] Cycloserine, which can be considered a cyclic amide, undergoes opening of its isoxazolidone ring due to hydrolysis in acidic media,[79] as shown in Scheme 14. Like loracarbef and ampicilllin, it also undergoes self-condensation.

The reactivity of these amides toward hydrolysis depends on the substituents R_1, R_2, and R_3 (Scheme 10), as shown in Table 3. The β-lactam antibiotics, including penicillins and cephalosporins, undergo surprisingly facile hydrolysis compared to other amides. The

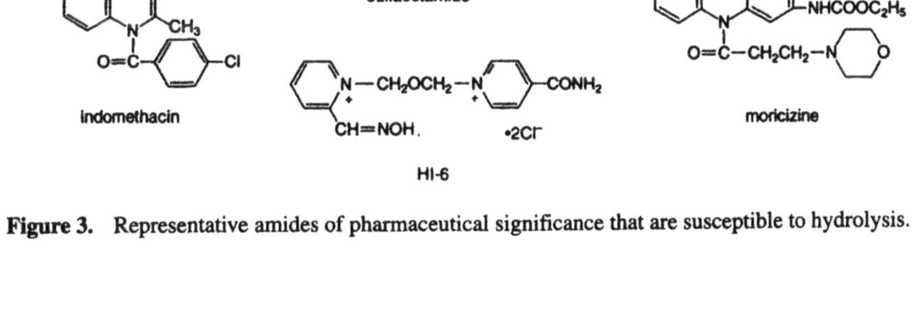

acetominophen

chloramphenicol

lincomycin

sulfacetamide

indomethacin

moricizine

HI-6

Figure 3. Representative amides of pharmaceutical significance that are susceptible to hydrolysis.

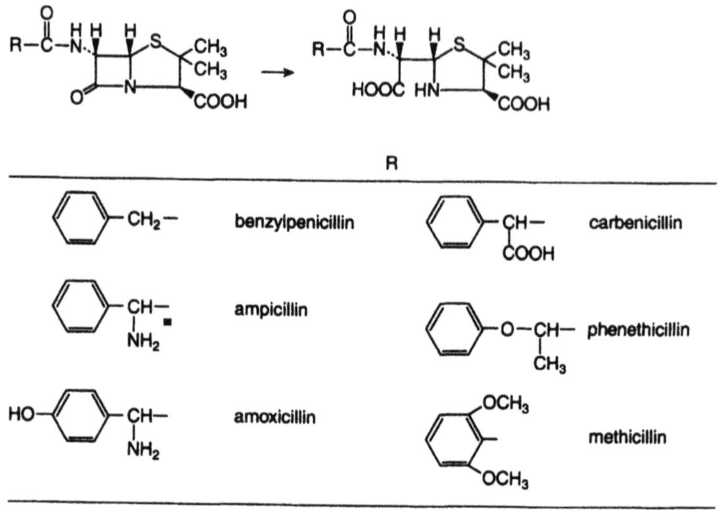

Scheme 11. Hydrolysis of β-lactam penicillins. This pathway is mostly seen in the neutral to alkaline pH range.

most likely contributors to this facile hydrolysis are electronic factors, the relief of ring strain (a four-membered ring coupled to a five- or six-membered ring), and the lower double bond character between the carbonyl carbon and the amide nitrogen.

2.1.1.3. *Barbiturates, Hydantoins, and Imides*

Barbiturates, hydantoins, and imides contain functional groups related to amides but tend to be more reactive. Barbituric acids such as barbital, phenobarbital, amobarbital, and metharbital undergo ring-opening hydrolysis, as shown in Scheme 15.[80,81] Decomposition products formed from these drug substances are susceptible to further decomposition reactions such as decarboxylation. The hydrolysis rates of these substances depend on the substituents R_1, R_2, and R_3. For some allylbarbituric acids, the effects of these substituents on hydrolysis rates can be explained in terms of Hammett's σ value.[82]

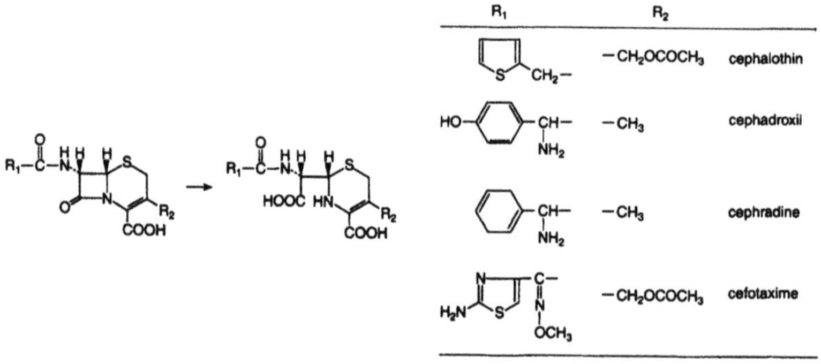

Scheme 12. Hydrolysis of β-lactam cephalosporins.

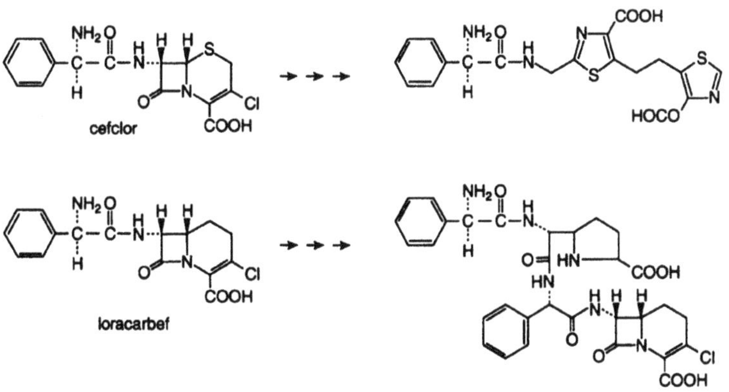

Scheme 13. Other degradation products of cefaclor and loracarbef.

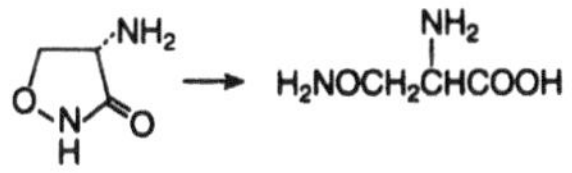

Scheme 14. Hydrolysis of cycloserine.

Table 3. Apparent Rate Constants of Hydrolysis of Various Amides under a Variety of pH and Temperature Conditions

	k (s^{-1}) (temperature)	pH	Reference
Benzylpenicillin	1.5×10^{-4} (25°C)	2.70	64
	3.9×10^{-6} (60°C)	6.75	63
Acetaminophen	1.0×10^{-9}(25°C)[b]	6	54
Indomethacin	2.2×10^{-4} (25.8°C)	11	58
	3.9×10^{-6} (60°C)	7	59
Cephalothin	5.6×10^{-5} (35°C)	9.84	70
Cefotaxime	2.4×10^{-5} (35°C)	8.94	73
Cephradine	2×10^{-5} (35°C)	10.00	70
Phenethicillin	3.3×10^{-6} (35°C)	7.4	68
Cefadroxil	2.1×10^{-6} (35°C)	7.20	71
carbenicillin	2.0×10^{-6} (35°C)	7.00	67
Amoxicillin	~1.1×10^{-6} (35°C)[a]	8.2	66
Ampicillin	2.5×10^{-7} (35°C)	7.11	65
Moricizine	8.2×10^{-6} (60°C)	6.0	61
Lincomycin	4.9×10^{-6} (70°C)	1	55
Chloramphenicol	6.0×10^{-6} (85, 36°C)	6.00	15
Sulfacetamide	9.3×10^{-6} (120°C)	6.91	60

[a]Value of k estimated from plots in Ref. 54.

[b]Value of k estimated using the reported value of the activation energy (E_a).

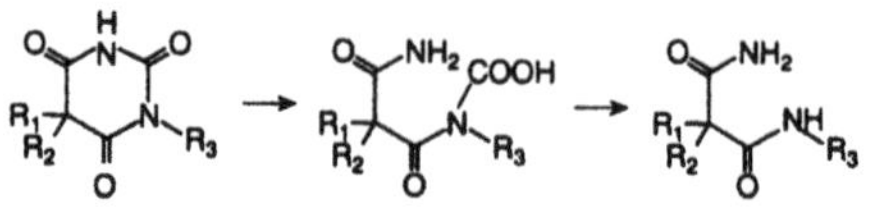

R_1	R_2	R_3	
C_2H_5-	C_2H_5-	H—	barbital
C_2H_5-	C_6H_5-	H—	phenobarbital
C_2H_5-	$(CH_3)_2CHCH_2CH_2-$	H—	amobarbital
C_2H_5-	C_2H_5-	CH_3-	metharbital

Scheme 15. Hydrolysis of barbituric acids.

As shown in Scheme 16, the hydantoin allantoin[83] is susceptible to hydrolysis, and the imide bonds in NSC-284356[84] and (+)-1,2-bis(3,5-dioxopiperazinyl-1-yl)propane (ICRF-187)[85,86] are hydrolyzed by parallel and successive reactions. In the case of ICRF-187, the reactivity of the imide groups is intramolecularly affected by the tertiary amine groups in its structure.[85,86] This conclusion was drawn from the observation that model compound A

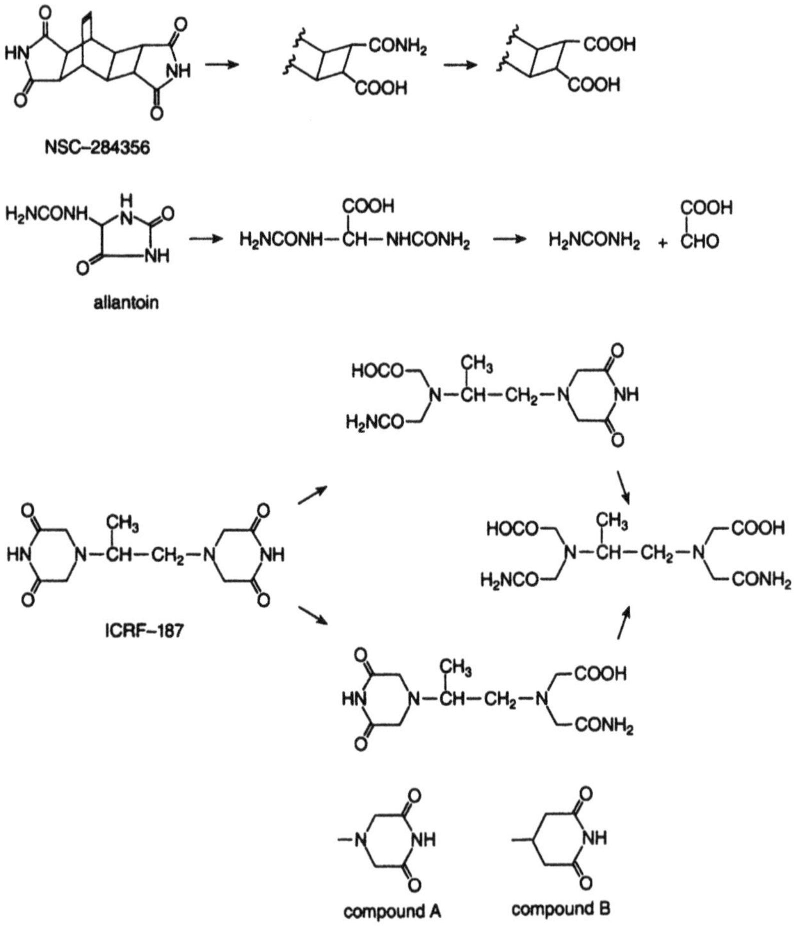

NSC-284356

allantoin

ICRF-187

compound A compound B

Scheme 16. Imides of pharmaceutical significance susceptible to hydrolysis.

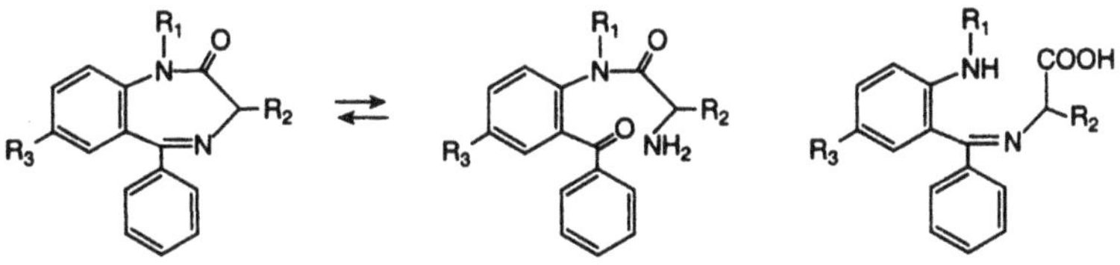

R_1	R_2	R_3	
$-CH_3$	$-H$	$-Cl$	diazepam
$-H$	$-OH$	$-Cl$	oxazepam
$-H$	$-H$	$-NO_2$	nitrazepam

Scheme 17. Hydrolysis of benzodiazepines.

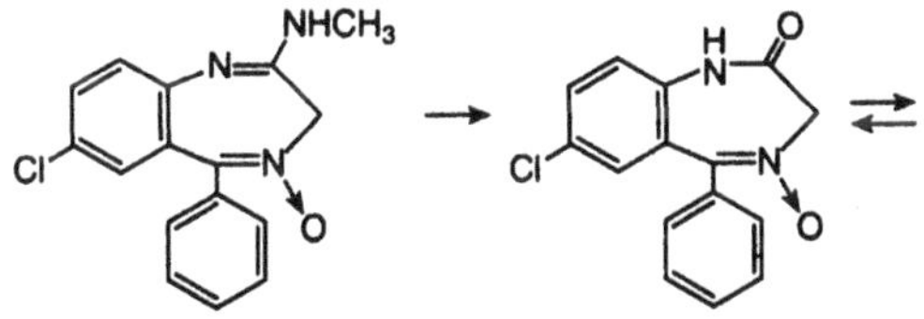

Scheme 18. Hydrolysis of chlordiazepoxide.

(Scheme 16) hydrolyzed, as expected, at approximately half the rate of ICRF-187, whereas the glutarimide, compound B, was significantly more stable.

2.1.1.4. Schiff Base and Other Reactions Involving Carbon–Nitrogen Bond Cleavage

Benzodiazepines such as diazepam,[87] oxazepam,[87] and nitrazepam[88,89] undergo ring opening due to reversible hydrolysis of the amide and azomethine bonds, as shown in Scheme 17. Chlordiazepoxide is converted to a lactam form, which is then similarly hydrolyzed (Scheme 18).[90,91] Triazolam, a triazole-condensed benzodiazepine, also undergoes ring opening due to hydrolysis, as shown in Scheme 19.[92] Benzodiazepinooxazoles (oxazole-condensed benzodiazepines) such as oxazolam,[93] flutazolam,[94] haloxazolam,[94] and cloxazolam[94] are not Schiff bases *per se* but undergo ring opening due to hydrolysis as shown

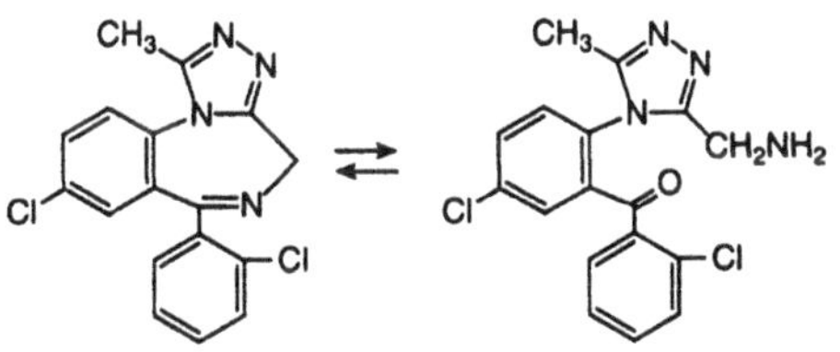

Scheme 19. Hydrolysis of triazolam.

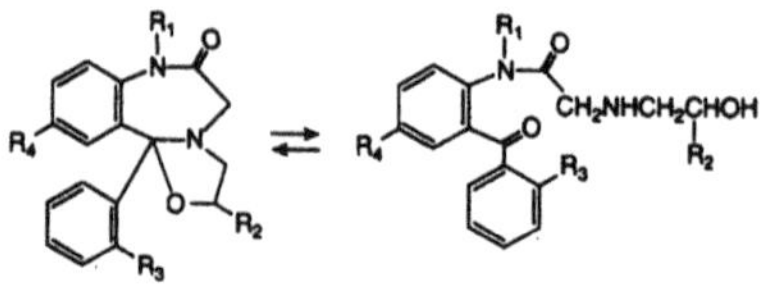

R_1	R_2	R_3	R_4	
H	CH_3	H	Cl	oxazolam
$-CH_2CH_2OH$	H	F	Cl	flutazolam
H	H	F	Br	haloxazolam
H	H	Cl	Cl	cloxazolam

Scheme 20. Hydrolysis of benzodiazepinooxazoles.

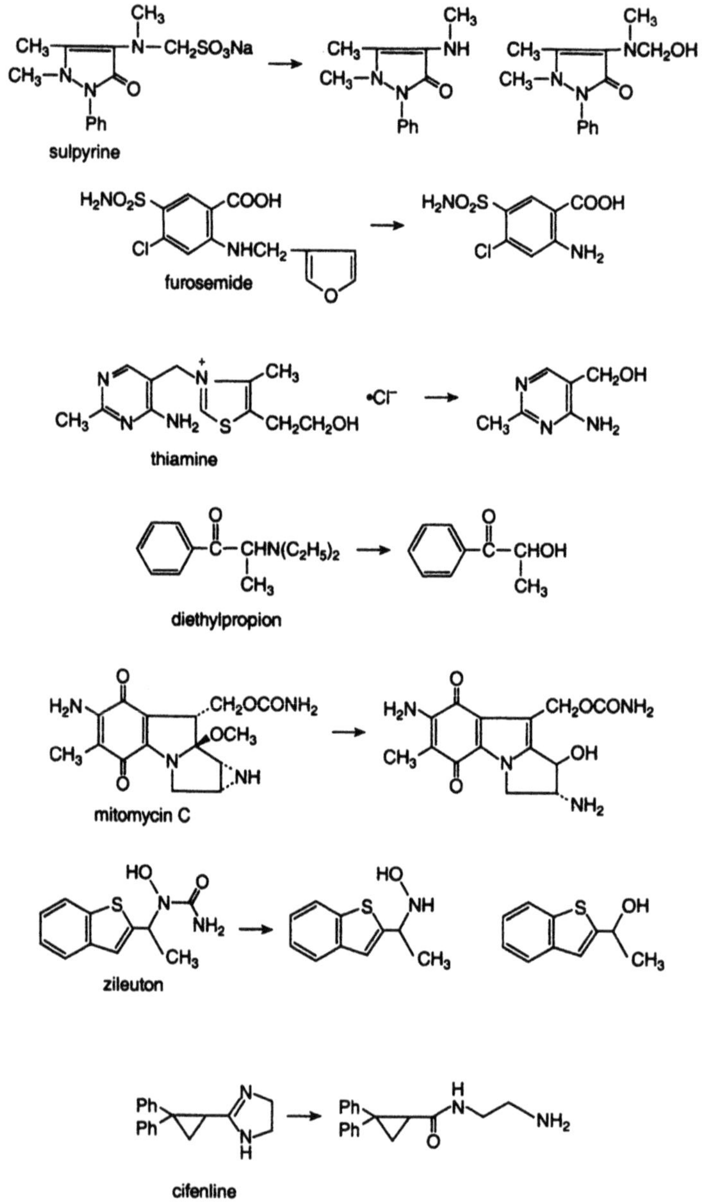

Scheme 21. Representative drug substances having a reactive nitrogen in their structure that are susceptible to hydrolysis.

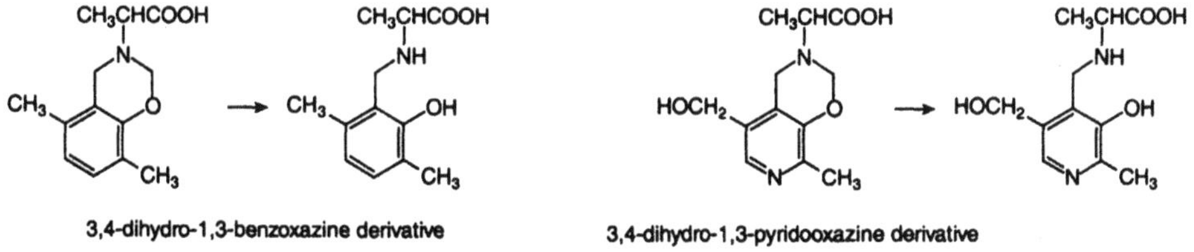

3,4-dihydro-1,3-benzoxazine derivative

3,4-dihydro-1,3-pyridooxazine derivative

Scheme 22. Hydrolysis of dihydrooxazines.

in Scheme 20. The oxazolidine ring of these drugs is known to exhibit an acid–base equilibrium reaction between ring opening and closing.[95–97]

Drug substances such as sulpyrine,[98,99] furosemide,[100,101] thiamine hydrochloride,[102] diethylpropion,[103] mitomycin C,[104,105] zileuton,[106] and cifenline[107] have reactive nitrogens in their molecular structure and undergo hydrolysis, as shown in Scheme 21. The derivatives of 3,4-dihydro-1,3-benzoxazine and 3,4-dihydro-1,3-pyridooxazine undergo ring opening due to hydrolysis accompanied by elimination of formaldehyde (Scheme 22).[108]

Nitrofurantoin[109,110] and rifampicin[111] undergo hydrolysis of the iminelike structure as shown in Scheme 23. Similarly, chlorothiazide[112] and hydrochlorothiazide[113,114] undergo ring opening due to hydrolysis by acid–base catalysis (Scheme 24). Nucleosides[115] such as 5-azacytidine[116,117] and cytarabine[118,119] form various hydrolysis products through different reactions (such as ring opening) depending on the conditions, as shown in Scheme 25.

2.1.1.5. Other Hydrolysis Reactions

Other drug substances susceptible to hydrolysis include chloramphenicol,[13] chlorambucil,[120,121] spirohydantoin mustard,[122] alkyl halides such as clindamycin,[123] azathioprine,[124] sulfides such as thimerosal,[125,126] and platinum compounds such as carboplatin,[127] as shown in Scheme 26.

Drug substances with carbohydrate moieties, such as digoxin,[128,129] eliminate the carbohydrate group(s) due to acid-catalyzed hydrolysis. Nucleosides such as 5-azacytidine

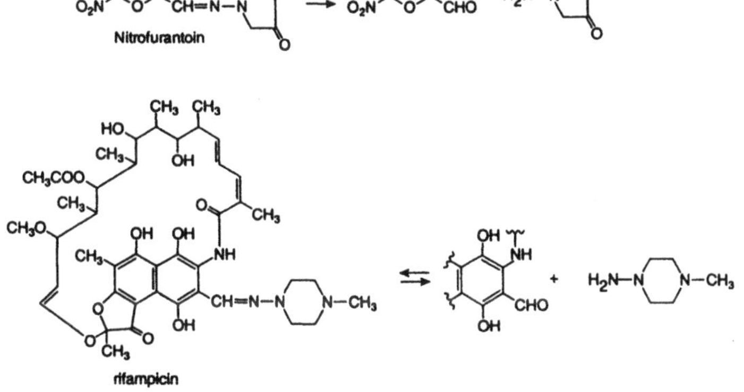

Nitrofurantoin

rifampicin

Scheme 23. Hydrolysis of other drug substances having a reactive nitrogen in their structure.

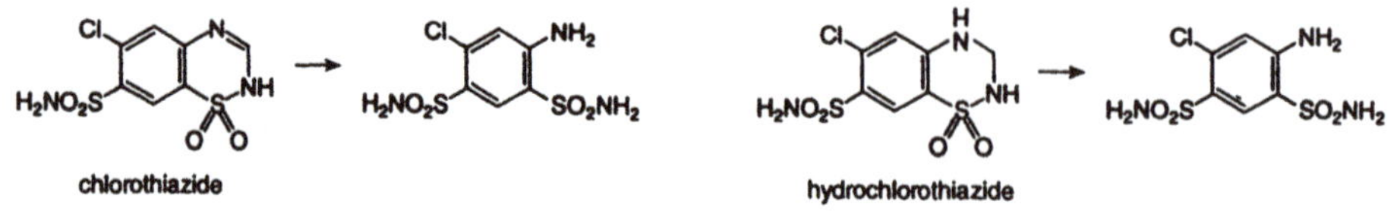

chlorothiazide

hydrochlorothiazide

Scheme 24. Hydrolysis of chlorothiazide and hydrochlorothiazide.

and cytarabine (Scheme 25) exhibit sugar-elimination reactions[130] in addition to the ring-opening reactions described earlier. Idoxuridine[131] and 2′,3′-dideoxyguanosine[132] undergo rapid hydrolysis in alkaline and acidic pH ranges, respectively (Scheme 27). 4′-Azidothymidine undergoes similar hydrolysis.[133] O^6-Benzylguanine hydrolyzes to benzyl alcohol and guanine in an acid-catalyzed reaction.[134]

2.1.2. Dehydration

Sugars such as glucose[135–137] (Scheme 28) and lactose[138,139] are known to undergo dehydration to form 5-(hydroxymethyl)furural. Erythromycin is susceptible to acid-catalyzed dehydration as shown in Scheme 29,[140,141] whereas prostaglandins E_1 and E_2 undergo dehydration followed by isomerization as shown in Scheme 30.[142–145] Batanopride undergoes an intramolecular ring-closure reaction in the acidic pH range due to dehydration (Scheme 31),[146] whereas streptovitacin A exhibits two successive acid-catalyzed dehydration reactions, as shown in Scheme 32.[147]

2.1.3. Isomerization and Racemization

Reported examples of isomerization of drug substances include *trans–cis* isomerization of amphotericin B (Scheme 33),[148] N,O-acyl rearrangement of cyclosporin A (Scheme 34),[149] and dienone–phenol rearrangement of steroids such as tirilazad (Scheme 35).[150]

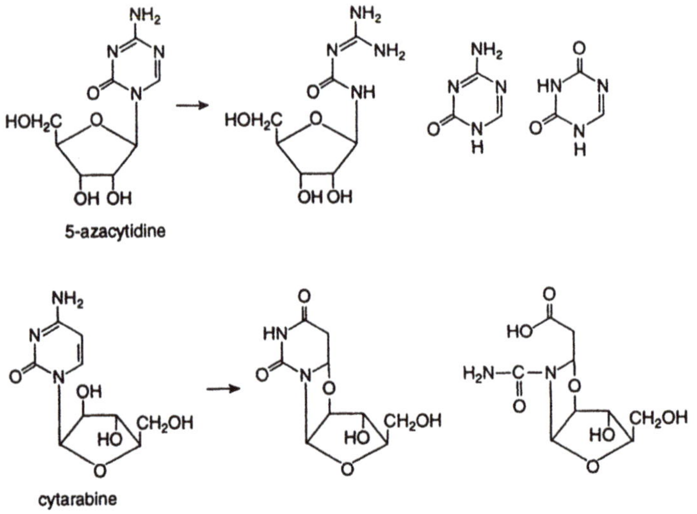

5-azacytidine

cytarabine

Scheme 25. Hydrolysis of 5-azacytidine and cytarabine.

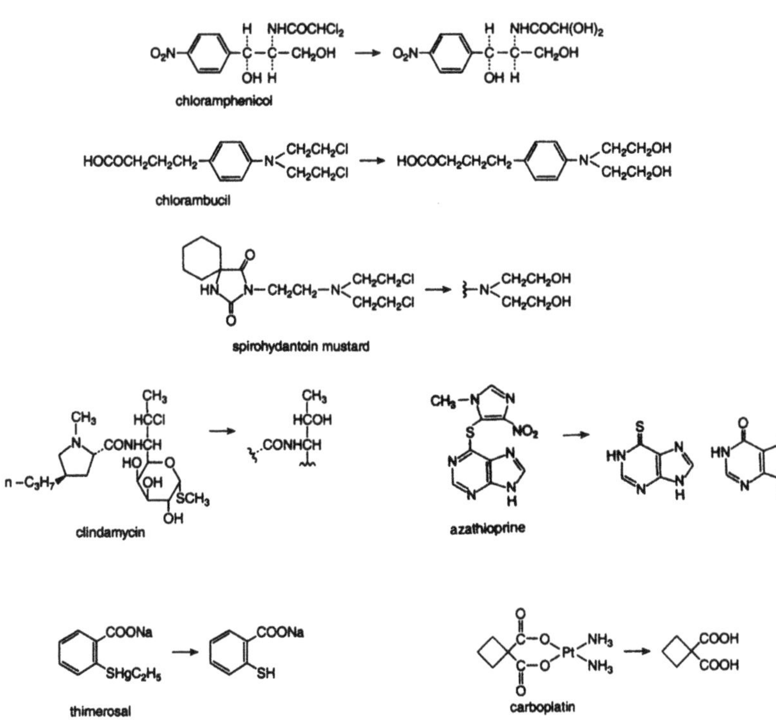

chloramphenicol

chlorambucil

spirohydantoin mustard

clindamycin

azathioprine

thimerosal

carboplatin

Scheme 26. Other drug substances that are susceptible to hydrolysis, including alkyl halides, sulfides, and platinum compounds.

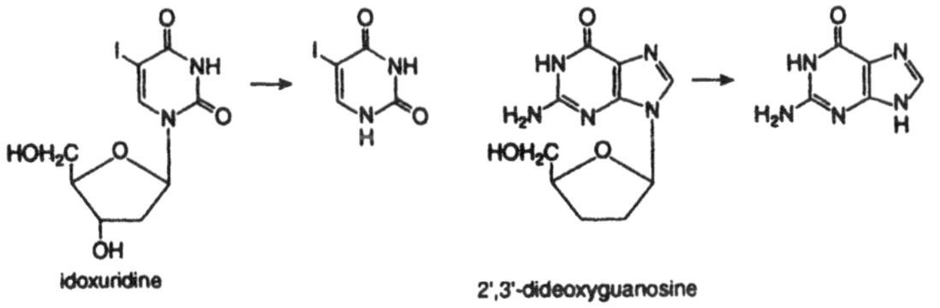

idoxuridine

2',3'-dideoxyguanosine

Scheme 27. Hydrolysis of nucleosides.

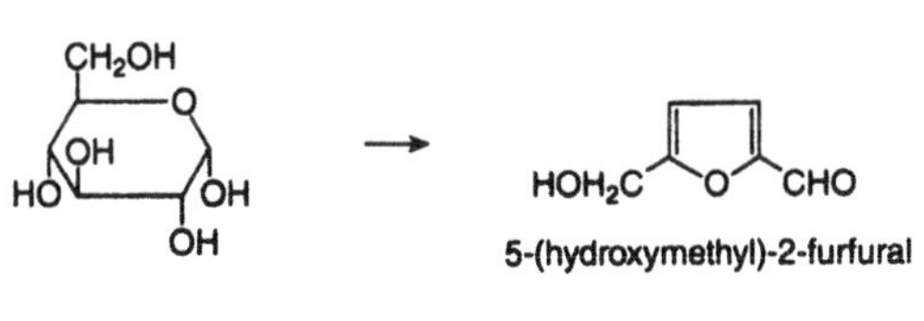

5-(hydroxymethyl)-2-furfural

Scheme 28. Dehydration of glucose.

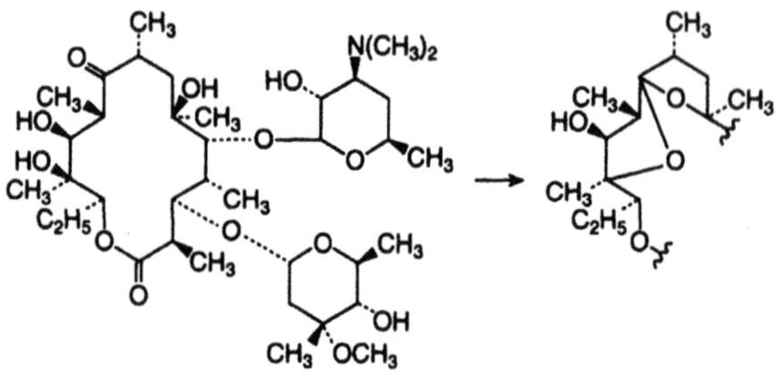

Scheme 29. Dehydration of erythromycin.

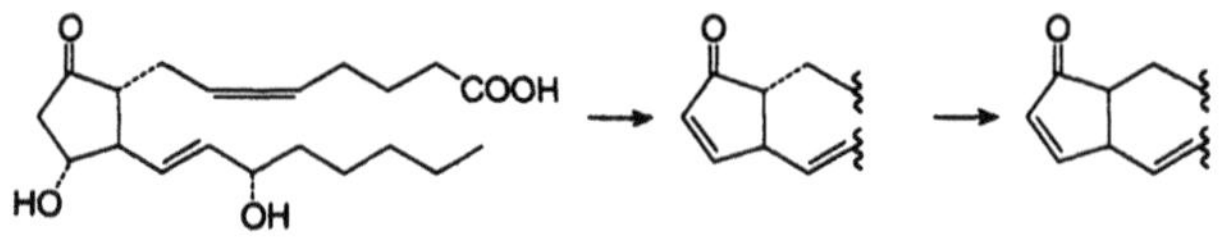

Scheme 30. Dehydration and isomerization of prostaglandin E_2.

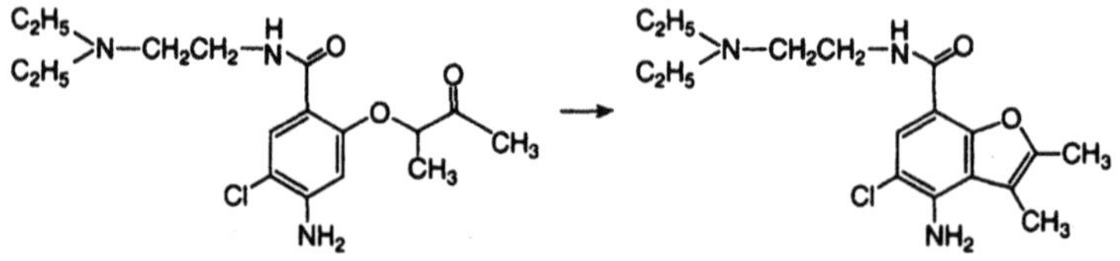

Scheme 31. Ring closure following dehydration of batanopride.

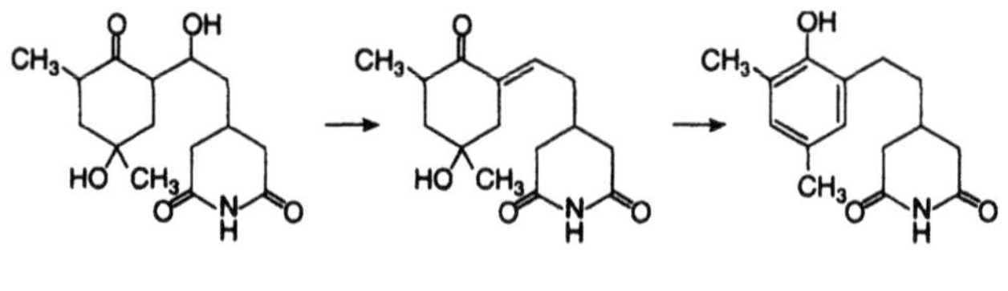

Scheme 32. Dehydration of streptovitacin A.

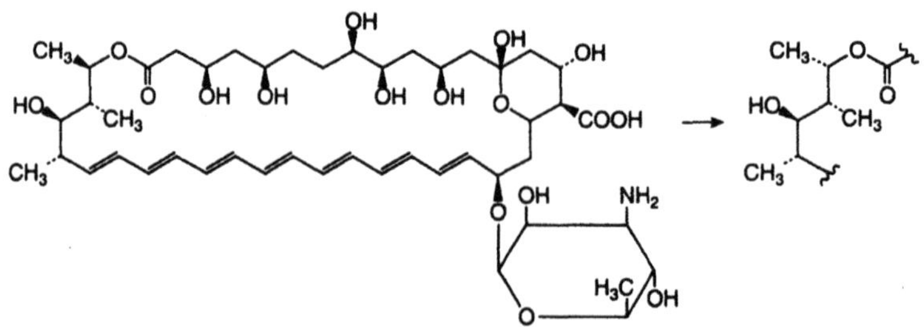

Scheme 33. *Trans–cis* Isomerization of amphotericin B.

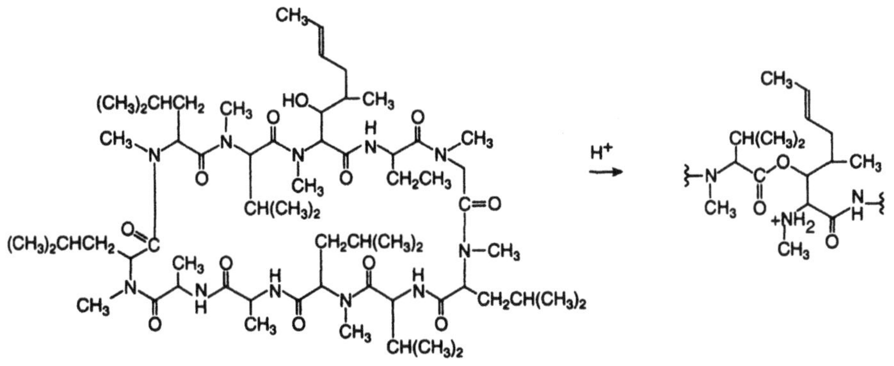

Scheme 34. N,O-Acyl rearrangement of cyclosporin A.

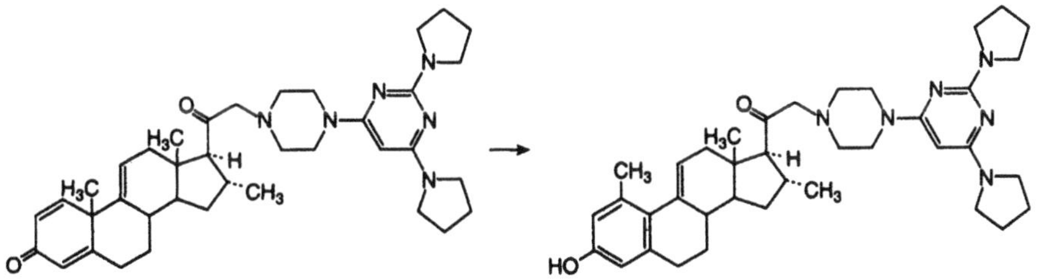

Scheme 35. Dienone–phenol rearrangement of tirilazad.

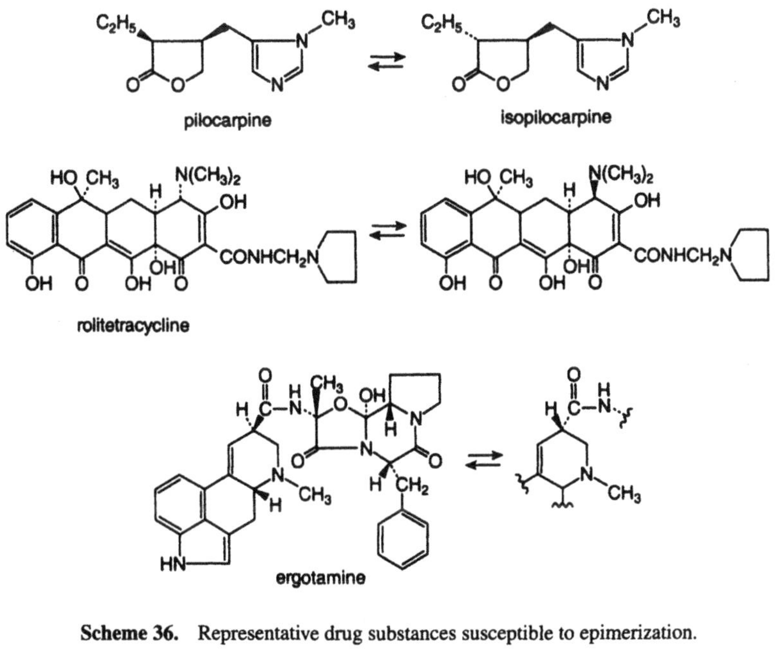

pilocarpine

isopilocarpine

rolitetracycline

ergotamine

Scheme 36. Representative drug substances susceptible to epimerization.

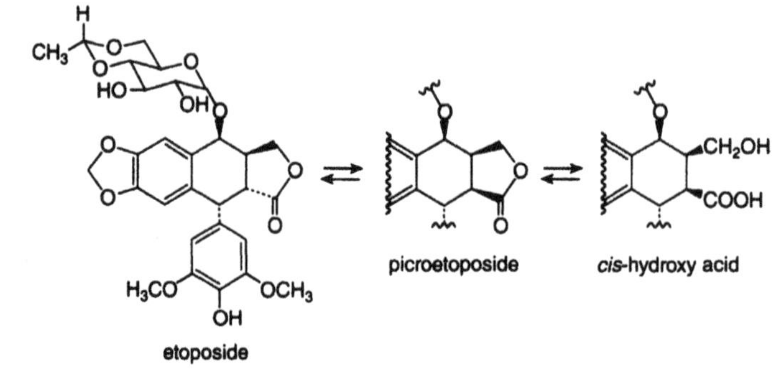

Scheme 37. Epimerization and hydrolysis of etoposide.

Racemization and epimerization, which are reversible conversions between optical isomers, have been reported for many drug substances. As shown in Scheme 36, pilocarpine undergoes epimerization by base catalysis,[36–38] whereas tetracyclines[151,152] such as rolitetracycline,[153,154] and ergotamine[155] exhibit epimerization by acid catalysis. Etoposide converts reversibly to picroetoposide, a *cis*-lactone, and then hydrolyzes to *cis*-hydroxy acid in the alkaline pH region, as shown in Scheme 37.[156,157] Epinephrine is oxidized (see Section 2.1.5) and undergoes racemization under strongly acidic conditions (Scheme 38).[158] Other drug substances susceptible to racemization include benzodiazepines, penicillins, and cephalosporins. Oxazepam undergoes racemization through a rapid equilibrium reaction in the neutral-to-alkaline pH region (Scheme 39).[159] Moxalactam exhibits epimerization of its side chain as well as hydrolysis of the β-lactam ring (Scheme 40).[160–162] Hetacillin exhibits epimerization of the lactam ring, hydrolysis of the side chain, and β-lactam ring cleavage (Scheme 41).[163] Similar racemization and hydrolysis have been reported for carbenicillin,[164,165] cefsulodin,[166] cefotaxime,[167] and dalvastatin.[39]

2.1.4. Decarboxylation and Elimination

Drug substances having a carboxylic acid group are sometimes susceptible to decarboxylation, as shown in Scheme 42. 4-Aminosalicylic acid is a good example.[168] Foscarnet also undergoes decarboxylation under strongly acidic conditions,[169] whereas etodolac is susceptible to decarboxylation by acid catalysis.[170]

Other elimination reactions have been reported for various drug substances, as shown in Scheme 43. Trimelamol eliminates its hydroxymethyl groups and forms formaldehyde.[171] Levothyroxine eliminates iodine.[172] ADD-17014, a derivative of triazoline, eliminates nitrogen and forms a derivative of aziridine.[173] Ditiocarb eliminates carbon disulfide.[174]

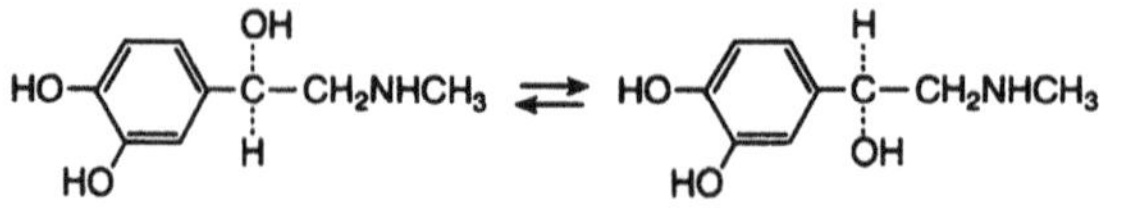

Scheme 38. Racemization of epinephrine.

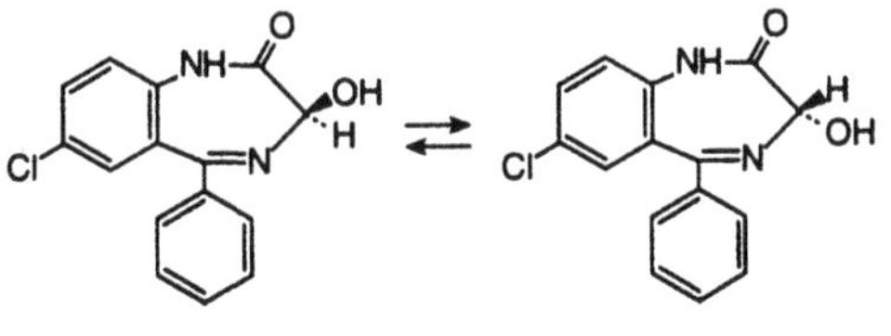

Scheme 39. Racemization of oxazepam.

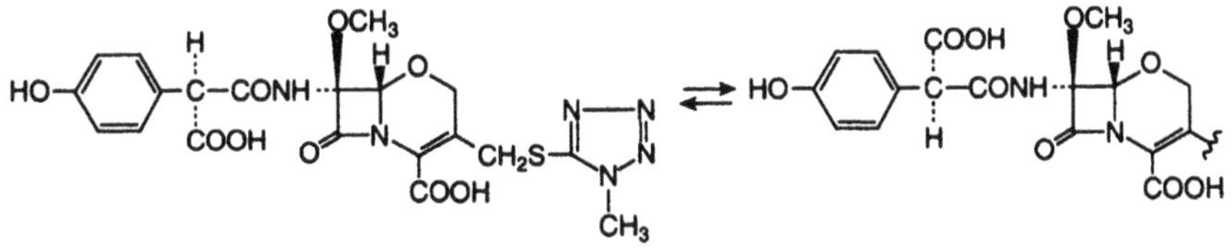

Scheme 40. Epimerization of moxalactam.

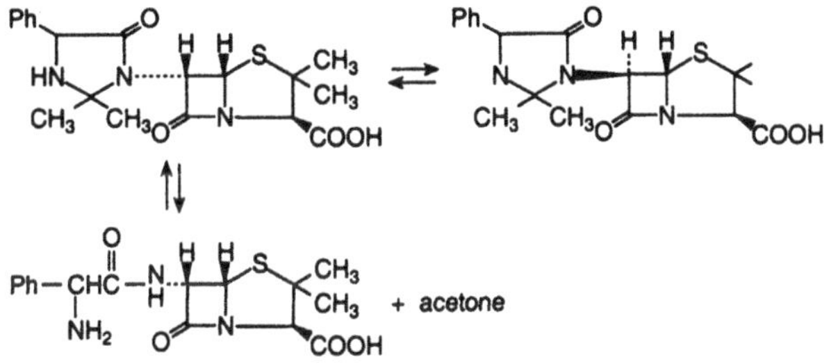

Scheme 41. Epimerization and hydrolysis of hetacillin.

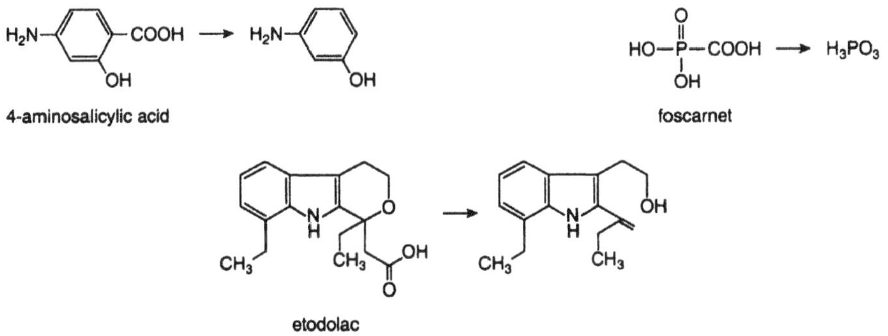

4-aminosalicylic acid

foscarnet

etodolac

Scheme 42. Representative drug substances that are susceptible to decarboxylation.

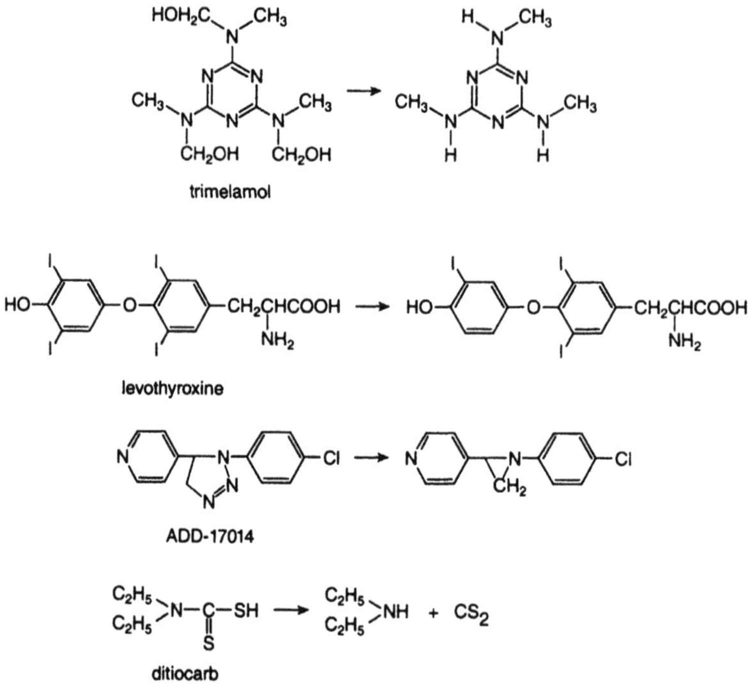

trimelamol

levothyroxine

ADD-17014

ditiocarb

Scheme 43. Other drug substances that are susceptible to elimination reactions.

2.1.5. Oxidation

Oxidation is a well-known chemical degradation pathway for pharmaceuticals. Oxygen, which participates in most oxidation reactions, is abundant in the environment to which pharmaceuticals are exposed, during either processing or long-term storage. Oxidation of ascorbic acid (Scheme 44) was reported as early as 1940,[175,176] and many factors affecting ascorbic acid oxidation have been discussed, including the role of metal ions.[177–179]

Oxidation mechanisms for drug substances depend on the chemical structure of the drug and the presence of reactive oxygen species or other oxidants. Catechols such as methyldopa[180] and epinephrine[181] are readily oxidized to quinones, as shown in Scheme 45. 5-Aminosalicylic acid undergoes oxidation and forms quinoneimine,[182] which is further degraded to polymeric compounds (Scheme 46).[183] Ethanolamines such as procaterol are oxidized to formyl compounds (Scheme 47),[184] whereas thiols such as 6-mercaptopurine,[185]

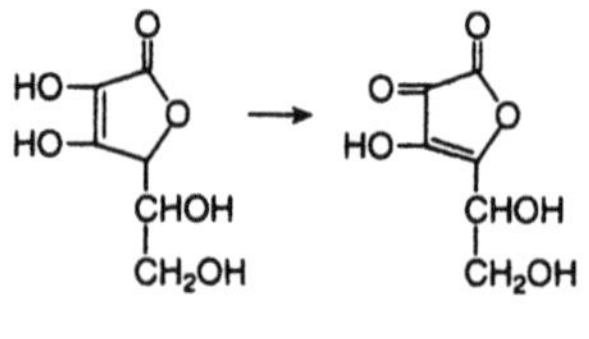

Scheme 44. Oxidation of ascorbic acid.

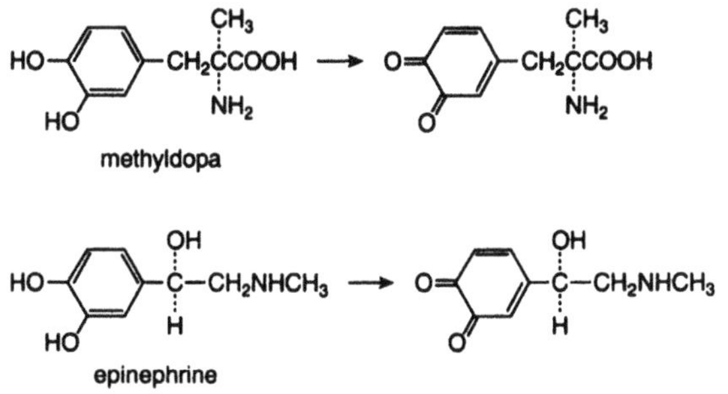

methyldopa

epinephrine

Scheme 45. Representative cathecol drug substances that are susceptible to oxidation.

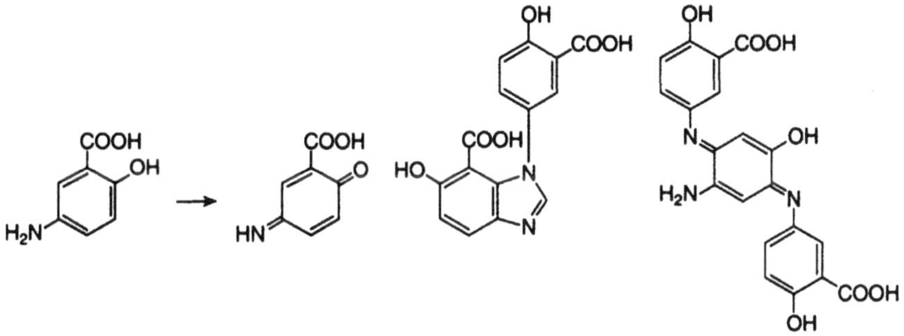

Scheme 46. Oxidation of 5-aminosalicylic acid.

captopril,[186] and NSC-629243 (a derivative of thiocarbamic acid)[187] are oxidized to disulfides (Scheme 48). Phenothiazines such as promethazine are oxidized via complex pathways and yield various products (Scheme 49).[188,189] As shown in Fig. 4, polyunsaturated molecules such as vitamin A,[190] as well as other polyenes such as ergocalciferol,[191,192] cholecalciferol,[192] fumagillin,[193] and filipin[194,195] are susceptible to oxidation. In additional, phenylbutazone,[196–199] sulpyrine,[200,201] morphine,[202] tetrazepam,[203] hydrocortisone,[204] and prednisolone[205] are oxidized to various products, as shown in Scheme 50. Spiradoline is susceptible to oxidative degradation, resulting in the formation of an imidazolidine ring in addition to hydrolysis of the amide bond (Scheme 51).[206] Sulfur atoms are becoming more common in new drug candidates and present a particular challenge owing to their propensity to oxidize to the corresponding sulfoxides and ultimately sulfones (Scheme 52).

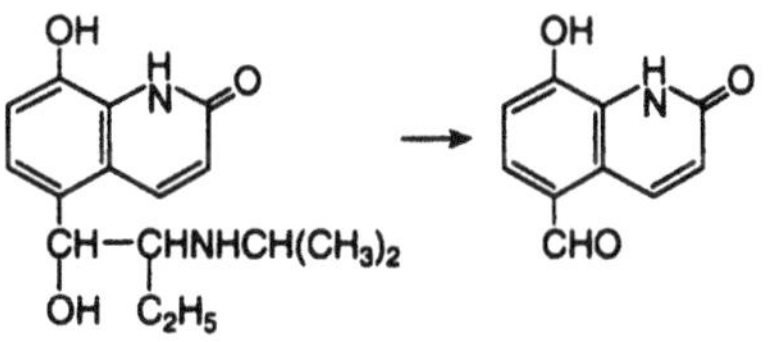

Scheme 47. Oxidation of procaterol.

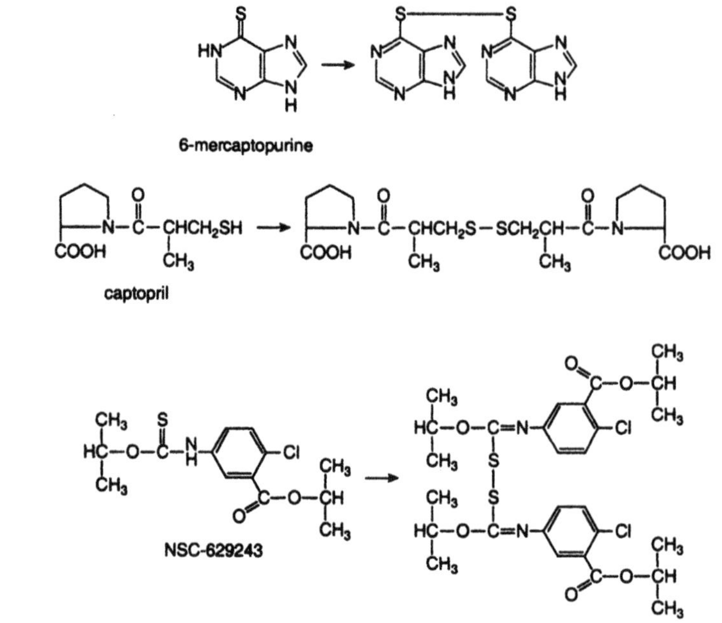

6-mercaptopurine

captopril

NSC-629243

Scheme 48. Representative thiol drug substances that are susceptible to oxidation.

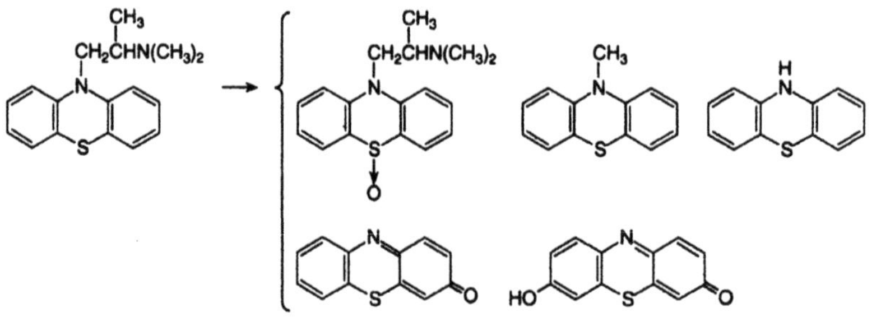

Scheme 49. Oxidation products of promethazine.

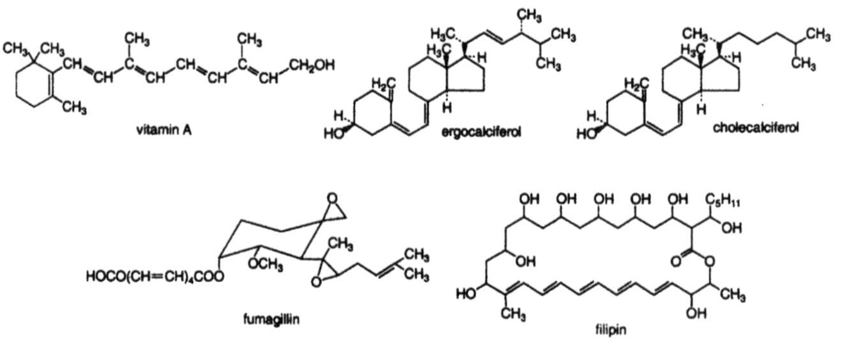

vitamin A

ergocalciferol

cholecalciferol

fumagillin

filipin

Figure 4. Representative polyene drug substances that are susceptible to oxidation.

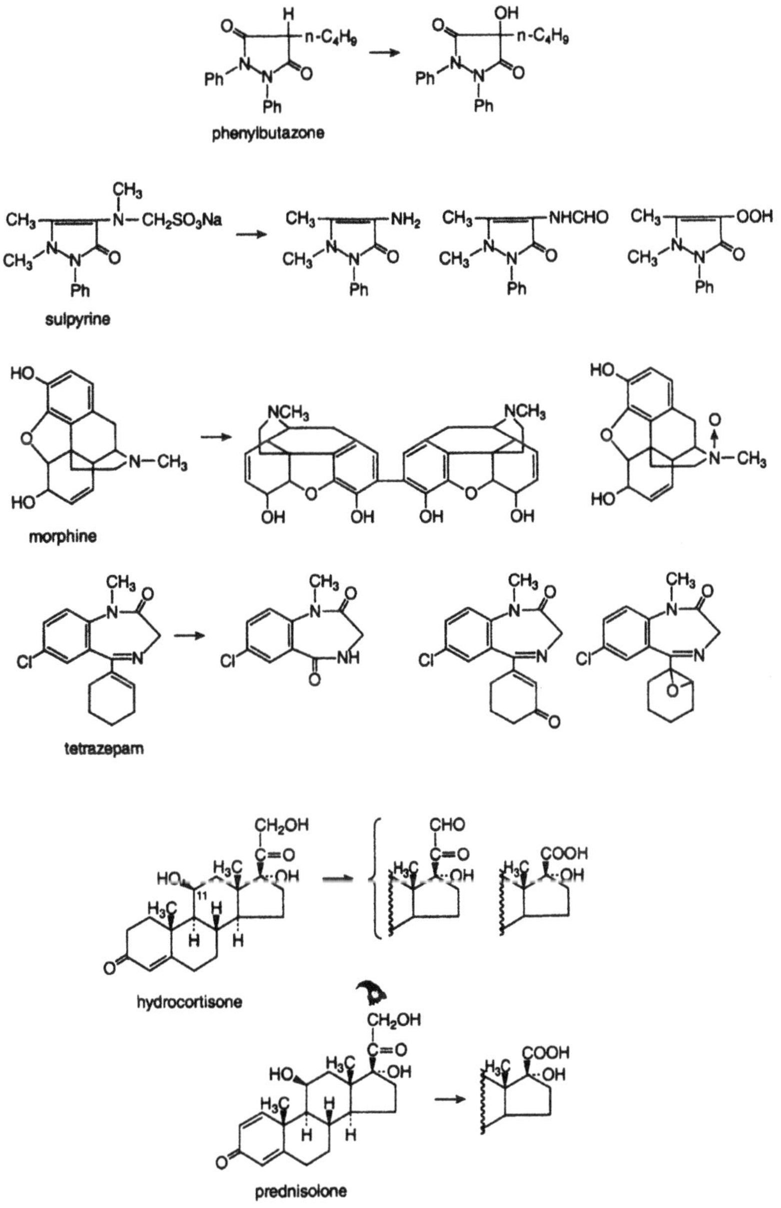

phenylbutazone

sulpyrine

morphine

tetrazepam

hydrocortisone

prednisolone

Scheme 50. Other drug substances that are susceptible to oxidation.

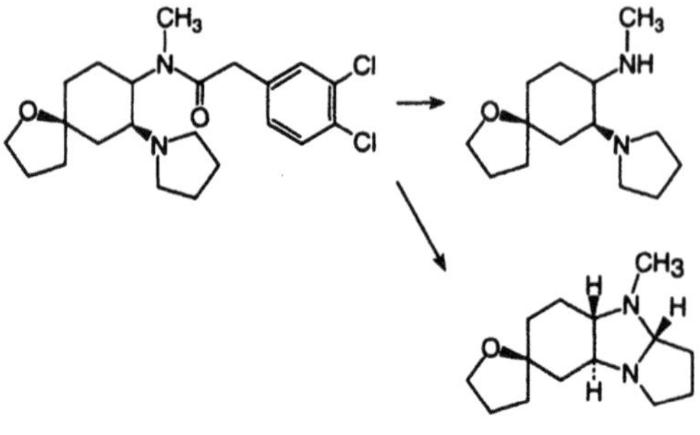

Scheme 51. Hydrolysis and oxidation of spiradoline.

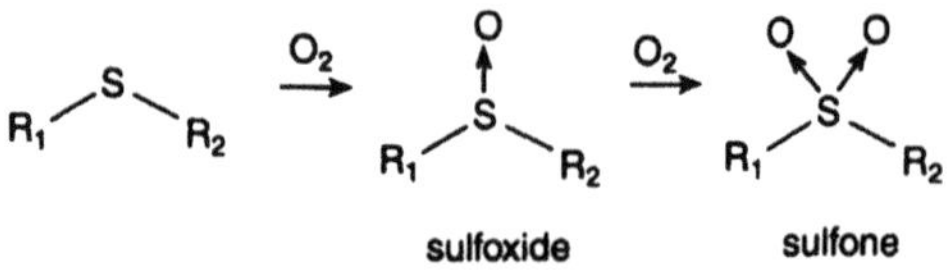

Scheme 52. Oxidation of dialkyl sulfides to sulfoxides and sulfones.

2.1.6. Photodegradation

Photodegradation has been reported for a large number of drug substances. The mechanisms for these reactions are generally very complex. As exemplified by chloroquine[207] and primaquine,[208] shown in Schemes 53 and 54, respectively, photodegradation

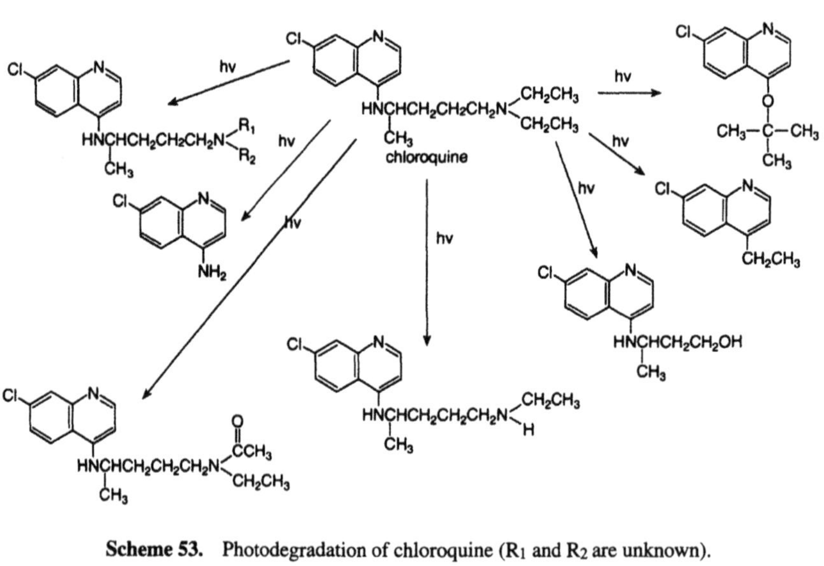

Scheme 53. Photodegradation of chloroquine (R_1 and R_2 are unknown).

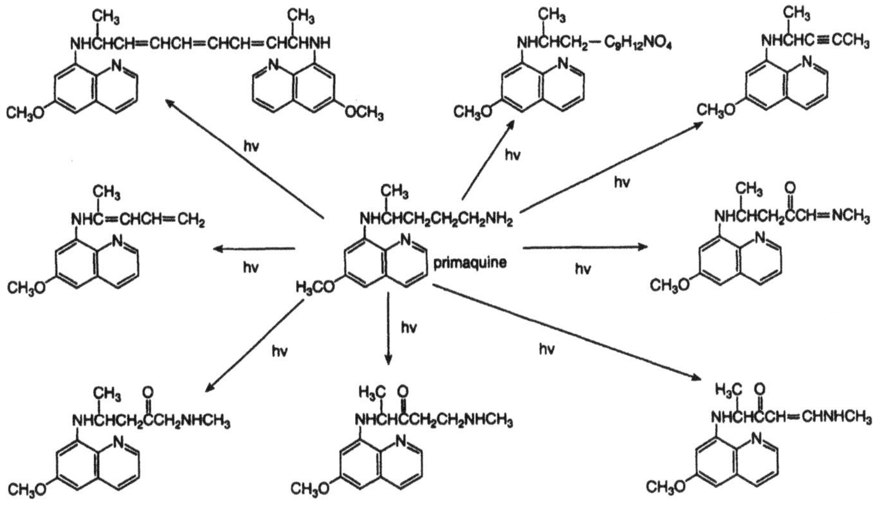

Scheme 54. Photodegradation of primaquine.

generally yields numerous products through complex pathways. Photodegradation is often accompanied by oxidation in the presence of oxygen. Thus, drug substances such as fumagillin,[209,210] phenothiazines,[211] and cholecalciferol,[192] whose oxidation was described in the previous section, are degraded to different products in the presence and absence of light.

Representative photodegradation routes for drug substances include dehydrogenation of nifedipine,[212–214] reserpine,[215] and nicardipine[216] (Scheme 55); dehydrogenation accompanied by transmutation of a nitro group in nimodipine[217] (Scheme 56); oxidation of a reactive methylene group to a carbonyl in 4-methoxy-2-(3-phenyl-2-propynyl)phenol (CO/1828),[218] tiaprofenic acid,[219] and KBT-3022 (a derivative of diphenylthiazole)[220,221] (Scheme 57); and rearrangement of chlordiazepoxide[222] (Scheme 58).

In addition, the following photoinduced degradation reactions have been reported: hydrolysis of mefloquine,[223] furosemide,[224] and LY277359 (a derivative of benzofuran carboxamide)[225] (Scheme 59); elimination of hydrogen halide from meclofenamic acid[226] (Scheme 60); oxidation of a hydroxyl group of 21-cortisol *tert*-butylacetate[227] and α-[(dibutylamino)methyl]-6,8-dichloro-2-(3′,4′-dichlorophenyl)-4-quinoline methanol[228] (Scheme 61); and rearrangement of benzydamine[229] (Scheme 62). Oxidation of menadione is enhanced by light (Scheme 63).[230]

2.1.7. Drug–Excipient and Drug–Drug Interactions

As will be seen in Chapter 4, drugs are rarely formulated as just the drug substance itself. Often, additives or excipients are present in the formulation. Quite often, reactions can occur between the drug and one or more additives. Similarly, two drugs might be formulated in the same product and react with each other.

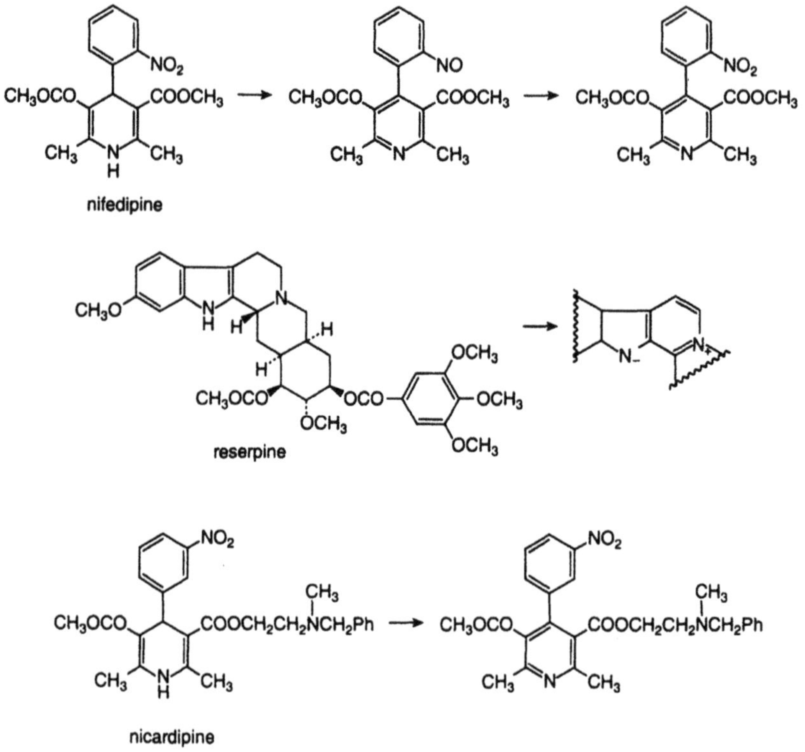

nifedipine

reserpine

nicardipine

Scheme 55. Photodegradation leading to dehydrogenation of nifedipine, reserpine, and nicardipine.

2.1.7.1. Reactions of Bisulfite, an Antioxidant

In the 1950s, it was reported that epinephrine, a catecholamine, undergoes displacement of its hydroxy group by bisulfite, as shown in Scheme 64.[231] Dexamethasone 21-phosphate, an α/β-unsaturated ketone, is known to undergo addition by bisulfite (Scheme 64).[232]

2.1.7.2. Reaction of Amines with Reducing Sugars

Reducing sugars readily react with primary amines, including those of amino acids, through the Maillard reaction. Drug substances with primary or secondary amine groups

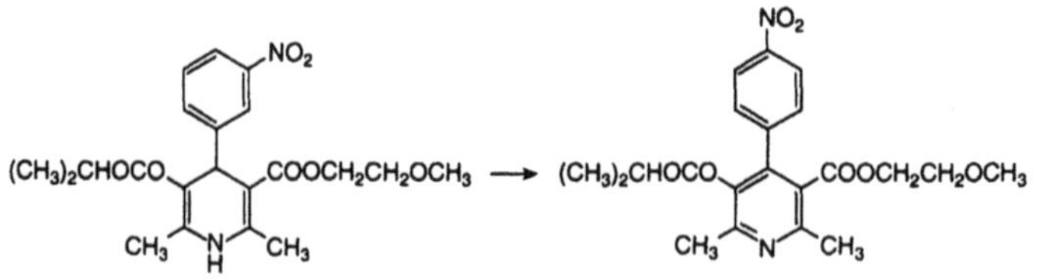

Scheme 56. Photodegradation of nimodipine.

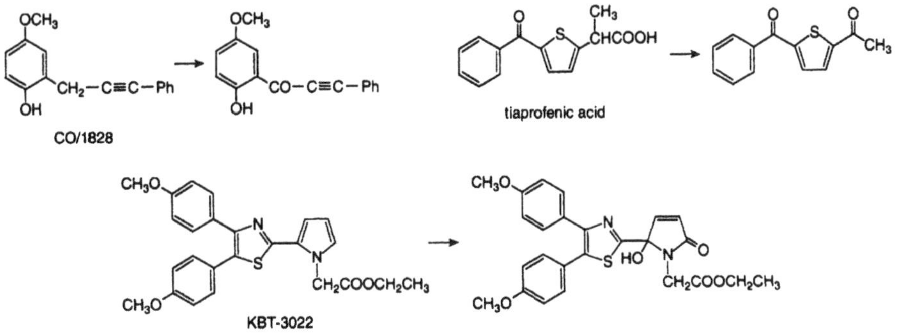

CO/1828

tiaprofenic acid

KBT-3022

Scheme 57. Some unusual photochemically induced reactions.

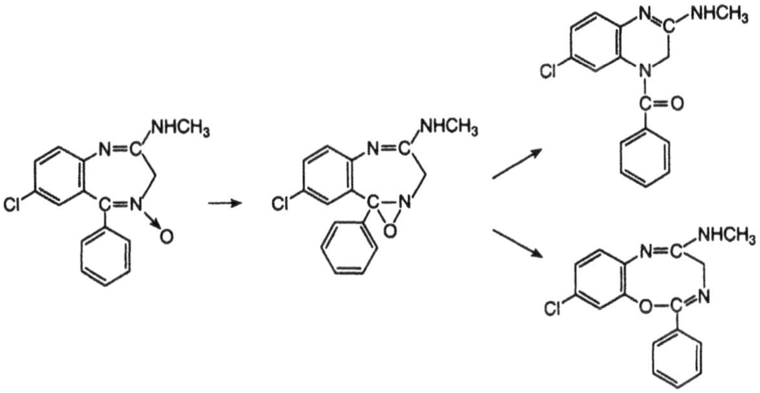

Scheme 58. Degradation of chlordiazepoxide.

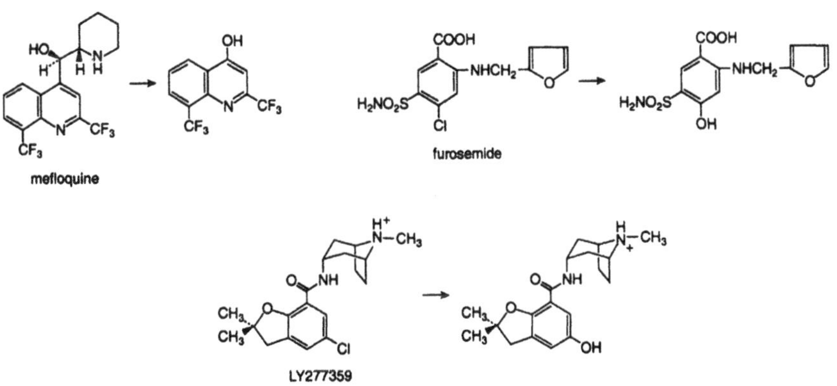

mefloquine

furosemide

LY277359

Scheme 59. Photochemically induced elimination/hydrolysis.

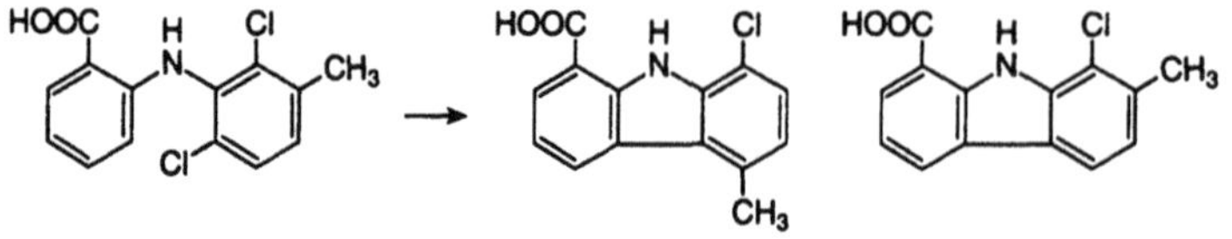

HOOC H Cl N CH3 Cl → HOOC H Cl N CH3 HOOC H Cl N CH3

Scheme 60. Photodegradation of meclofenamic acid.

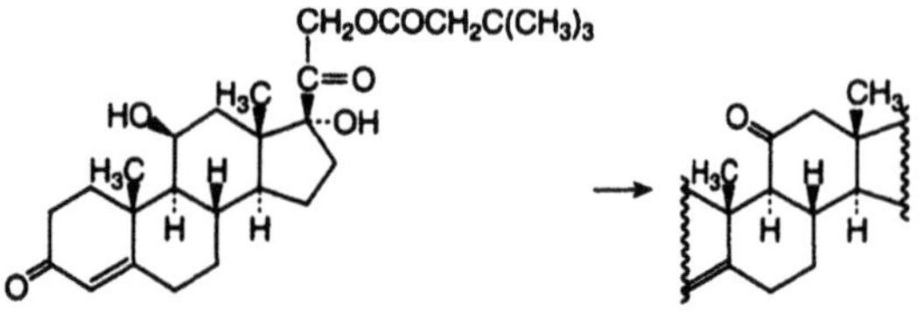

21-cortisol *tert*-butylacetate

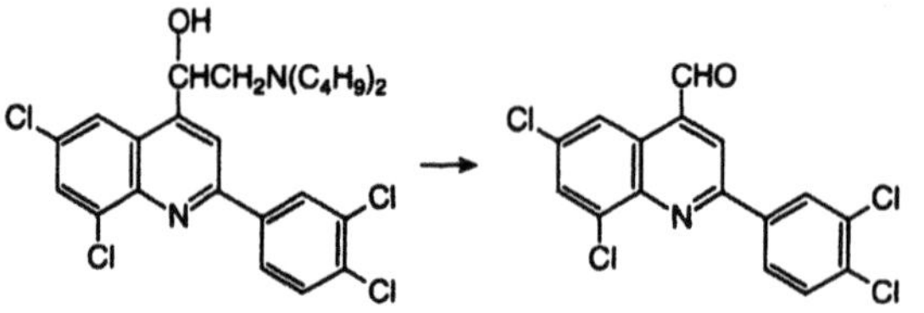

α-[(dibutylamino)methyl]-6,8-dichloro-2-(3',4'-dichlorophenyl)-4-quinoline methanol

Scheme 61. Photodegradation leading to oxidation of a hydroxy group.

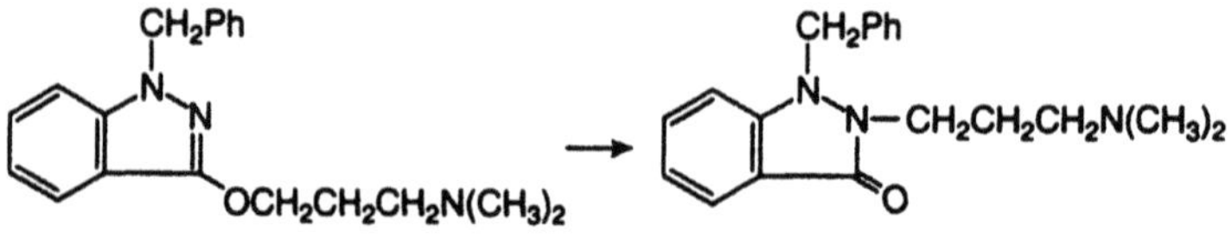

Scheme 62. Photodegradation of benzydamine.

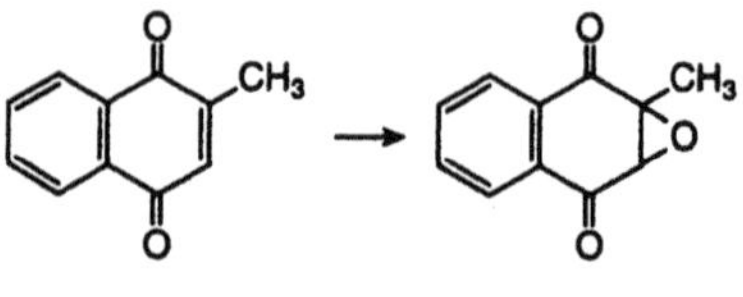

Scheme 63. Photooxidation of menadione.

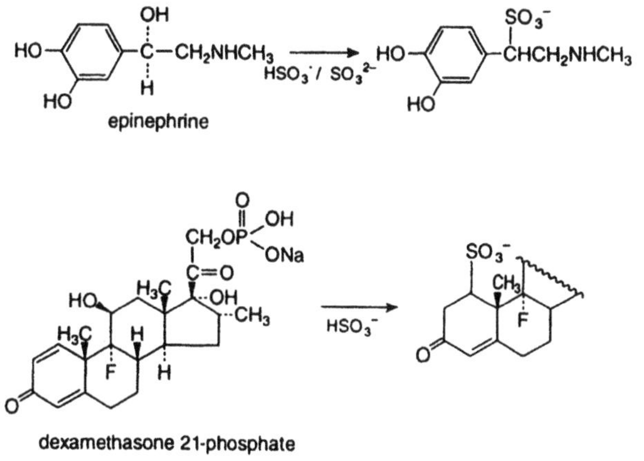

Scheme 64. Representative drug substances susceptible to substitution and addition reactions by bisulfite.

undergo this addition/rearrangement reaction, also called the "browning" reaction because of the resulting discoloration. Examples are the reaction of amphetamine,[233] isoniazid,[234] dextroamphetamine sulfate,[235,236] and norphenylephrine[237] with sugars such as lactose and the degradation products of sugars, such as 5-(hydroxymethyl)furfural. Sulpyrine forms an addition product with glucose.[238]

2.1.7.3. Transesterification Reactions

In the presence of drug substances with hydroxy groups, aspirin undergoes a reversible transacylation reaction to form salicylic acid, while acetylating the drug substance. For example, codeine[239] and sulfadiazine[240] are acetylated by aspirin, as shown in Scheme 65. Similar acetylation reactions with aspirin have been reported for acetaminophen[241] and the excipient polyethylene glycol.[242,243] Another example of transesterification is the reaction of benzocaine with polyvinyl acetate phthalate (Scheme 66).[244]

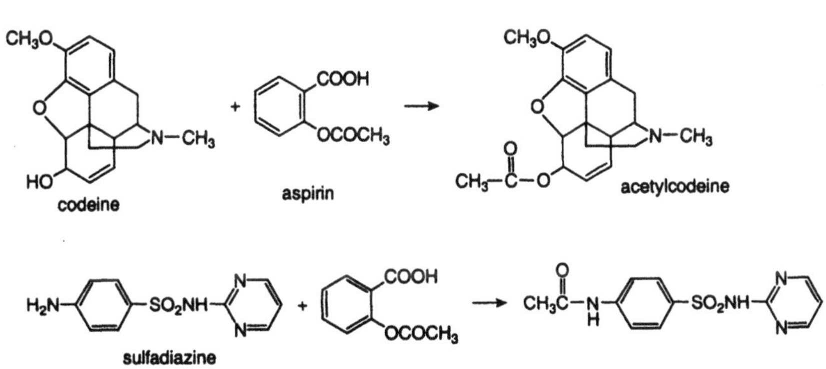

Scheme 65. Representative drug substances susceptible to acetylation by aspirin.

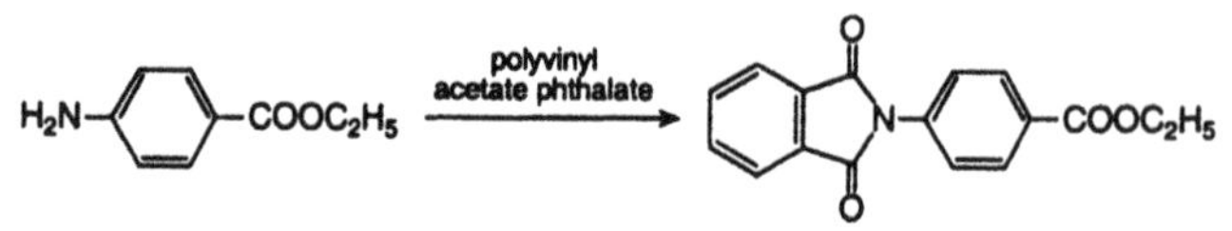

Scheme 66. Interaction of benzocaine and polyvinyl acetate phthalate.

2.2. Factors Affecting Chemical Stability

In the previous section, the chemical degradation pathways for many drug substances were classified according to various pathways and mechanisms. In this section, factors that affect the rates of chemical degradation are elaborated.

Factors determining the chemical stability of drug substances include intrinsic factors such as the molecular structure of the drug itself and environmental factors, such as temperature, pH, buffer species, ionic strength, light, oxygen, moisture, additives, and excipients. In the case of solid-state degradation, the solid-state properties of the drug such as melting point, crystallinity, and hygroscopicity are very important. In addition, mechanical forces such as pressure and grinding applied to drug substances may affect their chemical as well as physical stability. By applying well-established kinetic concepts, it is possible not only to summarize, numerically, the role that each variable might play in altering the kinetics of degradation but also to provide valuable insight into the mechanism(s) of degradation.

2.2.1. Basic Kinetic Principles

The simplest concept of chemical and physical reaction is the case of a drug D reacting to form a product P. This process is described by the following scheme:

$$\mathrm{D} \rightleftarrows \mathrm{P}$$

The extent to which D rearranges to P will depend on the free-energy differences between D and P. If P is of much lower free energy than D, then the reaction is better defined by

$$\mathrm{D} \rightarrow \mathrm{P}$$

Most drugs degrade by reactions that involve a so-called bimolecular reaction in which drug D collides with a reactant A to produce one or more products. This is illustrated in its simplest form by the following equation:

$$\mathrm{D} + \mathrm{A} \rightarrow \mathrm{P}$$

P will be formed if D and A collide with sufficient energy (and an appropriate orientation) to result in a molecular rearrangement to form P. In this simple case, the rate of loss of D, $-d[\mathrm{D}]/dt$, is said to be proportional to the activity (or, more simply, the concentration) of both D and A, as indicated by Eq. (2.1).

$$-\frac{d[\mathrm{D}]}{dt} \propto [\mathrm{D}][\mathrm{A}] \tag{2.1}$$

When the proportionality constant is included, the following equation is obtained:

$$-\frac{d[D]}{dt} = k[D][A] \tag{2.2}$$

where k is the proportionality constant, usually referred to as the rate constant. If k is large, the reaction is fast; if k is small, the reaction is slow. In this case, the reaction rate ($-d[D]/dt$) is said to be first-order in D and first-order in A.

If A is present in excess of D, that is, [A] >> [D], then even though some of A is consumed during the reaction, effectively only D is lost. Under these circumstances,

$$-\frac{d[D]}{dt} = (k[A])[D] = k_{obs}[D] \tag{2.3}$$

where k_{obs} is said to be the observed rate constant, a pseudo-first-order constant. In most studies of the stability of pharmaceuticals, especially in aqueous solution, the kinetics can often be simplified to pseudo-first-order conditions.

More generally, the degradation rate of a drug, D, depends on the drug concentration, [D], and the concentrations (more accurately, the activities) of chemical species participating in the reaction [A], [B], ... The rate at which [D] decreases, $-d[D]/dt$, is described by summing the terms for all the reactions that D might undergo:

$$\frac{-d[D]}{dt} = \{k_{0,AB\cdots}[A]^l[B]^m \cdots + k_{0,EF\cdots}[E]^o[F]^p \cdots + \}[D]^n \tag{2.4}$$

where $k_{0,AB\cdots}$ and $k_{0,EF\cdots}$etc., are rate constants of reactions in which species AB··· and species EF···, respectively, react with D. The terms n, l, m, o, and p are reaction orders for each species, and the sum of these is considered to be the overall reaction order. For example, assuming that an ester is hydrolyzed by both hydronium ion catalysis and water attack, the rate can described by Eq. (2.5). If additional species participate in the hydrolysis, their terms would be added to this expression.

$$\frac{-d[D]}{dt} = k_{H^+}[H^+][D] + k_{H_2O}[H_2O][D] \tag{2.5}$$

Assuming that degradation of drug D is a result of the direct reaction with a species A, the rate is proportional to the concentration of "activated complex," often referred to as the "transition state," $X^‡$, formed between D and A:

$$D + A \leftrightarrows X^‡ \xrightarrow{k'} P$$

$$-\frac{d[D]}{dt} = \frac{d[P]}{dt} = k'[X^‡]$$

$$= k''[D][A]f_D f_A / f_{X^‡}$$

$$= k[D][A] \tag{2.6}$$

In Eq. (2.6), $[X^‡]$ is the concentration of $X^‡$ and f_D, f_A, and $f_{X^‡}$ are the activity coefficients of D, A, and $X^‡$, respectively. The rate constant, k, can be described by the following equation, based on transition-state theory:

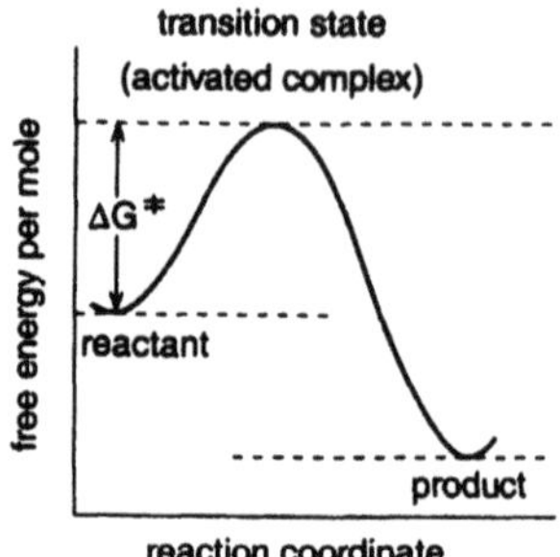

Figure 5. Free-energy diagram showing reactants proceeding to products through a transition state or activated complex.

$$k = \frac{\kappa T}{h} \exp\left(\frac{-\Delta G^{\ddagger}}{RT}\right) = \frac{\kappa T}{h} \exp\left(\frac{\Delta S^{\ddagger}}{R}\right) \exp\left(\frac{-\Delta H^{\ddagger}}{RT}\right) \tag{2.7}$$

where $\Delta G^{\ddagger}$, $\Delta S^{\ddagger}$, and $\Delta H^{\ddagger}$ are the free energy, entropy, and enthalpy of activation, respectively. $\Delta G^{\ddagger}$ is the difference in free energy between the reactant state and the activated complex, as shown in Fig. 5. The term κ is the Boltzmann constant, h is the Planck constant, and T is the temperature in degrees kelvin.

In descriptive terms, Eq. (2.7) essentially suggests that for chemical reaction to occur, molecules must first collide. The term $\kappa T/h$ represents a so-called universal collision number. Not only must the molecules collide, but they must collide with sufficient overall free energy for rearrangement of the molecules to occur. The term $e^{-\Delta G^{\ddagger}/RT}$ represents the fraction of molecules colliding with sufficient energy to overcome the free-energy barrier to reaction. This free-energy barrier is made up of both an enthalpic term ($\Delta H^{\ddagger}$) and an entropic term ($\Delta S^{\ddagger}$).

Other kinetic theories, such as the collision theory, were proposed earlier. In the collision theory, proposed by Lewis, the reaction rate, v, was given by

$$v = Z \exp\left(\frac{-E}{RT}\right) \tag{2.8}$$

where Z is the collision frequency, R is the gas constant, and E is the activation energy. Thereafter, Eyring developed the theory of absolute reaction rates by introducing the concept of the formation and breakdown of an activated complex. This so-called transition-state model, defined by the reaction illustrated in Fig. 5, is represented by

$$k = \frac{\kappa T}{h} \frac{Q^{\ddagger}}{Q_A Q_B} \exp\left(\frac{-E_0}{RT}\right) \tag{2.9}$$

where Q_A, Q_B, and $Q^{\ddagger}$ are the partition functions of A, B, and the activated complex, respectively, and E_0 is the energy required for the formation of 1 mol of the activated complex at 0°K. Replacing partition functions in Eq. (2.9) by thermodynamic functions yields Eq. (2.7).

Because chemical reaction rates depend not only on the concentrations (more accurately, the activities) of participating species but also on temperature and the free energy of reaction, $\Delta G^{\ddagger}$, as represented by Eqs. (2.4), (2.6), and (2.7), the reaction rate is strongly influenced by those factors affecting $\Delta G^{\ddagger}$.

With respect to degradation of drug substances in solutions, any observed rate or rate constant can be calculated according to Eqs. (2.4), (2.6), and (2.7), and factors affecting the degradation can be related to the terms in these equations. In other words, the stability of drug substances does not change unless the key parameters appearing in these equations change owing to changes in reaction conditions/media, etc. For a more comprehensive review of chemical kinetics in solution, the book by Connors[245] is highly recommended.

Equations (2.4), (2.6), and (2.7) are also applicable to the degradation of drug substances in the solid state. However, the factors affecting the reaction rates become more complex because reactions often proceed in heterogeneous physical states. For example, apparent reaction rates depend on solubility and dissolution rates of drug substances when degradation proceeds in water layers adsorbed on the surface of solid drugs. Therefore, these and other additional factors need to be considered.

In the following sections, factors affecting drug degradation in solutions are discussed, especially as they relate to basic kinetic concepts. Later sections discuss drug degradation in more complex heterogeneous states such as solid-state decomposition.

2.2.2. The Role of Molecular Structure

It has been noted earlier that the molecular structure of a drug substance determines its degradation mechanisms/pathways and that substituents around the reaction center can strongly influence its reactivity. For example, a drug substance having an electron-withdrawing functional group close to an ester bond will probably exhibit a higher propensity to nucleophilic attack by hydroxide ion than will a similar ester without that functional group. The increased rate of degradation can be explained by assuming that the electron-withdrawing group makes the carbonyl carbon of the ester group more susceptible to attack, as well as stabilizing the formed activated complex, thus decreasing $\Delta G^{\ddagger}$, [see Eq. (2.7)]. This increased propensity would manifest itself in the form of an increased rate. Similarly, the decreased hydrolysis rate of esters with bulky substituents near the ester group can be explained in terms of an increased $\Delta G^{\ddagger}$.

Steric factors can be significant for many chemical reactions. Consider the lactonization of isopilocarpic acid and pilocarpic acid to isopilocarpine and pilocarpine, respectively (Scheme 67). Isopilocarpic acid exhibits a rate of cyclization to its corresponding lactone that is 17.5 times larger than that of pilocarpic acid. Presumably, this is due to a smaller $\Delta G^{\ddagger}$ for the isopilocarpic acid reaction, resulting from differences in the three-dimensional structure of the lactone ring.[246] CI-988, whose structure is shown in Fig. 6, is a derivative of 4-amino-4-oxobutanoic acid that undergoes acid-catalyzed amide hydrolysis to the product in which R is $-NH_2$. This reaction appears to proceed via an intramolecular carboxyl-assisted mechanism. Replacement of the succinic acid amide by the phthalic acid amide (compound I in Fig. 6) accelerates the reaction by a factor of 100 (Fig. 6) because of the ease of formation of the ring-closed intermediate (the corresponding anhydride). The values of $\Delta H^{\ddagger}$ and $\Delta S^{\ddagger}$ estimated from the Eyring plots for the reaction indicate that $\Delta S^{\ddagger}$ for compound I is more favorable than that for CI-988.[247,248]

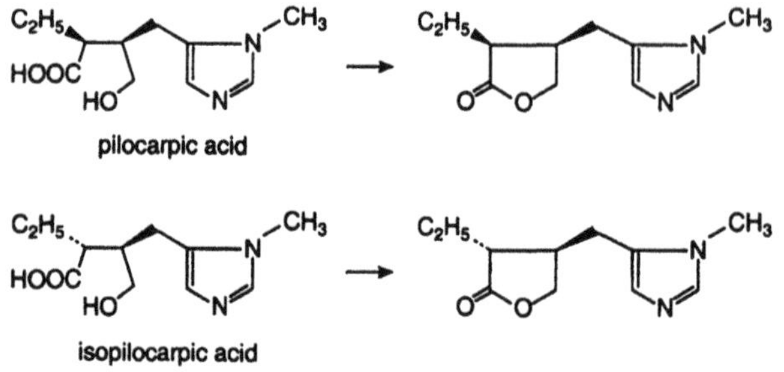

pilocarpic acid

isopilocarpic acid

Scheme 67. Cyclization of pilocarpic acid and isopilocarpic acid to pilocarpine and isopilocarpine, respectively.

2.2.3. Rate Equations and Kinetic Models

Drug substances undergo chemical degradation by various pathways and mechanisms, depending on their chemical structures. The rate of chemical degradation is determined by various factors contributing to the rate equations. Drug substances can be stabilized by inhibiting the degradation through control of these factors, as will be described in Section 2.3. Here, methods for describing the chemical degradation rate of drug substances on the basis of their kinetics are presented. Included are methods for kinetically analyzing an observed degradation curve under specific experimental conditions, obtaining rate constants, and predicting degradation rates under alternative conditions on the basis of this information. This section deals only with prediction of the chemical degradation rate of drug substances themselves. The prediction of the stability of dosage forms (for the purpose of estimating the shelf lives of drug products), in which physical degradation may also play a role, will be described in Chapter 4.

Obtaining reliable drug degradation data requires the development and validation of a stability-indicating assay. Once a stability study is initiated, one attempts to use a set of conditions that allows one to obtain a summary parameter, such as a rate constant, by kinetic analysis of a degradation versus time curve under these specific and controlled conditions. A kinetic model is selected to describe the degradation curve, and a rate constant is calculated by fitting the observed degradation curve to a suitable rate equation according to the assumed model. This section describes the selection of the kinetic model(s) and the calculation of a rate constant.

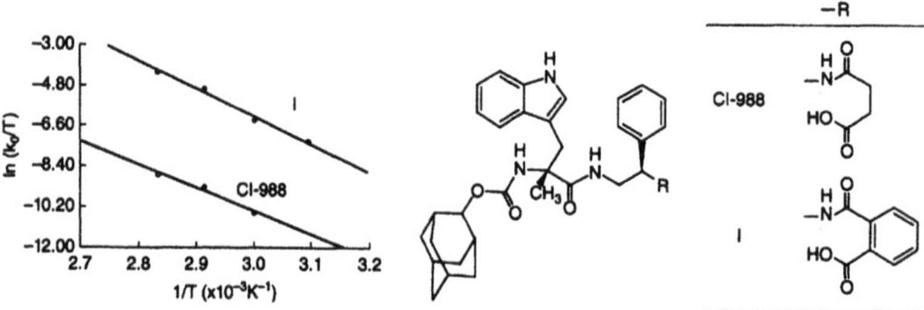

Figure 6. Eyring plots for the hydrolysis of CI-988 and analogue (compound I). (Reproduced from Ref. 248 with permission.)

2.2.3.1. Kinetic Models to Describe Drug Degradation in Solution

The generalized rate expression for drug degradation is represented by the rate equation that was given earlier (Eq. 2.4). When a drug substance, D, degrades via a certain mechanism in which reactants A, B, . . . participate, the degradation rate generally depends on the concentrations of the various reactants A, B, . . . and D according to Eq. (2.10), assuming that all the reactants are involved directly or indirectly in the rate-controlling step.

$$\frac{-d[\mathrm{D}]}{dt} = k_0[\mathrm{D}]^n[\mathrm{A}]^l[\mathrm{B}]^m \cdots \tag{2.10}$$

When the concentrations of A, B, . . . are maintained constant, that is, when the change in their concentrations during the reaction is negligible owing to their being present at much higher concentrations than drug D, or when these species are components that are maintained constant through the use of buffers, such as hydronium ion, the degradation rate is often described by

$$\frac{-d[\mathrm{D}]}{dt} = k[\mathrm{D}]^n \tag{2.11}$$

When n equals 0, 1, or 2, the reaction is said to be a pseudo-zero-, pseudo-first-, or pseudo-second-order reaction (pseudo-nth-order for higher order reactions), respectively. If the concentration of an additional reactant other than drug D is not constant during the reaction, the reaction order becomes $n + 1$.

Kinetic models generally used for drug stability prediction usually follow pseudo-zero-, pseudo-first-, or pseudo-second-order kinetics. Drug degradation higher than second order is rarely seen. Even complex degradation pathways involving multiple consecutive or parallel reactions can be represented by the combinations of zero-, first-, and second-order reactions. General kinetic models describing drug degradation are elaborated below.

2.2.3.1.a. Simple Pseudo-First-Order Reaction and Zero-Order Reaction

$$\mathrm{D} \xrightarrow{k} \mathrm{P}$$

The differential rate equation for a pseudo-first-order reaction is

$$\frac{-d[\mathrm{D}]}{dt} = k[\mathrm{D}] \tag{2.12}$$

The integrated form of this equation is

$$[\mathrm{D}] = [\mathrm{D}]_0 \exp(-kt) \tag{2.13}$$

where $[\mathrm{D}]_0$ is the initial concentration of the drug. From these equations, the degradation rate is seen to be proportional to drug concentration. Most drug degradation kinetics in solutions conform to apparent or pseudo-first-order kinetics, and the data are summarized by recording the apparent first-order rate constants, k.

The rate equations for pseudo-zero-order kinetics are

$$\frac{-d[\mathrm{D}]}{dt} = k \tag{2.14}$$

$$[\mathrm{D}] = [\mathrm{D}]_0 - kt \tag{2.15}$$

In this case, the drug degradation rate is independent of drug concentration. A specific example of pseudo-zero-order kinetics can be seen with drug degradation in suspensions. If a drug degrades in the solution phase of a suspension according to pseudo-first-order kinetics but is stable in the solid phase of the suspension, the degradation rate is proportional to the drug concentration in solution. Because the drug concentration in solution is given by the saturated solubility [S] and is maintained constant while drug in excess of its solubility is present, the amount of total drug remaining, M, decreases according to a pseudo-zero-order equation:

$$\frac{-dM}{dt} = k[\mathrm{S}] \tag{2.16}$$

Degradation of aspirin in suspension has been reported to follow zero-order kinetics (Fig. 7).[249]

2.2.3.1.b. *Simple Pseudo-Second-Order Reaction*

$$\mathrm{D} + \mathrm{A} \xrightarrow{k} \mathrm{P} \quad \text{or} \quad 2\mathrm{D} \xrightarrow{k} \mathrm{P}$$

The rate and concentration changes when drug D reacts with a reactant A are given by

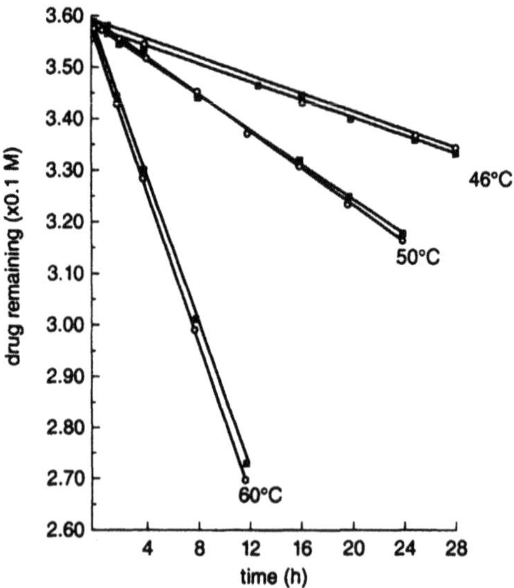

Figure 7. Time course of degradation of aspirin in suspension (pH 3.0), showing apparent zero-order behavior and a dependency on temperature but no dependency on particle size. Particle size: ■, 60 mesh; ○, 100 mesh. (Reproduced from Ref. 249 with permission of the American Pharmaceutical Association.)

$$\frac{-d[\mathrm{D}]}{dt} = k[\mathrm{D}][\mathrm{A}] \tag{2.17}$$

$$\ln\frac{[\mathrm{D}]}{[\mathrm{A}]} = ([\mathrm{D}]_0 - [\mathrm{A}]_0)kt + \ln\frac{[\mathrm{D}]_0}{[\mathrm{A}]_0} \tag{2.18}$$

where $[\mathrm{A}]_0$ is the initial concentration of A. Equations (2.19) and (2.20), result when drug D undergoes a bimolecular reaction with itself.

$$\frac{-d[\mathrm{D}]}{dt} = k[\mathrm{D}]^2 \tag{2.19}$$

$$[\mathrm{D}] = \frac{[\mathrm{D}]_0}{1 + [\mathrm{D}]_0 kt} \tag{2.20}$$

2.2.3.1.c. *Pseudo-First-Order Reversible Reaction*

$$\mathrm{D} \underset{k_2}{\overset{k_1}{\rightleftarrows}} \mathrm{P}$$

When drug D converts to product P according to reversible pseudo-first-order reactions, the rate is described by Eq. (2.21), which is integrated to Eq. (2.22) (Eq. 2.23) when the initial concentration of P, $[\mathrm{P}]_0$, equals zero.

$$\frac{-d[\mathrm{D}]}{dt} = k_1[\mathrm{D}] - k_2[\mathrm{P}] \tag{2.21}$$

$$[\mathrm{D}] = \frac{[\mathrm{D}]_0}{k_1 + k_2}\{k_1 \exp[-(k_1 + k_2)t] + k_2\} \tag{2.22}$$

$$\frac{[\mathrm{D}] - [\mathrm{D}]_e}{[\mathrm{D}]_0 - [\mathrm{D}]_e} = \exp[-(k_1 + k_2)t] \tag{2.23}$$

Hydrolysis of triazolam (see Scheme 19) and racemization of oxazepam (see Scheme 39) conform to this kinetic model, as shown in Figs. 8[250] and 9[159], respectively.

2.2.3.1.d. *Pseudo-Second- and Pseudo-First-Order Reversible Reactions*

$$\mathrm{D} + \mathrm{A} \underset{k_2}{\overset{k_1}{\rightleftarrows}} \mathrm{P}$$

When drug D reacts reversibly with A to form P according to a pseudo-second-order reaction, the rate expression for the loss of D is given by

$$\frac{-d[\mathrm{D}]}{dt} = k_1[\mathrm{D}][\mathrm{A}] - k_2[\mathrm{P}] \tag{2.24}$$

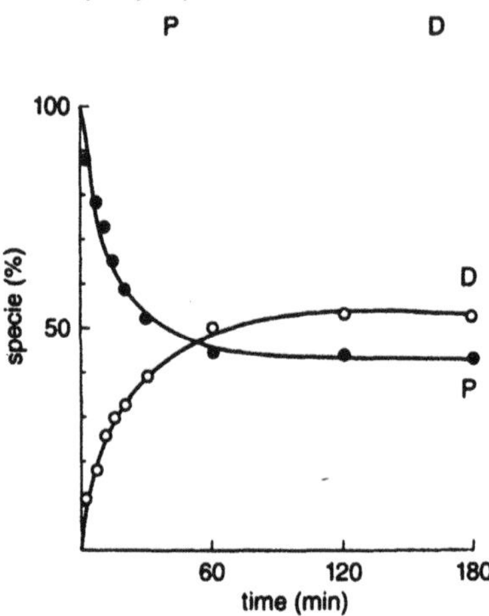

Figure 8. Time course of formation of triazolam from its hydrolysis product (pH 2.30, 37°C). (Reproduced from Ref. 250 with permission.)

When $[D]_0 = [A]_0$ and $[P]_0 = 0$ at $t = 0$, Eq. (2.24) is integrated to give

$$\ln \frac{([D]_0 - [D]_e)\{([D]_0 - [D]_e)[D] + [D]_e [D]_0\}}{[D]_0^2([D] - [D]_e)} = \frac{[D]_e(2[D]_0 - [D]_e)}{[D]_0 - [D]_e} kt \tag{2.25}$$

where $[D]_e$ is the concentration of D at equilibrium. Equation (2.25) can be simplified to

$$\ln \frac{[P]_e([D]_0^2 - [P]_e[P])}{[D]_0^2([P]_e - [P])} = \frac{[D]_0^2 - [P]_e^2}{[P]_e} kt \tag{2.26}$$

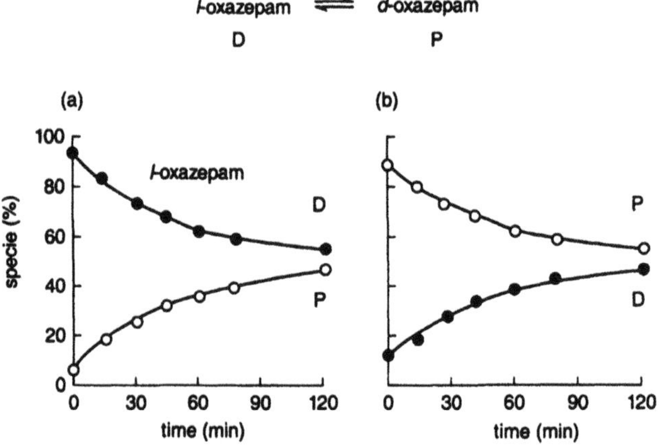

Figure 9. Time course of racemization of oxazepam (pH 12, 0°C). (a) Racemization of *l*-oxazepam; (b) racemization of *d*-oxazepam. (Reproduced from Ref. 159 with permission.)

where $[P]_e$ is the concentration of product formed at equilibrium. A similar equation is derived for the case of $[D]_0 = /[A]_0$ and is used to describe the interaction of isoniazid and reducing sugars, as shown in Fig. 10.[251]

2.2.3.1.e. Pseudo-First- and Pseudo-Second-Order Reversible Reactions

$$D \underset{k_2}{\overset{k_1}{\rightleftarrows}} P_1 + P_2$$

Equation (2.27) represents the rate of reversible conversion of drug D to products P_1 and P_2. When $[P_1]_0 = [P_2]_0 = 0$ at $t = 0$, Eq. (2.27) can be integrated to give Eq. (2.28).

$$\frac{-d[D]}{dt} = k_1[D] - k_2[P_1][P_2] \tag{2.27}$$

$$\ln \frac{[D]_0([D]_0 - [D]_e) + [D]_e([D]_0 - [D])}{[D]_0([D] - [D]_e)} = \frac{[D]_0 + [D]_e}{[D]_0 - [D]_e} kt \tag{2.28}$$

The loss of hydrochlorothiazide follows this model,[252] although a complicated mechanism including multiple reaction steps has been proposed for its degradation.[253]

2.2.3.1.f. Pseudo-First-Order Consecutive Reactions

$$D \overset{k_1}{\rightarrow} P_1 \overset{k_2}{\rightarrow} P_2$$

Equations (2.29) and (2.30) represent the case when drug D converts to P_1, which is subsequently converted to P_2 according to consecutive pseudo-first-order reactions.

$$\frac{-d[D]}{dt} = k_1[D] \qquad \frac{d[P_1]}{dt} = k_1[D] - k_2[P_1] \tag{2.29}$$

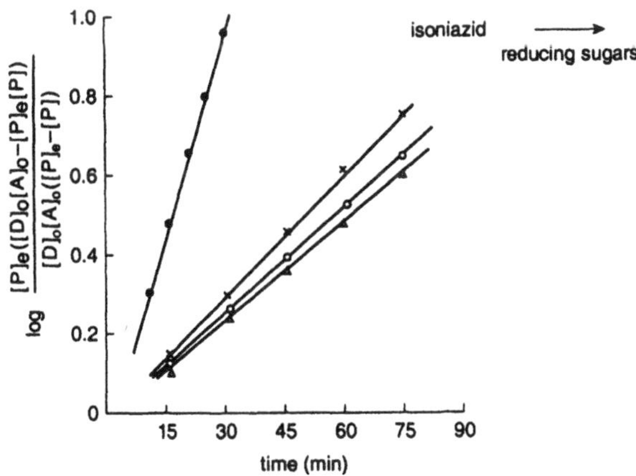

Figure 10. Time course of reaction of isoniazid with various reducing sugars under second-order reaction conditions (pH 1.8, 37°C). ●, Galactose; ×, lactose; ○, glucose; △, maltose. (Reproduced from Ref. 251 with permission.)

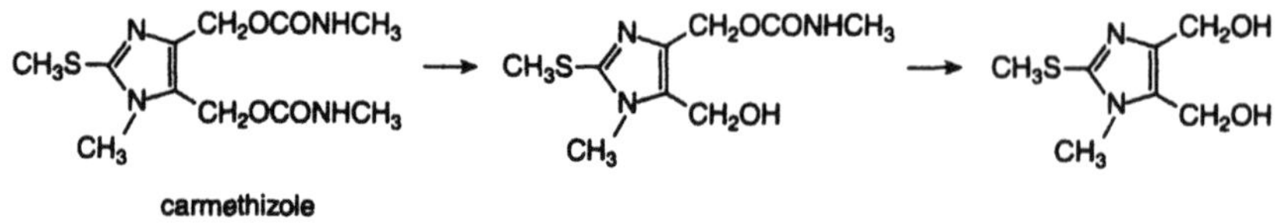

CH_3S–(imidazole)($CH_2OCONHCH_3$)($CH_2OCONHCH_3$)(N–CH_3) → CH_3S–(imidazole)($CH_2OCONHCH_3$)(CH_2OH)(N–CH_3) → CH_3S–(imidazole)(CH_2OH)(CH_2OH)(N–CH_3)

carmethizole

Scheme 68. Consecutive loss of carbamate groups in carmethizole (NSC-602668). The reactivity of one carbamate group is greater than that of the other. (Reproduced from Ref. 46 with permission.)

$$[\mathrm{D}] = [\mathrm{D}]_0 \exp(-k_1 t)$$

$$[\mathrm{P}_1] = \frac{k_1}{k_2 - k_1} [\mathrm{D}]_0 [\exp(-k_1 t) - \exp(-k_2 t)] \qquad (2.30)$$

A good example of consecutive reactions is the degradation of carmethizole (NSC-602668), an experimental cytotoxic agent (Scheme 68).[44] The hydrolysis of hydrocortisone hemisuccinate (Fig. 11)[30] and alkaline epimerization followed by hydrolysis of etoposide (Fig. 12)[156,157] fit this mathematical model even though their degradation pathways are more complex.

2.2.3.1.g. Pseudo-First-Order Reversible and Consecutive Reactions

$$\mathrm{D} \underset{k_2}{\overset{k_1}{\rightleftarrows}} \mathrm{P}_1 \underset{k_4}{\overset{k_3}{\rightleftarrows}} \mathrm{P}_2$$

When drug D is reversibly converted to P_1, which is reversibly converted to P_2 by apparent first-order kinetics, Equations (2.31) and (2.32) represent the rate and integrated expressions, respectively.

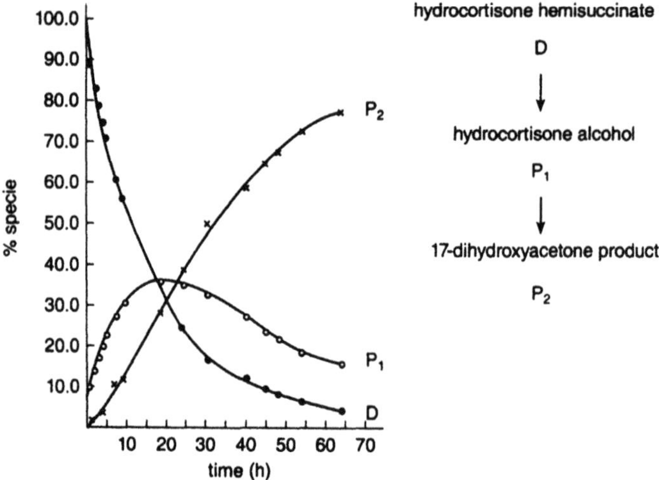

Figure 11. Time course of hydrolysis of hydrocortisone hemisuccinate (pH 6.9, 70°C). (Reproduced from Ref. 30 with permission.)

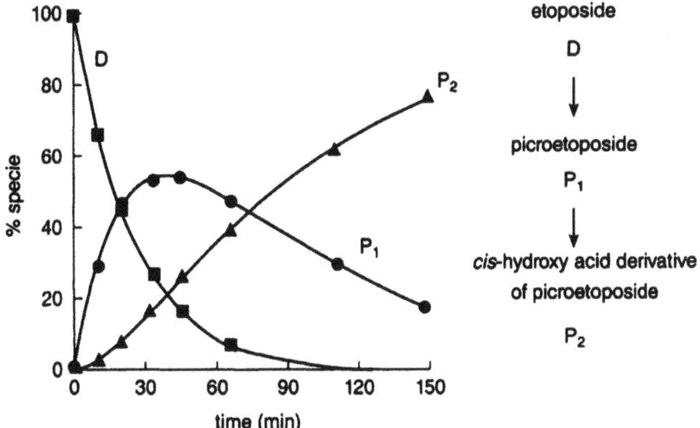

Figure 12. Time courses of epimerization and hydrolysis of etoposide (pH 10, 37°C). (Reproduced from Ref. 157 with permission.)

$$\frac{-d[\mathrm{D}]}{dt} = k_1[\mathrm{D}] - k_2[\mathrm{P}_1]$$

$$\frac{d[\mathrm{P}_1]}{dt} = k_1[\mathrm{D}] - (k_2 + k_3)[\mathrm{P}_1] + k_4[\mathrm{P}_2]$$

$$\frac{d[\mathrm{P}_2]}{dt} = k_3[\mathrm{P}_1] - k_4[\mathrm{P}_2] \tag{2.31}$$

$$[\mathrm{D}] = [\mathrm{D}]_0\left[\frac{k_1k_3 + k_1k_4 + k_1\gamma_1}{(\gamma_2 - \gamma_1)\gamma_1}\exp(\gamma_1 t) + \frac{k_1k_3 + k_1k_4 + k_1\gamma_2}{(\gamma_1 - \gamma_2)\gamma_2}\exp(\gamma_2 t) + \frac{k_2k_4}{\gamma_1\gamma_2}\right]$$

$$[\mathrm{P}_2] = [\mathrm{D}]_0\left[\frac{k_1k_3}{(\gamma_1 - \gamma_2)\gamma_1}\exp(\gamma_1 t) + \frac{k_1k_3}{(\gamma_2 - \gamma_1)\gamma_2}\exp(\gamma_2 t) + \frac{k_1k_3}{\gamma_1\gamma_2}\right]$$

$$[\mathrm{P}_1] = [\mathrm{D}]_0 - [\mathrm{D}] - [\mathrm{P}_2]$$

$$= [\mathrm{D}]_0 k_1\left[\frac{k_4}{\gamma_1\gamma_2} + \frac{k_4 + \gamma_1}{(\gamma_1 - \gamma_2)\gamma_1}\exp(\gamma_1 t) + \frac{k_4 + \gamma_2}{(\gamma_2 - \gamma_1)\gamma_2}\exp(\gamma_2 t)\right] \tag{2.32}$$

where

$$\gamma_1 = -\frac{1}{2}(k_1 + k_2 + k_3 + k_4) + \frac{1}{2}\sqrt{(k_1 + k_2 + k_3 + k_4)^2 - 4(k_1k_3 + k_2k_4 + k_1k_4)}$$

$$\gamma_2 = -\frac{1}{2}(k_1 + k_2 + k_3 + k_4) - \frac{1}{2}\sqrt{(k_1 + k_2 + k_3 + k_4)^2 - 4(k_1k_3 + k_2k_4 + k_1k_4)}$$

2.2.3.1.h. *Pseudo-First-Order Parallel Reactions*

$$P_2 \xleftarrow{k_2} D \xrightarrow{k_1} P_1$$

Equations (2.33) and (2.34) pertain when drug D converts to P_1 and P_2 via independent first-order pathways.

$$\frac{-d[D]}{dt} = (k_1 + k_2)[D]$$

$$\frac{d[P_1]}{dt} = k_1[D] \tag{2.33}$$

$$[D] = [D]_0 \exp[-(k_1 + k_2)t]$$

$$[P_1] = \frac{k_1}{k_1 + k_2}[D]_0\{1 - \exp[-(k_1 + k_2)t]\} \tag{2.34}$$

Many drugs degrade to more than one product so this scheme is quite common. Often, more than two products are formed.

2.2.3.1.i. *Pseudo-First-Order Reversible and Parallel Reactions*

$$P_2 \underset{k_4}{\overset{k_3}{\rightleftarrows}} D \underset{k_2}{\overset{k_1}{\rightleftarrows}} P_1$$

When both P_1 and P_2 are capable of being converted back to D, Eqs. (2.35) and (2.36) adequately describe the kinetics.

$$\frac{-d[D]}{dt} = k_1[D] - k_2[P_1] + k_3[D] - k_4[P_2]$$

$$\frac{d[P_1]}{dt} = k_1[D] - k_2[P_1]$$

$$\frac{d[P_2]}{dt} = k_3[D] - k_4[P_2] \tag{2.35}$$

$$[P_1] = [D]_0 k_1\left[\frac{\gamma_1 + k_4}{(\gamma_1 + \gamma_2)\gamma_1}\exp(\gamma_1 t) + \frac{\gamma_2 + k_4}{(\gamma_2 - \gamma_1)\gamma_2}\exp(\gamma_2 t) + \frac{k_4}{\gamma_1\gamma_2}\right]$$

$$[P_2] = [D]_0 k_3\left[\frac{\gamma_1 + k_2}{(\gamma_1 - \gamma_2)\gamma_1}\exp(\gamma_1 t) + \frac{\gamma_2 + k_2}{(\gamma_2 - \gamma_1)\gamma_2}\exp(\gamma_2 t) + \frac{k_2}{\gamma_1\gamma_2}\right]$$

$$[D] = 1 - [P_1] - [P_2] \tag{2.36}$$

where

$$\gamma_1 = -\frac{1}{2}(k_1 + k_2 + k_3 + k_4) + \frac{1}{2}\sqrt{(k_1 + k_2 + k_3 + k_4)^2 - 4(k_4k_1 + k_3k_2 + k_4k_2)}$$

$$\gamma_2 = -\frac{1}{2}(k_1 + k_2 + k_3 + k_4) - \frac{1}{2}\sqrt{(k_1 + k_2 + k_3 + k_4)^2 - 4(k_4k_1 + k_3k_2 + k_4k_2)}$$

Degradation of pilocarpine in the neutral pH region appears to conform to this model (Fig. 13).[254] Technically, however, isopilocarpine can also degrade to isopilocarpic acid; therefore, a more complete scheme for the degradation of polocarpine is

$$\begin{array}{c} P_2 \rightleftarrows P_3 \\ \uparrow\downarrow \\ D \rightleftarrows P_1 \end{array}$$

However, within the limits of the experimental conditions, this more complex scheme reduces to that defined by Eq. (2.35).

2.2.3.1.j. *Pseudo-First- and Psuedo-Second-Order Reversible and Parallel Reactions*

$$P_3 \xleftarrow{k_3} D \underset{k_2}{\overset{k_1}{\rightleftarrows}} P_1 + P_2$$

A reaction pathway similar to, but more complicated than, that considered above is one in which the conversion of D to P_1 and P_2 is reversible. Epimerization and hydrolysis of hetacillin apparently conform to this model, as shown in Fig. 14,[163] even though epihetacillin can also dissociate and the isomerization of hetacillin to epihetacillin should be considered reversible.

2.2.3.1.k. *Pseudo-First-Order Parallel and Consecutive Reactions*

$$P_2 \xleftarrow{k_2} D \xrightarrow{k_1} P_1 \xrightarrow{k_3} P_3$$

When P_1 subsequently converts to P_3 according to a first-order reaction in a pseudo-first-order parallel reaction, Eqs. (2.37) and (2.38) represent the rate and integrated expressions, respectively.

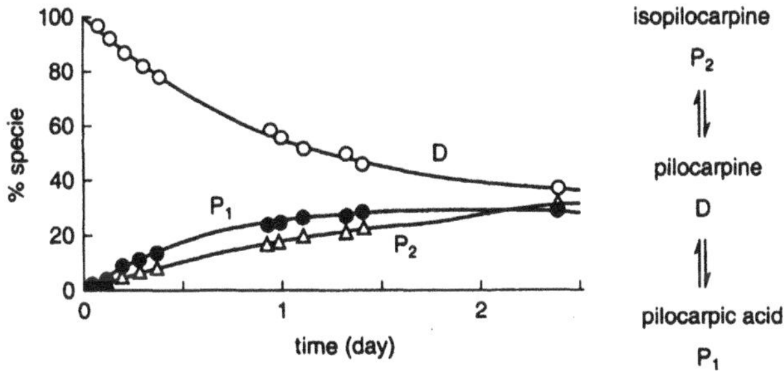

Figure 13. Time course of degradation of pilocarpine (pH 6.0, 80°C). (Reproduced from Ref. 254 with permission.)

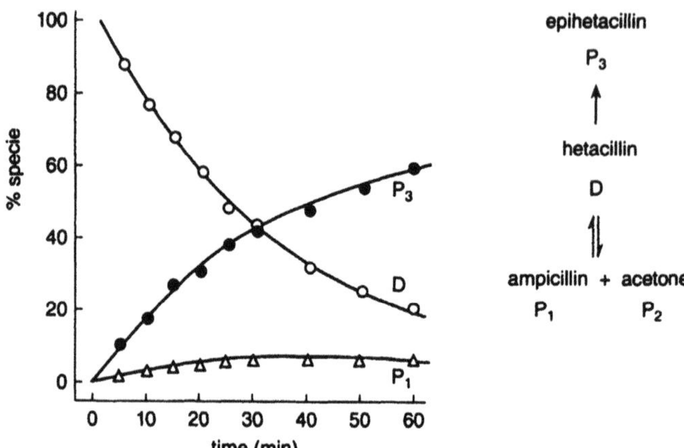

Figure 14. Time courses of epimerization and hydrolysis of hetacillin (pD 10.6, 35°C). (Reproduced from Ref. 163 with permission.)

$$\frac{-d[\mathrm{D}]}{dt} = k_1[\mathrm{D}] + k_2[\mathrm{D}]$$

$$\frac{d[\mathrm{P}_1]}{dt} = k_1[\mathrm{D}] - k_3[\mathrm{P}_1] \tag{2.37}$$

$$[\mathrm{D}] = [\mathrm{D}]_0 \exp[-(k_1 + k_2)t]$$

$$[\mathrm{P}_1] = \frac{k_1}{k_1 + k_2 - k_3}[\mathrm{D}]_0\{\exp(-k_3 t) - \exp[-(k_1 + k_2)t]\} \tag{2.38}$$

The alkaline degradation of cefixime[255] and pilocarpine[38] appears to conform to this model, even though the actual pathways/mechanisms may be more complex than indicated by this scheme.

2.2.3.1.1. Pseudo-First-Order Reversible, Parallel and Consecutive Reactions

$$\mathrm{P}_2 \xleftarrow{k_3} \mathrm{D} \underset{k_2}{\overset{k_1}{\rightleftarrows}} \mathrm{P}_1 \xrightarrow{k_4} \mathrm{P}_3$$

When P_1 is in equilibrium with D, Eqs. (2.39) and (2.40) can describe this model.

$$\frac{-d[\mathrm{D}]}{dt} = (k_1 + k_3)[\mathrm{D}] - k_2[\mathrm{P}_1]$$

$$\frac{d[\mathrm{P}_1]}{dt} = k_1[\mathrm{D}] - (k_2 + k_4)[\mathrm{P}_1]$$

$$\frac{d[\mathrm{P}_2]}{dt} = k_3[\mathrm{D}] \tag{2.39}$$

$$[\mathrm{D}] = [\mathrm{D}]_0 \left[\frac{\gamma_1 + k_2 + k_4}{\gamma_1 - \gamma_2} \exp(\gamma_1 t) - \frac{\gamma_2 + k_2 + k_4}{\gamma_1 - \gamma_2} \exp(\gamma_2 t) \right]$$

$$[\mathrm{P}_1] = \frac{k_1}{\gamma_1 - \gamma_2} [\mathrm{D}]_0 [\exp(\gamma_1 t) - \exp(\gamma_2 t)]$$

$$[\mathrm{P}_2] = k_3 [\mathrm{D}]_0 \left\{ - \frac{\gamma_1 + k_2 + k_4}{\gamma_1(\gamma_1 - \gamma_2)} [1 - \exp(\gamma_1 t)] + \frac{\gamma_2 + k_2 + k_4}{\gamma_2(\gamma_1 - \gamma_2)} [1 - \exp(\gamma_2 t)] \right\} \quad (2.40)$$

where

$$\gamma_1 = -\frac{1}{2}(k_1 + k_2 + k_3 + k_4) + \frac{1}{2}\sqrt{(k_1 + k_2 + k_3 + k_4)^2 + 4(k_1 k_4 + k_2 k_3 + k_3 k_4)}$$

$$\gamma_2 = -\frac{1}{2}(k_1 + k_2 + k_3 + k_4) - \frac{1}{2}\sqrt{(k_1 + k_2 + k_3 + k_4)^2 - 4(k_1 k_4 + k_2 k_3 + k_3 k_4)}$$

Isomerization and hydrolysis of chlorphenesin carbamate under strongly alkaline pH conditions[256] and epimerization and hydrolysis of carumonam (Fig. 15)[257] and moxalactam[160,161] all appear to conform to this model. Hydrolysis of chlorothiazide, under alkaline pH conditions,[258] is explained by this model when k_3 is set to zero (Fig. 16).

A more recent example is the aqueous degradation of the neuraminidase inhibitor prodrug GS-4104, which, like chlorphenesin carbamate, undergoes an acyl migration and hydrolysis.[259] In the case of GS-4104, products P_2 and P_3 are also capable of interconverting as in the scheme below.

$$\begin{array}{ccc} \mathrm{D} & \rightleftarrows & \mathrm{P}_1 \\ \downarrow & & \downarrow \\ \mathrm{P}_2 & \rightleftarrows & \mathrm{P}_3 \end{array}$$

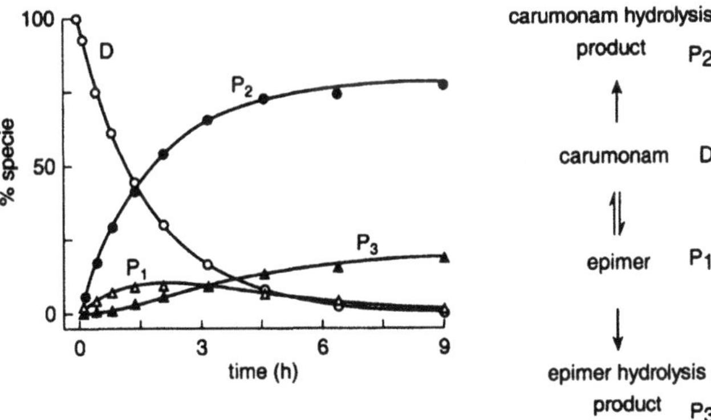

Figure 15. Time courses of epimerization and hydrolysis of carumonam (pH 9.50, 35°C). (Reproduced from Ref. 257 with permission.)

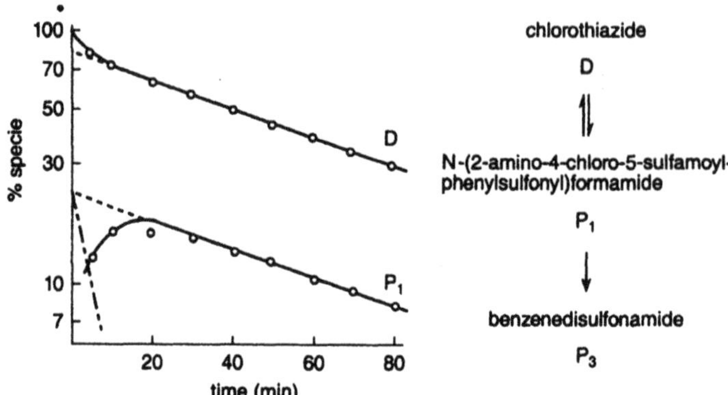

Figure 16. Time course of hydrolysis of chlorothiazide ($[OH^-] = 0.1N$, 80°C). (Reproduced from Ref. 258 with permission.)

2.2.3.1.m. Pseudo-First- and Pseudo-Second-Order Parallel Reactions

$$D \xrightarrow{k_1} P \qquad 2D \xrightarrow{k_2} P$$

When a reaction pathway involves both pseudo-first and pseudo-second-order pathways, Eqs. (2.41) and (2.42) adequately describe the kinetics.

$$\frac{-d[D]}{dt} = k_1[D] + 2k_2[D]^2 \tag{2.41}$$

$$\ln\left\{\frac{[D]_0(k_1 + 2k_2[D])}{[D](k_1 + 2k_2[D]_0)}\right\} = k_1 t \tag{2.42}$$

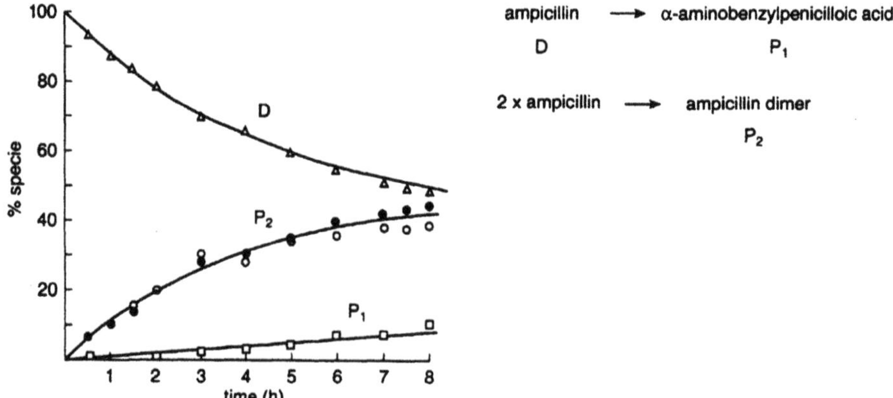

Figure 17. Time course of degradation of ampicillin (pH 8.50, 35°C). P_2 was determined by two assay methods (○, ●). (Reproduced from Ref. 260 with permission.)

Under neutral-to-alkaline pH conditions, the degradation of ampicillin (Fig. 17),[260] amoxicillin,[261,262] cefaclor,[263] and cefatrizine[263] can be reasonably described by this model.

2.2.3.1.n. Equilibrium, Pseudo-First-Order Parallel Reactions

$$P_1 \xleftarrow{k_1} D + A \overset{K}{\rightleftarrows} DA \xrightarrow{k_2} P_2$$

This case obtains when a drug, D, forms a complex (DA) with A, which is defined by the equilibrium constant, K, and both D and DA are capable of undergoing independent pseudo-first-order reactions. When the concentration of A is significantly higher than that of D, the kinetics can be described by Eqs. (2.43) and (2.44).[264]

$$\frac{-d([\mathrm{D}]+[\mathrm{DA}])}{dt} = k_1[\mathrm{D}] + k_2[\mathrm{DA}] = k_1[\mathrm{D}] + k_2K[\mathrm{D}][\mathrm{A}] \tag{2.43}$$

$$[\mathrm{D}] = \frac{[\mathrm{D}]_0}{1+K[\mathrm{A}]}\exp(-k_{\mathrm{obs}}t) \tag{2.44}$$

where

$$k_{\mathrm{obs}} = \frac{k_1 + k_2K[\mathrm{A}]}{1+K[\mathrm{A}]}$$

or

$$\frac{1}{k_1 - k_{\mathrm{obs}}} = \frac{1}{(k_1-k_2)K[\mathrm{A}]} + \frac{1}{k_1-k_2}$$

The degradation of carbenicillin in the presence of human serum albumin[165] conforms to this model, as does the degradation of drugs in the presence of cyclodextrins (see Section 2.3.3 for a more complete discussion).

2.2.3.1.o. Product Catalysis

$$\mathrm{D} \underset{\mathrm{P}}{\overset{k}{\rightarrow}} \mathrm{P}$$

Equation (2.45) describes the degradation rate of drug D catalyzed by its product P, which is continuously changing as the reaction proceeds.

$$\frac{-d[\mathrm{D}]}{dt} = k[\mathrm{D}]([\mathrm{D}]_0 - [\mathrm{D}] + [\mathrm{P}]_0) \tag{2.45}$$

When $[\mathrm{D}]_0 >> [\mathrm{P}]_0$, Eqs. (2.46) and (2.47) can be used to describe the kinetics.

$$\frac{-d[\mathrm{D}]}{dt} = k[\mathrm{D}][\mathrm{P}] \tag{2.46}$$

$$\ln\frac{[\mathrm{D}]}{[\mathrm{P}]} = [\mathrm{D}]_0kt + \ln\frac{[\mathrm{D}]_0}{[\mathrm{P}]_0} \tag{2.47}$$

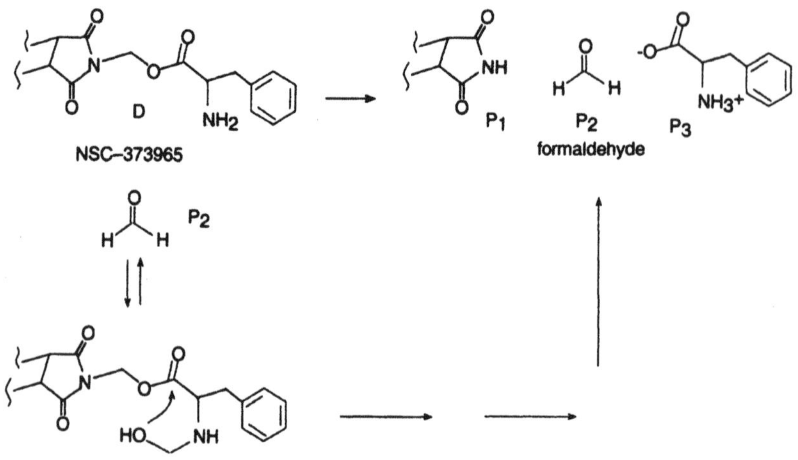

Scheme 69. Degradation of NSC-373965 catalyzed by its degradant, formaldehyde. (Reproduced from Ref. 265 with permission.)

An example that closely follows this scheme is the hydrolysis of NSC-373965 (Scheme 69), a water-soluble prodrug of NSC-284356, where P, in this case, is formaldehyde.[265] Initial hydrolysis of D in this example to generate formaldehyde was not dependent on formaldehyde being present in the starting solution.

2.2.3.2. *Kinetic Models Describing Chemical Drug Degradation in the Solid State*

The rate equations used to describe drug degradation in solution can be derived theoretically on the basis of the proposed degradation mechanisms. Data can then be tested to see if they conform to the proposed scheme. When the scheme is validated, the appropriate rate constant can be calculated and used to further refine the model. Similar theoretical rate equations for drug degradation in the solid state have been derived. Because drug degradation in the solid state generally occurs in a heterogeneous system where the physical state of the drug and other components varies with time, the rate equations describing solid-state degradation are much more complicated than those for degradation in solution.

A descriptor rate constant for solid-state degradation can be obtained once a theoretical rate equation has been derived and the data have been tested to see if they conform to the proposed model. However, for solid-state degradation in which the factors affecting the degradation mechanism have not been elucidated, because of the complexity involved, often an apparent constant (or constants) obtained by fitting the observed degradation curve to an empirical equation or equations is utilized. Such constants and the empirical relationships themselves can sometimes be used for stability prediction purposes. This section first discusses various theoretical equations used to describe the solid-state stability of drugs and introduces an empirical equation that can often describe the data adequately.

2.2.3.2.a. Diffusion-Controlled Reaction—The Jander Equation (Three-Dimensional Diffusion). For a model in which a sphere of a reactant B exists in another reactant A (Fig. 18), a rate equation for reaction between A and B at the interface was derived by Jander in

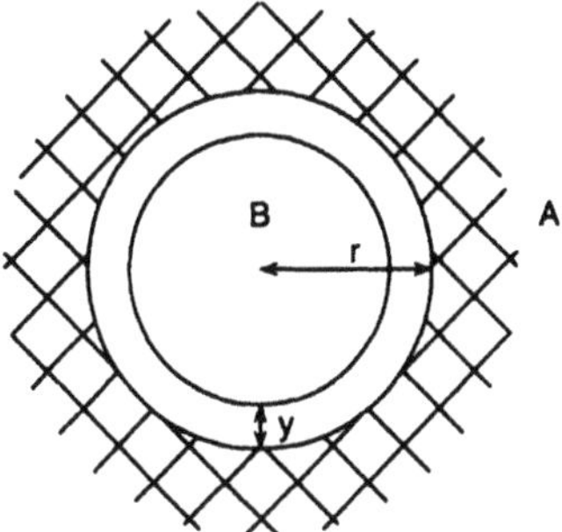

Figure 18. Schematic representation of the Jander model for solid-state degradation.

the 1920s. When a reaction starts at the interface of a sphere B, with radius r, and proceeds inside the sphere forming a reaction product phase with a thickness y, the fractional decomposition, x, is given by Eq. (2.48), and the value of y is given by Eq. (2.49).

$$x = \frac{\frac{4}{3}\pi r^3 - \frac{4}{3}\pi(r-y)^2}{\frac{4}{3}\pi r^3} \tag{2.48}$$

$$y = r[1-(1-x)^{1/3}] \tag{2.49}$$

Assuming that the rate at which the thickness of the product phase, y, increases is proportional to the diffusion rate of A into B yields Eq. (2.50), which is integrated to give Eq. (2.51).

$$\frac{dy}{dt} = \frac{kD([\mathrm{A}]_\mathrm{A} - [\mathrm{A}]_\mathrm{B})}{y} \tag{2.50}$$

$$y^2 = 2kD([\mathrm{A}]_\mathrm{A} - [\mathrm{A}]_\mathrm{B})t \tag{2.51}$$

In these equations, $[\mathrm{A}]_\mathrm{A}$ is the concentration of A in the A phase, $[\mathrm{A}]_\mathrm{B}$ is the concentration of A in the interfacial area, and D is the diffusion constant. The Jander equation (Eq. 2.52) is derived from Eqs. (2.50) and (2.51).

$$[1-(1-x)^{1/3}]^2 = \frac{2kD([\mathrm{A}]_\mathrm{A} - [\mathrm{A}]_\mathrm{B})}{r^2}t \tag{2.52}$$

Jander reported that the decarboxylation of inorganic carbonate can be described by Eq. (2.52), as shown in Fig. 19.[266,267]

$$CaCO_3 + MoO_3 \infty CaMoO_4 + CO_2$$

$$BaCO_3 + SiO_2 \infty BaSiO_2 + CO_2$$

Assuming that the rate at which the thickness of the product phase, y, increases (Eq. 2.49), independent of y, as represented by Eq. (2.53), yields Eq. (2.54). Prout and Tompkins reported that the decarboxylation of mercuric oxalate ($Hg_2C_2O_4$) conforms to Eq. (2.54).[268]

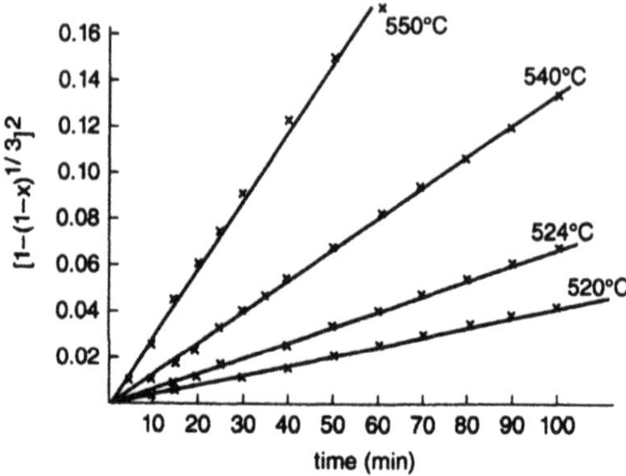

Figure 19. Decarboxylation of inorganic carbonate in the solid state as described by the Jander equation. (Reproduced from Ref. 266 with permission.)

$$\frac{dy}{dt} = k, \qquad y = kt \tag{2.53}$$

$$1 - (1 - x)^{1/3} = \frac{k}{r}t \tag{2.54}$$

The Jander equation has been applied to the degradation kinetics of various pharmaceuticals. For example, the degradation of freeze-dried thiamine diphosphate (Fig. 20)[269,270] and the degradation of propantheline bromide in the presence of aluminum hydroxide gel (Fig. 21)[271] have been described by the Jander equation.

2.2.3.2.b. Autocatalytic Reactions Controlled by Formation and Growth of Reaction Nuclei. When a reaction proceeds with a growing number of nuclei or imperfection sites, it tends to exhibit an S-shaped degradation curve, where the rate of product formation depends on the rate of nuclei formation and growth, as represented generally by Eq. (2.55), where l, m, and n are constants. And x is as defined in Eq. (2.48).

$$\frac{dx}{dt} = kx^l(1 - x)^m t^n \tag{2.55}$$

Equation Describing the Initial Stage of an Autocatalytic Reaction. The rate of an autocatalytic reaction is proportional to the number of nuclei (N). Nuclei represent imperfection sites in the crystal where it is assumed that chemical reactions can take place. It is also assumed that as chemical reactions proceed, strain is placed on the crystal, resulting in more imperfections. The rate at which N increases at the initial stage of this autocatalytic process is described by Eq. (2.56). Equation (2.57) then describes the reaction rate.

$$\frac{dN}{dt} = At^q \tag{2.56}$$

$$\frac{dx}{dt} = kN = kt^n \tag{2.57}$$

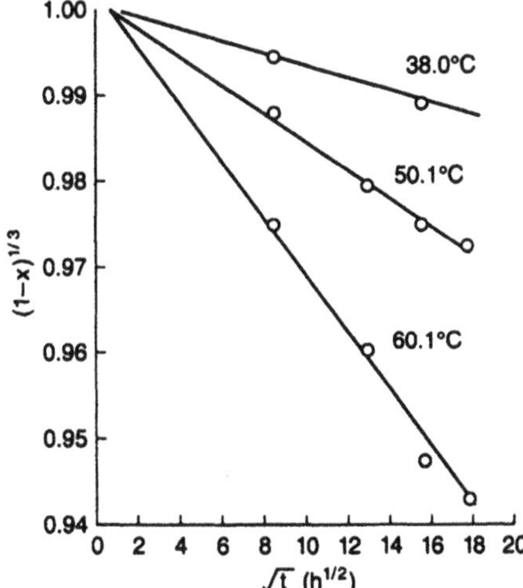

Figure 20. Degradation of freeze-dried thiamine diphosphate plotted according to the Jander equation (11% RH). (Reproduced from Ref. 270 with permission.)

Equation (2.57) is equivalent to Eq. (2.55) when l and m equal zero. Equation (2.58), obtained by integrating Eq. (2.57), describes the initial hydrolysis rate of meclofenoxate hydrochloride in the solid state.[272,273] Similarly, this equation describes the hydrolysis of propantheline bromide under high-humidity conditions, as shown in Fig. 22.[274] Equation (2.58) has also been applied to the initial degradation of aspirin derivatives in the solid state (Fig. 23).[275]

$$x = kt^{n'} \tag{2.58}$$

The Prout–Tompkins Equation. It can be argued that, as more stress/nuclei appear, the rate of nuclei formation will eventually decrease. That is, when termination of the nuclei

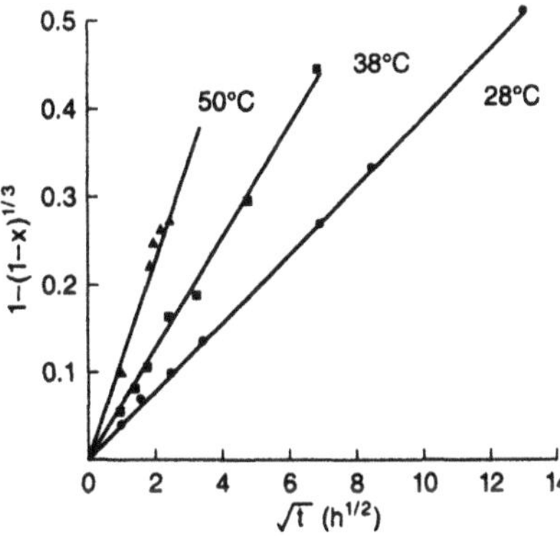

Figure 21. Degradation of propantheline bromide in the presence of aluminum hydroxide gel plotted according to the Jander equation (75% RH). Drug: aluminum hydroxide gel = 50:4950 (weight ratio). (Reproduced from Ref. 271 with permission.)

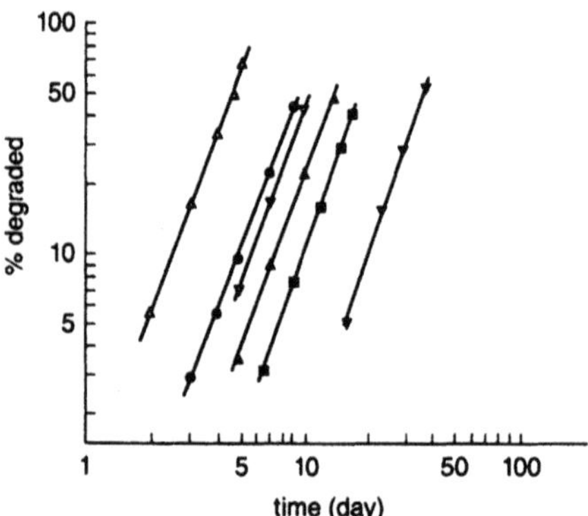

Figure 22. Degradation of propantheline bromide represented by Eq. (2.58) (equation for an initial autocatalytic reaction). △, 90°C, 78.3% RH; ●, 80°C, 79.5% RH; ▽, 80°C, 65.5% RH; ▲, 80°C, 51.0% RH; ■, 70°C, 80.0% RH; ▼, 60°C, 80.5% RH. (Reproduced from Ref. 274 with permission.)

occurs in addition to their propagation, the number of nuclei is determined by the probability of propagation (α) and the probability of termination (β), as described by Eq. (2.59). Equation (2.60) represents the reaction rate and is equivalent to Eq. (2.55) when l, m, and n are equal to 1, 1, and 0, respectively.

$$\frac{dN}{dt} = (\alpha - \beta)N \tag{2.59}$$

$$\frac{dx}{dt} = kx(1 - x) \tag{2.60}$$

Prout and Tompkins reported that the thermal degradation of potassium permanganate can be described by Eq. (2.61), which is obtained by integrating Eq. (2.60).[276] Equation (2.60) adequately describes the degradation of solid aspirin in the presence of limited moisture, as shown in Fig. 24.[277]

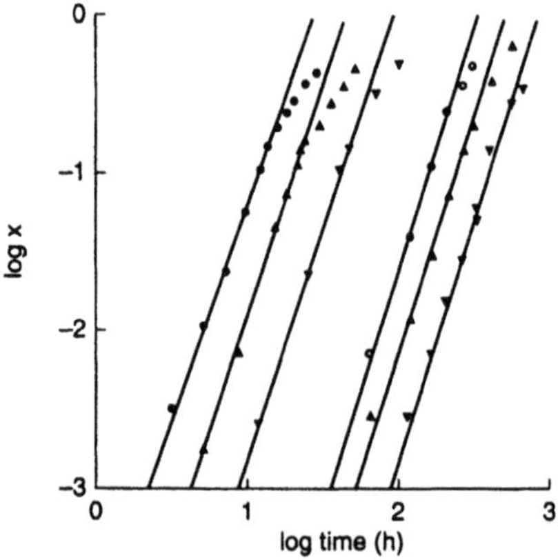

Figure 23. Degradation of acetyl-5-nitrosalicylic acid represented by Eq. (2.58).[275] ●, 90°C, 78% RH; ▲, 90°C, 51% RH; ▼, 90°C, 23% RH; ○, 70°C, 80% RH; △, 70°C, 51% RH; ▽ 70°C, 22% RH.

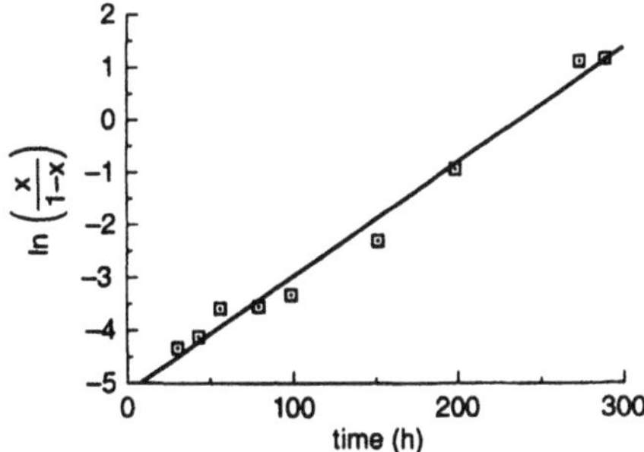

Figure 24. Degradation of aspirin plotted according to the Prout–Tompkins equation (62.5°C, water content: 10%). (Reproduced from Ref. 277 with permission.)

$$\ln\left(\frac{x}{1-x}\right) = kt + C \tag{2.61}$$

The Kawakita Equation. Because the value of l in Eq. (2.55) depends on the catalytic effect of the degradation product, a more general form of Eq. (2.60) is Eq. (2.62). Kawakita reported that the reduction of ferric oxide can be described by Eq. (2.62) with various values of l, depending on temperature.[278]

$$\frac{dx}{dt} = kx^{l}(1-x) \tag{2.62}$$

The Avrami Equation. Equation (2.55) yields Eq. (2.63) when $l = 0$ and $m = 1$.

$$-\ln(1-x) = kt^{n'} \tag{2.63}$$

Equation (2.63) describes the reaction between ZnO and $BaCO_3$. It is also used to describe the rate of some polymorphic transitions[279,280] described in Chapter 3.

2.2.3.2.c. Reaction Forming a Liquid Product—the Bawn Equation. A rate equation proposed by Bawn is applicable to reactions of solids forming gaseous and liquid products. In the latter case, the reaction rate is described by summing the reaction rates in the solid state and those in the solution formed by the liquid product:

$$\frac{dx}{dt} = k_s(1 - x - Sx) + k_1 Sx \tag{2.64}$$

where S is the solubility of the drug in the liquid formed, and k_s and k_l are the rate constants in the solid state and in solution, respectively. In Eq. (2.64), Sx, the product of the fraction degraded, x, and the solubility S, represents the molecular fraction of drug in solution, and $(1 - x - Sx)$ represents the fraction in the solid state. Integrating Eq. (2.64) gives Eq. (2.65), which has been used to describe the decarboxylation of various benzoic acid derivatives.[281] Decarboxylation of various alkoxyfuroic acids such as 5-(tetradecyloxy)-2-furoic acid[282] and octyloxy furanoic acid[283,284] was adequately described by Eq. (2.65), as shown in Figs. 25 and 26, respectively.

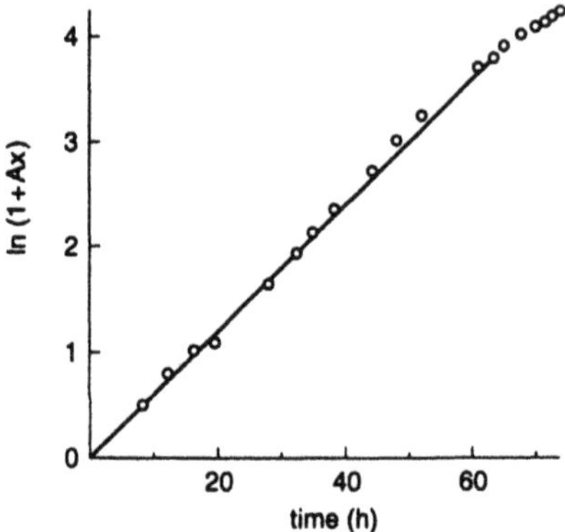

Figure 25. Decarboxylation of 5-(tetradecyloxy)-2-furoic acid plotted according to the Bawn equation (90°C). $A = (Sk_1 - Sk_s - k_s)/k_s$. (Reproduced from Ref. 282 with permission.)

$$\ln\left(1 + \frac{Sk_1 - Sk_s - k_s}{k_s} x\right) = (Sk_1 - Sk_s - k_s)t \tag{2.65}$$

2.2.3.2.d. Reaction Controlled by an Adsorbed Moisture Layer—The Leeson–Mattocks Equation. Leeson and Mattocks proposed a model in which drug degradation occurs in an adsorbed moisture layer. An S-shaped degradation curve observed for aspirin in the solid state (Fig. 27) was explained by this model.[285,286] Aspirin was assumed to be dissolved rapidly in an adsorbed moisture layer so as to form a saturated solution in which decomposition occurs. It was assumed that the decomposition was catalyzed by hydronium ion, as described by Eq. (2.66), and that the rate increased as the amount of salicylic acid formed (x) increased, thus yielding an S-shaped curve.

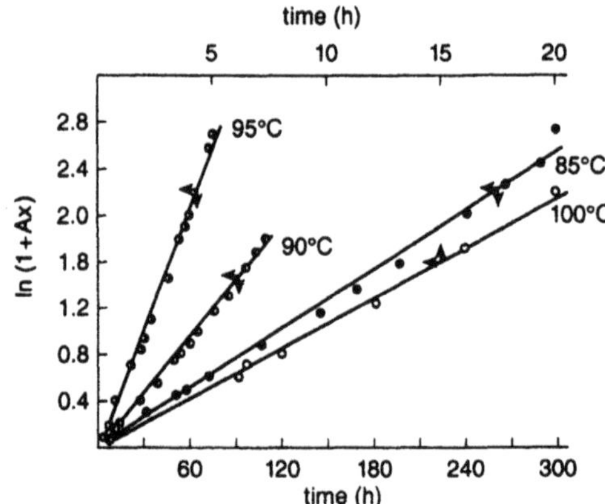

Figure 26. Decarboxylation of octyloxy furanoic acid plotted according to the Bawn equation. $A = (Sk_1 - Sk_s - k_s)/k_s$. (Reproduced from Ref. 283 with permission.)

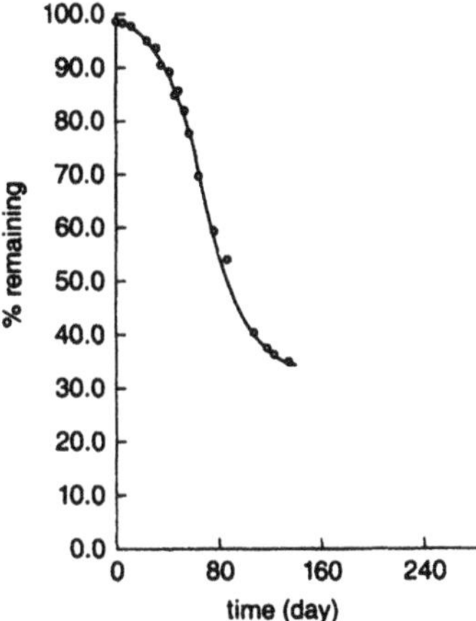

Figure 27. Time course of degradation of solid aspirin explained by the Leeson–Mattocks equation (60°C, 80.6% RH). (Reproduced from Ref. 285 with permission of the American Pharmaceutical Association.)

$$\frac{dx}{dt} = k[\mathrm{D}][\mathrm{H}^+] = k[\mathrm{D}]\sqrt{KVx} \tag{2.66}$$

In the above equation, V is the volume of the layer of adsorbed moisture, and K is the ionization constant of the degradant. Subsequent studies showed that the reaction was not affected by the formed salicylate, thus invalidating the model that led to Eq. (2.66).[275]

If it is assumed that the degradation rate of aspirin in the adsorbed moisture layer is determined by aspirin concentration, [D], the amount of moisture, $[H_2O]$, the volume of the adsorbed moisture layer, V, and a rate constant, k, then the reaction should be described by Eq. (2.67).

$$\frac{dx}{dt} = k(t) \cdot V(t) \cdot [\mathrm{D}(t)] \cdot [\mathrm{H_2O}(t)] \tag{2.67}$$

Each of these parameters was experimentally established by Carstensen and co-workers and utilized to predict the degradation curve.[277,287,288] The predicted degradation curve diverged from the observed curve, indicating that the adsorbed moisture layer theory cannot fully explain the degradation of aspirin (Fig. 28).

For reactions in which the catalytic effect of degradation products is negligible and the volume of the adsorbed moisture layer and the drug solubility can be regarded as constant, the rate should be described by Eq. (2.68) according to the adsorbed moisture layer, and the degradation should conform to apparent zero-order kinetics.

$$\frac{dx}{dt} = kV[\mathrm{D}] = kVS \tag{2.68}$$

Oxidative degradation of solid sulpyrine in the presence of moisture was described by Eq. (2.68).[289] This equation was also used to describe decarboxylation of 4-aminosalicylic acid.[290,291]

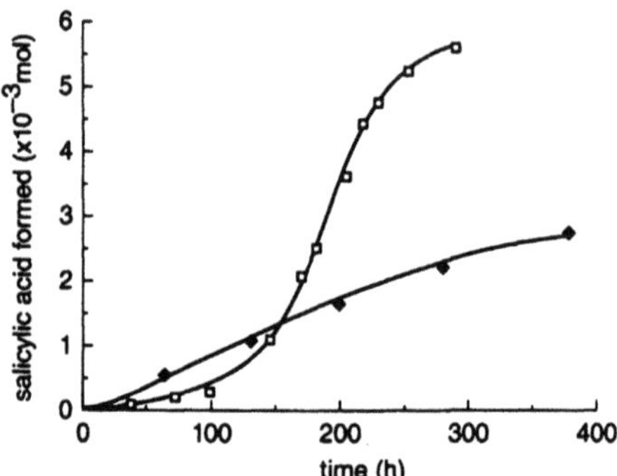

Figure 28. Aspirin degradation curve (62.5°C,water content: 10%). □, Observed; ♦, predicted from Eq. (2.67). (Reproduced from Ref. 277 with permission.)

2.2.3.2.e. Use of Empirical Expressions Such as the Weibull Equation. It is often difficult to derive the rate equation for degradation of drugs in the solid state because the physical state of the system varies with time in complex ways. This is especially true for degradation in the presence of moisture, such that derivation of adequate rate equations is often impossible. Empirical equations such as the Weibull equation (Eq. 2.69) have been used to describe this kind of solid drug degradation. Degradation curves for various drugs have been fitted to the Weibull equation without significant deviations (Fig. 29).[292] Weighted least-squares analysis of data fit to the Weibull equation has resulted in smaller deviations.[293]

$$\ln \ln\left(\frac{1}{1-x}\right) = \ln k + m \ln t \tag{2.69}$$

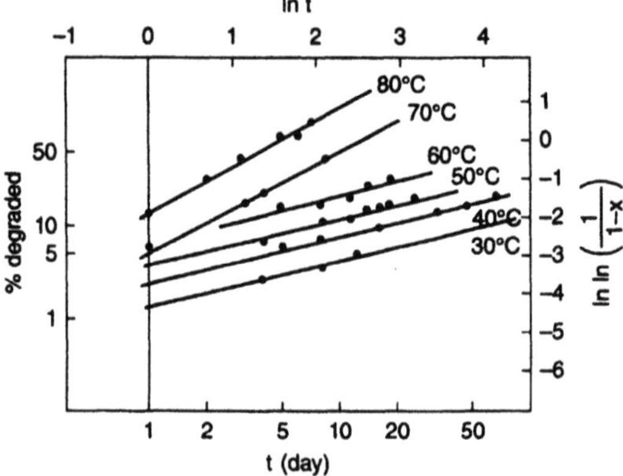

Figure 29. Degradation of ascorbic acid in the presence of mannitol ([ascorbic acid]:[mannitol] = 1:9) plotted according to the Weibull equation. (water content: 2%). (Reproduced from Ref. 292 with permission.)

2.2.3.3. *Calculation of Rate Constants by Fitting to Kinetic Models*

A rate constant for drug degradation can be obtained by fitting an observed degradation curve to a suitable kinetic model chosen on the basis of a proposed mechanism. Since any experimental data include errors, least-squares regression analysis is usually carried out to test the validity of the model and to calculate the apparent rate constant(s).

The concentration of remaining drug, [D], and the fraction of degradation, x, are generally represented by a linear equation for zero-order kinetics or linearized forms of various equations, e.g., log [D], versus time for first-order kinetics. Recently, with the advent of computers, nonlinear regression analysis has become popular. Various methods for obtaining accurate estimates utilizing linear regression analysis were developed,[294] and some of these are still used as a method of analysis or to obtain initial values needed in nonlinear regression analysis. Various programs are available for the preliminary estimation of parameters for drug degradation kinetics of various orders.[295] Computer programs such as MULTI,[296] NONLIN, and other commercial software programs are also used, especially spreadsheet programs such as EXCEL, which allow one to perform nonlinear regression analyses very easily.

In the estimation of rate constants through the fitting of degradation data to a kinetic model, the validity of the model and the reliability of the estimated rate constant should be evaluated, taking into account experimental errors. Additional data are sometimes required to obtain accurate estimates. For example, in the case of consecutive reactions, the time courses for both the parent drug and the intermediate are required to estimate the pseudo-first-order rate constant for the formation and loss of the intermediate (see Section 2.2.3.7.f), especially when k_2/k_1 is larger than 0.5.[297]

2.2.4. Temperature

2.2.4.1. *General Principles*

Temperature is one of the primary factors affecting drug stability. The rate constant/temperature relationship has traditionally been described by the Arrhenius equation,

$$k = A \exp\left(\frac{-E_a}{RT}\right) \tag{2.70}$$

where E_a is the activation energy and A is the frequency factor. This equation is a variant of the equation describing the effect of temperature on equilibrium processes that was developed by van't Hoff in 1887. Arrhenius applied his equation to various reaction processes.[298] Comparison of the empirical Arrhenius equation to the Eyring equation, Eq. (2.7), shows some similarities and some differences. The frequency factor A in the Arrhenius equation corresponds to the product of the universal collision and entropy terms in Eq. (2.7) while the E_a term in Eq. (2.70) is related to the enthalpy term in Eq. (2.7).[299] Because a plot of the logarithm of k against the reciprocal of absolute temperature generally yields a linear relationship (Arrhenius plots), the frequency factor and activation energy are regarded as independent of temperature, and the activation energy, E_a, is used as a measure of the temperature dependence of the rate constant. However, the Eyring equation suggests that both A and E_a should be temperature-dependent. E_a can be shown to be related to $\Delta H^{\ddagger}$ as indicated in Eq. (2.71). Observed linear Arrhenius plots can be explained by the much larger temperature dependency of the exponential term in Eq. (2.70) as compared to that of A. In theory, however, Arrhenius plots should not be linear.

$$E_a = \Delta H^{\ddagger} + RT \tag{2.71}$$

Nevertheless, Arrhenius plots have been traditionally used to describe the temperature dependency for various chemical reactions by regarding A and E_a as independent of temperature. A prerequisite for the application of Eq. (2.70) [and Eq. (2.7)] is that the degradation mechanism does not change in the temperature range of interest. E_a values of about 10–30 kcal/mol (40–130 kJ/mol) are generally observed in the degradation of drug substances. Table 4 shows E_a values for degradation of representative drug substances. The values of E_a are presented in units of calories per mole rather than kilojoules per mole because those were the units reported in the original reference sources (1 kcal/mol = 4.18 kJ/mol).

As an alternative to Arrhenius plots, the data can be fitted to the Eyring equations:

$$\frac{k}{T} = \left(\frac{\kappa}{h}\exp(\Delta S^{\ddagger}/R)\right)\exp(-\Delta H^{\ddagger}/RT) \tag{2.72}$$

$$\ln k/T = \ln \kappa/h + \Delta S^{\ddagger}/R - \Delta H^{\ddagger}/RT \tag{2.73}$$

A plot of ln k/T versus $1/T$ is linear. Thus, plots of either k versus $1/T$ or k/T versus $1/T$ are usually plotted by taking either the natural logarithm (ln) or the logarithm to the base 10 (log) of k. The slopes of plots of log k or log k/T versus $1/T$ are $-E_a/2.303RT$ and $-\Delta H^{\ddagger}/2.303RT$, respectively.

Temperature is obviously an important parameter because most reactions proceed faster at elevated temperatures than at lower temperatures. The terms E_a and $\Delta H^{\ddagger}$ are a measure of how sensitive the degradation rate of a drug is to temperature changes. Table 5 shows the effect of a 10°C change in temperature on the rate constant. If the E_a for a degradation process is only 10 kcal/mol, this temperature change results in only a 1.76-fold change in drug reactivity. However, if the E_a is 30 kcal/mol, a 10°C increase in temperature results in about a 5.5-fold increase in the degradation rate.

2.2.4.2. *Quantitation of the Temperature Dependency of Degradation Rate Constants*

Estimation of an appropriate rate or rate constant for drug degradation is an important step in predicting the stability of pharmaceuticals. Knowing how such a rate or rate constant changes with temperature in a quantitative way may allow one to predict the stability at other temperatures. Even if a rate or rate constant cannot be estimated by fitting the data to a theoretical or empirical equation, constants such as time required for 10% degradation (t_{90}) can be utilized instead of rate constants. Stability prediction is possible, for example, from the relationship between the reciprocal of t_{90} and temperature.

In the previous section, the Arrhenius equation was described. The Arrhenius equation was applied to the prediction of drug degradation in the 1940s and 1950s. Taking the logarithm of both sides of Eq. (2.70) yields

$$\ln k = \frac{-E_a}{RT} + \ln A \tag{2.74}$$

Table 4. Activation Energy Values for Degradation of Representative Drug Substances

Drug hydrolysis	E_a (kcal/mol)	pH[a]	Reference
Hydrolysis			
Ampicillin	9.2[b]	9.78	65
	18.3[b]	4.93	65
Cefotaxime	9.4	8.94	73
	24.7	5.52	73
Echothiophate iodide	10	OH^-	49
	23	H^+	49
Indomethacin	10.1	11	58
Methylphenidate	12.35	OH^-	19
	15.98	H^+	19
Oxazolam	12.4	8.0	93
Succinylcholine chloride	13.09	OH^-	32
	17.23	H^+	32
Haloxazolam	15.2	1.3	94
Mitomycin C	16.2[b,c]	5.20	105
Procaine	16.8	H^+	10
Hydrocortisone sodium phosphate	17.0	H^+	47
Diazepam	17.2	10.18	87
Atropine	17.2	H^+	16
	20.9	H^+	26
Acetaminophen	17.42	6	53
Phenethicillin	17.6	6.7	68
Amobarbital	18.6[b]	10.12	80
Benzocaine	18.6	H^+	10
Ethylparaben	18.7	9.16	22
Meperidine	18.77	6	28
Nitrofurantoin	18.9	H^+	109
Rifampicin	19.2	H^+	111
Oxazepam	20.0	3.24	87
	26.1	8.52	87
Benzylpenicillin	20.3	2.7	64
Thiamine hydrochloride	21	9.90	102
	29	1.70	102
Cefadroxil	21.4	7.00	71
Carmethizole	21.5	9.90	44
Chlorphenesin carbamate	21.8	H^+	43
Cephalothin	22.6	5.00	70
	28.5	10.00	70
Sulfacetamide	22.9[d]	7.40	60
Carbenicillin	23.4	4.45	67
	27.45	10.47	67
Furosemide	23.5	H^+	100
Chloramphenicol	24.0	6.00	15
Chlorambucil	24.4[e]	7.00	120
Pilocarpine	25.02	OH^-	37
Cocaine	26.2	6.25	34
Moricizine	27.5	6.0	61
Clindamycin	38.0	1.10	123

(continued)

Table 4. Continued

Drug hydrolysis	E_a (kcal/mol)	pH[a]	Reference
Racemization and epimerization			
Etoposide	18.2[f]	11.04	156
Hetacillin	21.2	OH^-	163
Epinephrine	23.19	H^+	158
Cefsulodin	27.0	9.0	166
Pilocarpine	28.48	OH^-	37
Dehydration			
Prostaglandin E_1	18	1.2	142
	24	8	142
Isomerization			
Amphotericin B	16.4	7	148
Prostaglandin E_1	20	8	142
Decarboxylation			
Etodolac	26	H^+	170
Oxidation			
Ascorbic acid	7.8	6.60	179
	12.2	3.52	179
Morphine	22.8	6	202
Procaterol	34.5	6.0	184

[a]H^+, Hydronium ion-catalyzed reaction; OH^-, hydroxide ion-catalyzed reaction.
[b]Activation enthalpy $\Delta H^‡$.
[c]Value reported in the reference (68 kJ/mol) has been converted to kilocalories per mole.
[d]Value reported in the reference (95.9 kJ/mol) has been converted to kilocalories per mole.
[e]Value reported in the reference (102 kJ/mol) has been converted to kilocalories per mole.
[f]Value reported in the reference (76 kJ/mol) has been converted to kilocalories per mole.

So-called Arrhenius plots in which the logarithm of rate (rate constant) is plotted against the reciprocal of absolute temperature (degrees kelvin) have been used to validate the conformity of the degradation rates of various drugs to this equation. Linear Arrhenius plots were reported for the degradation of penicillin G (Fig. 30),[300] the discoloration of a liquid

Table 5. Effect of a 10°C Temperature Change (20°C to 30°C) on the Relative Degradation Rate Constants for Drugs Having Different E_a Values

E_a (cal/mol)	k_{T2}/k_{T1} ($T1$ = 20°C, $T2$ = 30°C)
10,000	1.76
15,000	2.34
20,000	3.11
25,000	4.12
30,000	5.48

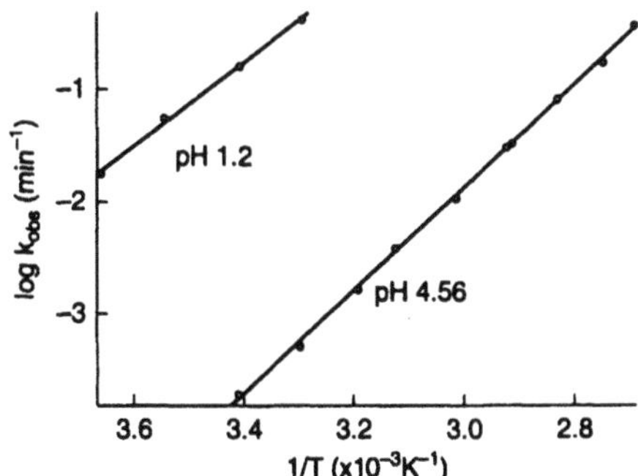

Figure 30. Arrhenius plots for the degradation of penicillin G at pH 1.2 and pH 4.56. Note: The authors plotted the values of $1/T$ from high to low values (rather than from low to high values as is normally done). (Reproduced from Ref. 300 with permission.)

multisulfa preparation (Fig. 31),[301] and the degradation of liquid multivitamin preparations (Fig. 32),[302,303] to give a few examples. A linear Arrhenius plot indicates that, once degradation rates are obtained at several temperature levels, the degradation rate at some other specific temperature can be estimated. Thus, the Arrhenius equation has been successfully applied to the prediction of the stability of various pharmaceuticals: the degradation of vitamin A dosage forms,[304] the change in appearance of tablets and powders,[305,306] the degradation of multivitamin tablets,[307] and many other examples too numerous to cite here.

The Arrhenius equation is valid in the temperature range where both A and E_a in Eq. (2.70) can be regarded as constant. Changes in the degradation mechanism with temperature may result in nonlinear Arrhenius plots. A very low degradation rate of phenylbutazone tablets was observed at 50°C in contrast to a high rate at 60°C, suggesting a change in the degradation mechanism.[308] The Arrhenius plots for the hydrolysis of soybean phosphatidylcholine exhibited different slopes at temperatures above and below its phase-transition temperature (Fig. 33).[309] The slopes also changed with changes in the pH of the solution.

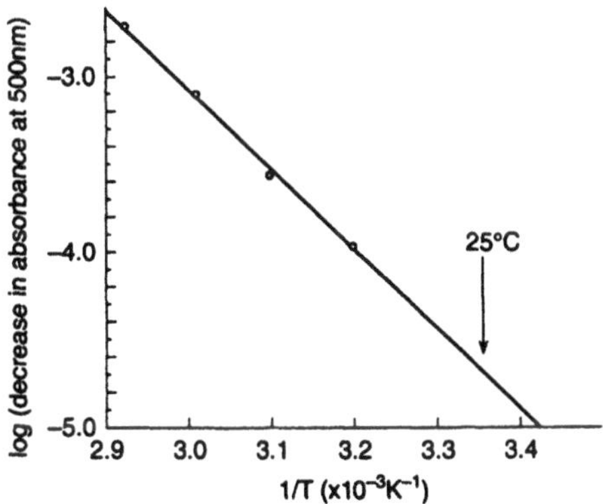

Figure 31. An Arrhenius plot for the discoloration of a liquid multisulfa preparation (Reproduced from Ref. 301 with permission of the American Pharmaceutical Association.)

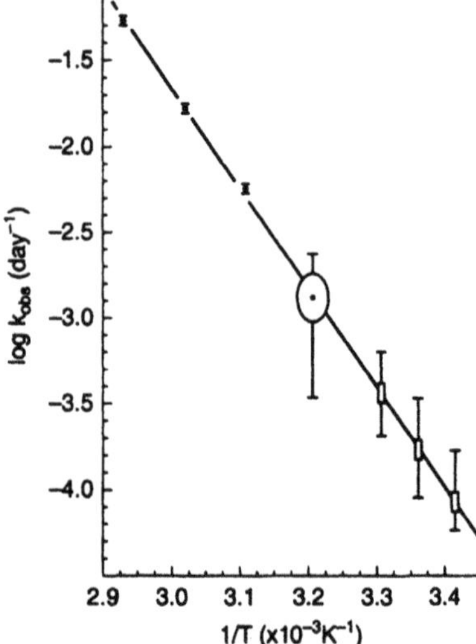

Figure 32. An Arrhenius plot for the degradation of thiamine hydrochloride in a liquid multivitamin preparation (Reproduced from Ref. 303 with permission of the American Pharmaceutical Association.)

On the other hand, the degradation rate of amorphous indomethacin at a temperature slightly lower than its melting point yielded the same linear Arrhenius plots as obtained for the molten substance (Fig. 34).[310]

A recent paper proposed that the hydrolysis rate of aspirin could not be predicted on the basis of the Arrhenius plots, owing to changes in the activation energy in the temperature range 30–70°C.[311] The authors proposed that the structures of icebergs formed near the hydrophobic groups in the activated complex changed between the temperature ranges below 42°C, 42–58°C, and above 58°C. However, this interpretation has been questioned based on the statistical uncertainty of the observed differences in the activation energy.[312,313]

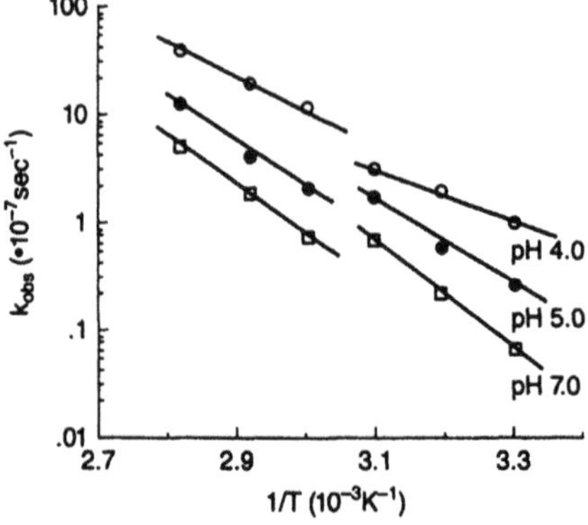

Figure 33. Arrhenius plots describing the degradation of soybean phosphatidylcholine at three pH values. Phosphatidylcholine concentration: 0.05*M*. (Reproduced from Ref. 309 with permission.)

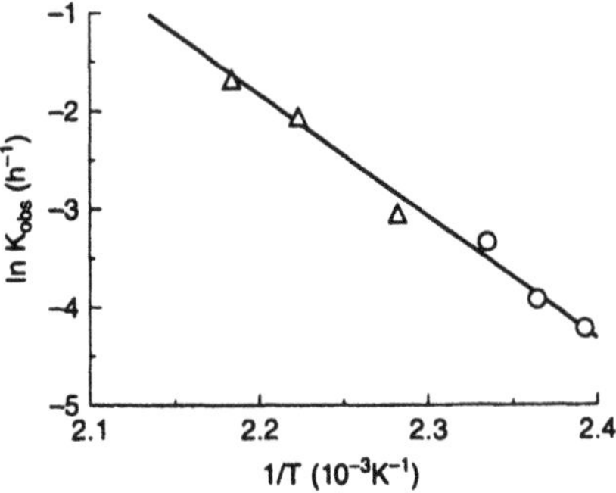

Figure 34. An Arrhenius plot for the degradation of indomethacin in amorphous (○) and molten (△) states. (Reproduced from Ref. 310 with permission.)

2.2.4.2.a. Prediction of Degradation Rate by Linear Regression Analysis of the Arrhenius Equation. Drug degradation rates at room temperature can be estimated by using Arrhenius plots of the rate constants calculated at each of several temperatures, as described above. This regression method using multiple rate constants calculated separately at different temperatures has been called "classical Arrhenius analysis" and is still used as a basic method for the prediction of drug stability. However, this classical analysis may result in large errors in the estimated rate constant. Slater *et al.*[314] determined the degradation rate of vitamin A in multivitamin tablets at 40, 45, 50, and 55°C and obtained the Arrhenius plot shown in Fig. 35. Estimation of the rate at 25°C according to the regression equation derived from this plot resulted in a calculated degradation rate curve that deviated significantly from that observed experimentally, as shown in Fig. 36.[314]

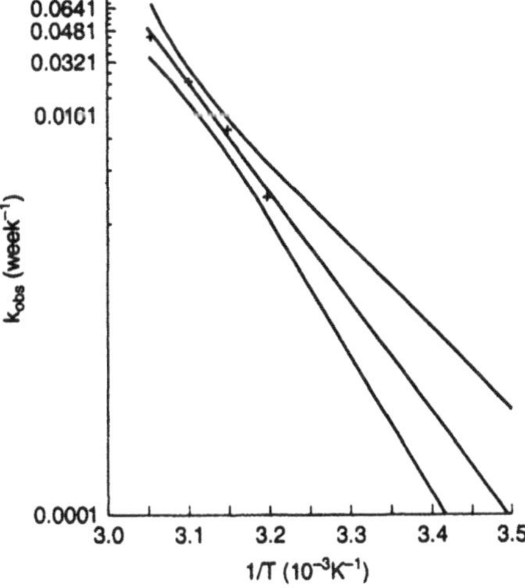

Figure 35. An Arrhenius plot for the degradation of vitamin A in multivitamin tablets, obtained by classical analysis. The upper and lower curves represent 95% significance limits. (Reproduced from Ref. 314 with permission.)

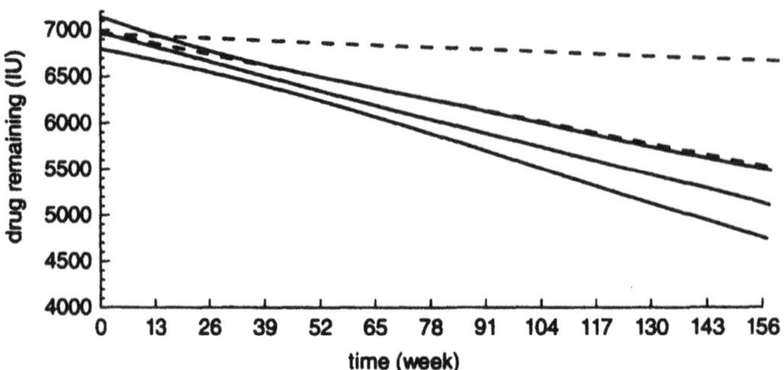

Figure 36. Time course of degradation of vitamin A at 25°C. ---, Estimated using upper and lower predicted k values from the Arrhenius plot, —— experimentally observed (upper and lower curves represent 95% significance limits. (Reproduced from Ref. 314 with permission.)

In classical Arrhenius analysis, the rate constants are calculated by fitting normal degradation versus time data, which have inherent experimental errors, and then fitting the calculated rate constants again to the Arrhenius equation. The errors included in the original data points are thus not directly reflected in the Arrhenius plots. Therefore, the weight given in the analysis to a single data point varies if the number of data points differs among the different temperature levels. Furthermore, the variation of estimates of rate constant obtained from the Arrhenius plots becomes larger owing to the reduced number of degrees of freedom. This occurs because the number of data points used in the Arrhenius regression analysis (the number of calculated rate constants) is smaller than the number of original data points used to estimate the original experimental rate constants.

The Arrhenius regression analysis can also be performed by using the multiple rate constants calculated from each of the drug concentration/time data points observed at a single temperature level. As shown in Fig. 37, the data at several temperature levels are fitted to Eq. (2.75) (for first-order degradation kinetics), which is obtained by replacing the rate constant k of Eq. (2.70) with the concentration of remaining drug, [D]:

$$\ln \frac{[\mathrm{D}]}{[\mathrm{D}]_0} = -tA \exp\left(\frac{-E_a}{RT}\right) \tag{2.75}$$

where $[\mathrm{D}]_0$ is initial drug content. This "modified Arrhenius analysis"[314–316] uses the same weight for each data point and provides an estimate of the rate constant at room temperature with a smaller 95% confidence interval owing to the larger number of degrees of freedom (the larger number of data points used). Nevertheless, in the analysis of the degradation of vitamin A, this method provided a regression curve similar to that obtained by classical Arrhenius analysis, resulting in a calculated room-temperature degradation curve which deviated significantly from that observed experimentally.

The large deviation between the estimated rate constants for a given temperature extrapolated from classical (Fig. 35) and modified Arrhenius regression analysis (Fig. 37) and the experimental rate constants determined at the actual temperature is due to the linear regression analysis method used. That is, the values of the logarithms of the rate constants do not reflect directly the errors in the experimental data. As Bentley pointed out,[317] when the error term ε is added to the logarithm of k, in accordance with Eq. (2.76), in the linear

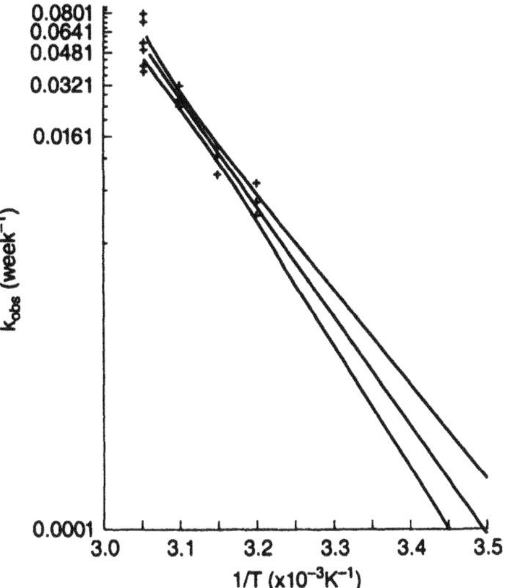

Figure 37. Arrhenius plot for the degradation of vitamin A obtained by a modified analysis according to Eq. (2.75). The upper and lower curves represent 95% significance limits. (Reproduced from Ref. 314 with permission.)

least-squares regression analysis of the rate constant k_T at a temperature T, this results in a larger contribution from the data obtained at lower temperatures than from those obtained at higher temperatures (Fig. 38). To avoid this problem, Bentley proposed a weighted least-squares analysis in which the error term ε is added to k as described by Eq. (2.77) rather than Eq. (2.76).

$$\ln k_T = \ln A - \frac{E_a}{RT} + \varepsilon \tag{2.76}$$

$$k = A \exp\left(\frac{-E_a}{RT}\right) + \varepsilon \tag{2.77}$$

Weighted least-squares analysis provides the variance of experimental errors on the basis of which the linearity of the Arrhenius plots can be assessed.

2.2.4.2.b. Prediction of Degradation Rate by Nonlinear Regression Analysis of the Arrhenius Equation. Although linear regression analysis provides biased results and weighted least-squares analysis is required to improve the estimates, nonlinear regression analysis does not suffer from the same problem. Representing Eq. (2.75) in a nonlinear manner yields Eq. (2.78) for first-order degradation kinetics.[318] Similarly, Eq. (2.79) is obtained for zero-order degradation kinetics.

$$[\mathrm{D}] = [\mathrm{D}]_0 \exp\left[-tA \exp\left(\frac{-E_a}{RT}\right)\right] \tag{2.78}$$

$$[\mathrm{D}] = [\mathrm{D}]_0 - tA \exp\left(\frac{-E_a}{RT}\right) \tag{2.79}$$

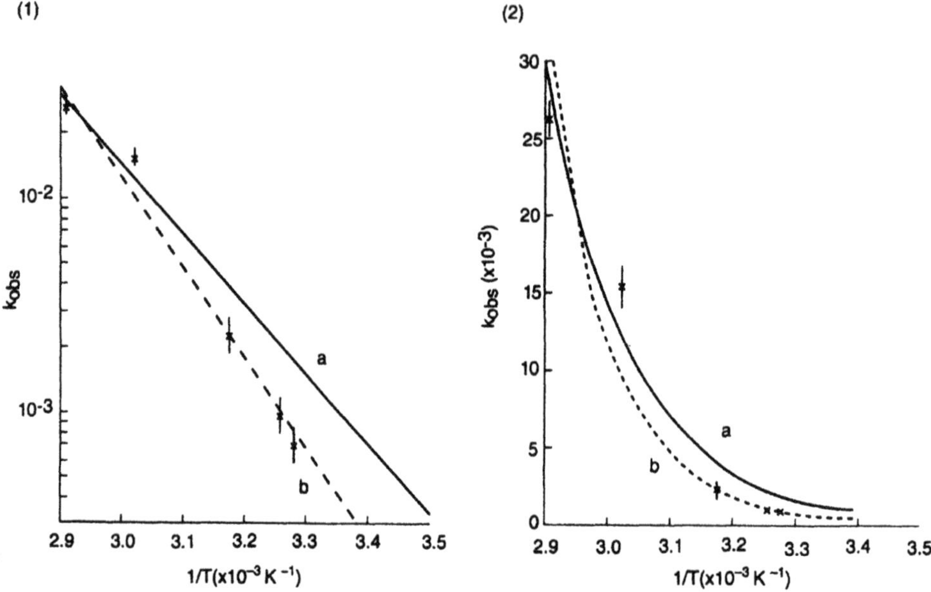

Figure 38. Linear Arrhenius regression by weighted least-squares analysis (a) and least-squares analysis (b). *y*-axis: (1) logarithmic scale, (2) arithmetical scale. (Reproduced from Ref. 317 with permission.)

The frequency factor A corresponds to the rate constant at infinite temperature. Replacing A with k_{298}, which has a more practical meaning than A [rate constant of degradation at 25°C represented by Eq. (2.80)], yields Eq. (2.81).

$$A = k_{298} \exp\left(\frac{E_a}{298 \cdot R}\right) \tag{2.80}$$

$$[\mathrm{D}] = [\mathrm{D}]_0 \exp\left\{-k_{298} t \exp\left[\frac{E_a}{R}\left(\frac{1}{298} - \frac{1}{T}\right)\right]\right\} \tag{2.81}$$

Again, replacing k_{298} in Eq. (2.81) with the shelf life of a pharmaceutical, t_{90} (time required for 10% degradation), yields Eq. (2.82). Similarly, Eq. (2.83) is obtained for zero-order degradation kinetics.

$$[\mathrm{D}] = [\mathrm{D}]_0 \exp\left\{\frac{-0.1054t}{t_{90}} \exp\left[\frac{E_a}{R}\left(\frac{1}{298} - \frac{1}{T}\right)\right]\right\} \tag{2.82}$$

$$[\mathrm{D}] = [\mathrm{D}]_0 \left\{1 - \frac{0.1t}{t_{90}} \exp\left[\frac{E_a}{R}\left(\frac{1}{298} - \frac{1}{T}\right)\right]\right\} \tag{2.83}$$

The estimates of t_{90} and E_a can be obtained directly by the nonlinear regression analysis of the observed degradation versus time data according to Eq. (2.82) or Eq. (2.83). A Monte Carlo simulation study performed by King *et al.*[319] suggested that nonlinear regression analysis could provide more reliable estimates, with smaller deviations and biases, than does

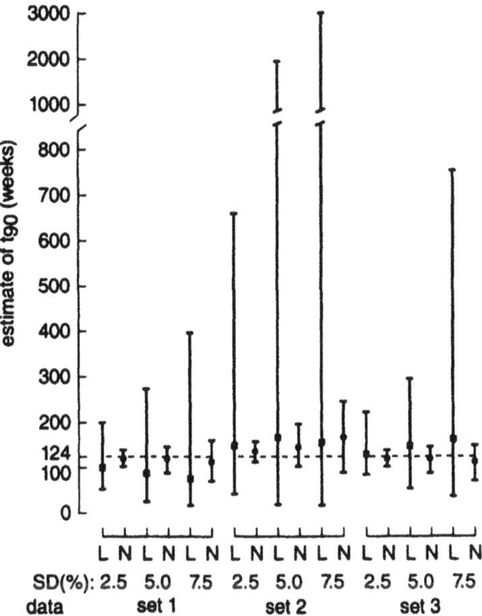

Figure 39. Estimates of t_{90} obtained by linear (L) and nonlinear (N) regression analysis. Theoretical value of t_{90}: 124 weeks. (Reproduced from Ref. 319 with permission.)

linear analysis. As shown in Fig. 39, nonlinear analysis produced an estimate of t_{90} closer to the theoretical value with smaller and more symmetrical 95% confidence limits than did linear analysis.

2.2.4.2.c. Nonisothermal Prediction of Degradation Rate

Rate Equation for Nonisothermal Prediction. In the previous sections, the estimation of rate constants at room temperature by linear and nonlinear Arrhenius regression of the degradation data determined at several fixed temperatures was described. The method of nonisothermal prediction of degradation rate was developed to reduce the experimental effort required by allowing kinetic parameters to be estimated from a single set of drug concentration versus time data obtained while the temperature is changed as a function of time according to some algorithm. For the nonisothermal prediction of degradation rate, the rate constant k is represented by Eq. (2.84), in which the term, T, or temperature, in the Arrhenius equation is replaced by a temperature time function, $G(t)$.

$$k = A \exp\left[\frac{-E_a}{R \cdot G(t)}\right] \tag{2.84}$$

The basic theory for nonisothermal prediction of degradation rate was established in the 1950s.[320,321] Its application to the stability prediction of pharmaceuticals was reported by Rogers[322] and extended by Eriksen and Stelmach.[323] Initial temperature programs or algorithms used relationships that could easily be integrated when inserted into Eq. (2.84).

For example, the following relationships were employed by Rogers[322] and Eriksen and Stelmach,[323] respectively:

$$\frac{1}{T}=\frac{1}{T_0}-b\ln(1+t) \tag{2.85}$$

$$\frac{1}{T}=\frac{1}{T_0}-at \tag{2.86}$$

In these equations, T_0 and T are the temperatures at time zero and time t, respectively, and b and a are constants. Assuming that the temperature changes according to Eq. (2.86), a rate equation obtained by combining the first-order rate equation, Eq. (2.12), with Eq. (2.84) can be integrated to give Eq. (2.87). The estimates of E_a and k_{T_0} (rate constant at T_0) can be obtained by fitting the drug concentration versus time data to the following equation:

$$\ln\left(\ln\frac{[\mathrm{D}]}{[\mathrm{D}]_0}\right)=\frac{aE_a}{R}t+\ln\left(\frac{Rk_{T_0}}{aE_a}\right) \tag{2.87}$$

Estimations using Eq. (2.87) and variants of this equation were performed manually with limited temperature programs and were not generally applicable to drug degradation studies.[324,325]

New analysis methods using flexible temperature programs have been reported with the increasing availability of computers to facilitate subsequent calculations. Zoglio and co-workers[326–328] proposed a method for obtaining optimal kinetic parameters. They described the degradation versus time curve as a function of E_a by using the arithmetic mean of an individual rate constant at time t as the mean rate constant and by representing temperature change in terms of a linear or polynomial expression.[326–328] Kay and Simon performed this estimation using an analog computer.[329] Edel and Baltzer applied a stepped heating program to this method.[330]

In contrast to these approximate methods, the nonlinear regression methods reported by Madsen *et al.*[331] and Tucker and Owen[332] utilized numerical integration of Eq. (2.88), which is obtained from the general rate equation (Eq. 2.11) and Eq. (2.84). Equation (2.88) becomes Eq. (2.89) for first-order degradation kinetics.

$$\frac{-d[\mathrm{D}]}{[\mathrm{D}]^n}=A\exp\left[\frac{-E_a}{R\cdot G(t)}\right]dt \tag{2.88}$$

$$[\mathrm{D}]=[\mathrm{D}]_0\exp(-AI)$$

$$I=\int\exp\left[\frac{-E_a}{R\cdot G(t)}\right]dt \tag{2.89}$$

Hempenstall *et al.*[333] reported a calculation method that can be performed by simple computers, whereby the degradation curve is represented by a polynomial equation in order to easily obtain a rate constant k_T at a temperature T. Using Eq. (2.90), k_T can be represented by Eq. (2.91) in the case of first-order degradation kinetics. Inserting the coefficients a_0, a_1, . . . , a_n, [which are obtained by fitting the drug concentration versus time data to Eq. (2.90)] into Eq. (2.91) yields individual k_T values at a series of temperatures, T, which are then used to estimate kinetic parameters according to the Arrhenius equation (Eq. 2.70).

$$\ln [D] = a_0 + a_1 t + a_2 t^2 + \cdots + a_n t^n \tag{2.90}$$

$$k_T = \frac{-d(\ln[D])}{dt} = -a_1 - 2a_2 t - 3a_3 t^2 - \cdots - n a_n t^{n-1} \tag{2.91}$$

Yoshioka *et al.*[334] reported a method for obtaining the estimates of t_{90} and E_a directly from the concentration versus time data using an equation obtained by replacing the frequency factor A with t_{90}. Equation (2.92) was used to describe the process for first-order degradation kinetics.

$$[D] = [D]_0 \exp\left[-0.1054 \exp\left(\frac{E_a}{R \cdot 298}\right)\right] \cdot \frac{I}{t_{90}} \tag{2.92}$$

where

$$I = \int \exp\left[\frac{-E_a}{R \cdot G(t)}\right] dt$$

In addition, many papers have reported various methods for nonisothermal estimation of kinetic parameters and its application to stability prediction of pharmaceuticals.[335–347]

Reliability of Kinetic Parameters Estimated by Nonisothermal Prediction. The reliability of the kinetic parameters estimated by fitting degradation data observed under varying temperature conditions to a nonisothermal rate equation depends largely on the temperature program used. The accuracy and precision of kinetic parameter estimation using various nonisothermal temperature programs were studied by using a series of Monte Carlo simulations.[334,348] Three hundred sets of experimental data with specified temperature and assay errors were generated using the 12 linear temperature programs shown in Fig. 40. These data sets were then analyzed according to Eq. (2.92), resulting in estimates of t_{90} and E_a values. All of the temperature programs used provided estimates with similar means, which were close to the theoretical values, but with significantly different variations. For example, a

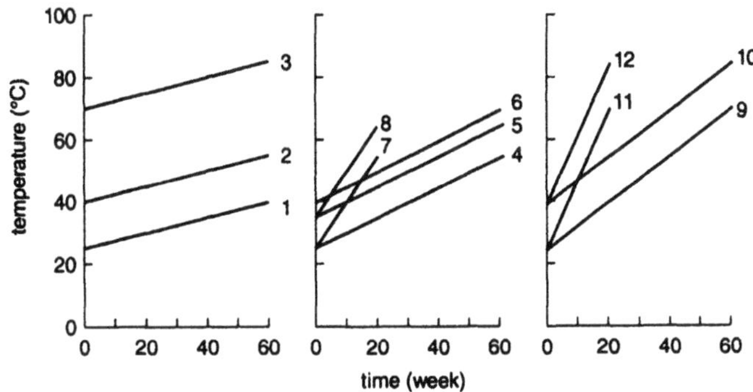

Figure 40. Twelve different patterns of temperature programs used to study the effect of temperature programs on kinetic parameters estimated by nonisothermal prediction. (Reproduced from Ref. 334 with permission.)

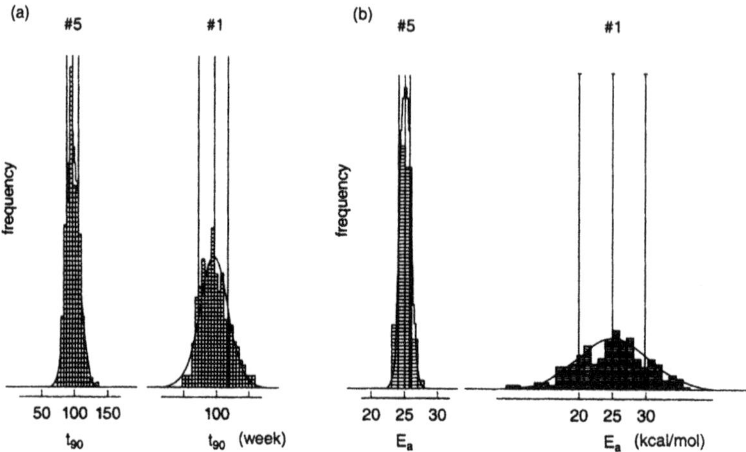

Figure 41. Effect of linear temperature programs on t_{90} (a) and E_a (b) estimates and variations. Theoretical value of t_{90}, 100 weeks: theoretical value of E_a, 25.00 kcal/mol; assay error (standard deviation): 2%. (Reproduced from Ref. 334 with permission.)

marked difference in the variations of the t_{90} and E_a estimates was observed between temperature programs 1 and 5 when the theoretical values of t_{90} and E_a were 100 weeks and 25 kcal/mol, respectively (see Fig. 41). The variation of the estimates depended on the temperature variation range and the final percentage of degraded drug, so that estimates with smaller variations were obtained by using a program providing a larger temperature range and a higher percentage degraded (see Fig. 42). The t_{90} estimate is affected by the number of observations at temperatures near room temperature; thus, estimates with smaller variations are obtained by using a program including temperatures near room temperature. For example, when an E_a value of 10 kcal/mol and programs 3 and 6 were used, estimates with enormous variations were obtained.

Although temperature programs providing a higher final percentage degraded and a larger temperature range and those including temperatures close to room temperature are needed to obtain estimates with small variations, optimization of temperature programs is not easily achieved due to the interrelationship of these factors. As shown in Fig. 43, nonlinear temperature programs with different patterns (Fig. 44) provided estimates with different variations even when the final degradation ratios were the same. The Monte Carlo simulation study suggested that estimates with relatively small variations for a pharmaceutical having a t_{90} of approximately 3 years can be obtained by using the program $T(°C) = 25 + 1.56 \times 10^{-5}t^4$ (week) for a 40-week experiment and the program $T(°C) = 25 + 4t$ (week) for a 10-week experiment.

Advantages of Nonisothermal Prediction. In both isothermal and nonisothermal predictions of degradation rate at room temperature from data obtained at elevated temperatures, degradation data that include higher overall degradation yield more reliable estimates. Therefore, the difference in the precision of the estimates between isothermal and

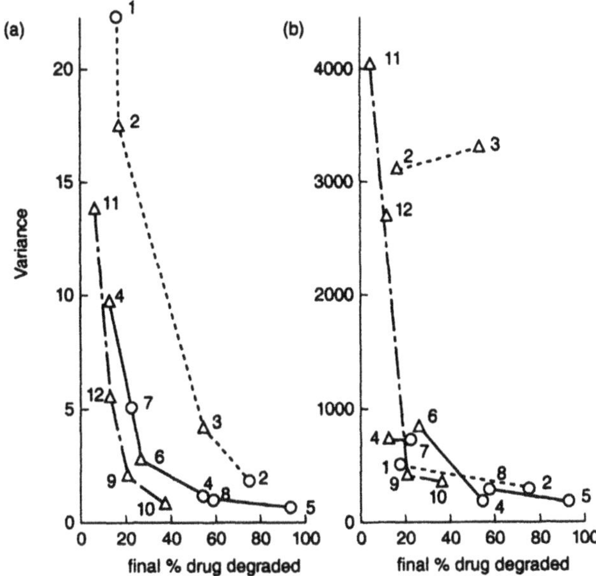

Figure 42. Effects of final degradation ratio and temperature variation range on E_a (a) and t_{90} (b) estimates. Temperature variation range: ----, 15°C; ——, 30°C; —·—, 45°C. ○, E_a=25 kcal/mol; △, E_a = 10 kcal/mol. (Reproduced from Ref. 334 with permission.)

nonisothermal predictions is due to the difference in the final percentage degradation achieved.

In the prediction of degradation rate according to the Arrhenius equation for degradations exhibiting slightly curved Arrhenius plots, nonisothermal prediction may provide an

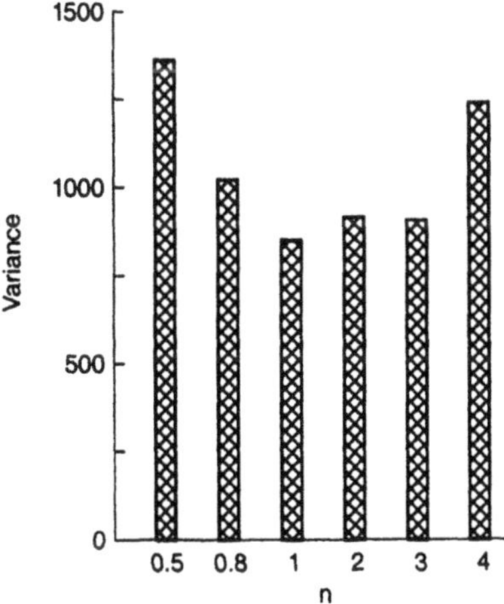

Figure 43. Effect of nonlinear temperature programs on the variance of t_{90} estimate (effect of n in the nonlinear temperature program, T (°C) = $25 + kt^n$). Assay error (standard deviation): 2%; temperature error (standard deviation), 0.5°C. Theoretical value of t_{90}, 156 weeks; E_a, 25 kcal/mol. (Reproduced from Ref. 348 with permission.)

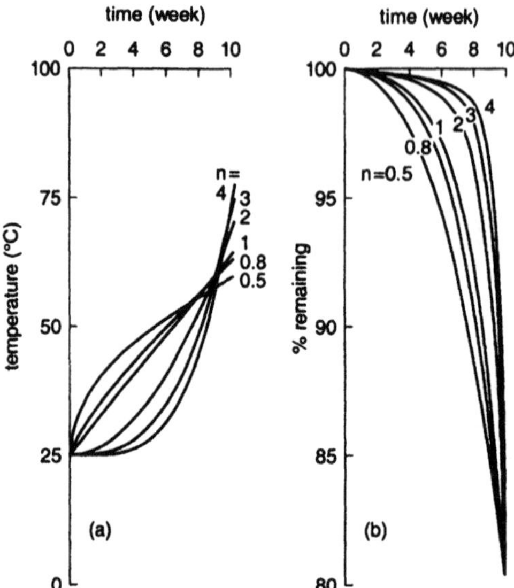

Figure 44. Nonlinear temperature programs (a) and degradation curves (b) used to study the effect of temperature programs on kinetic parameters estimated by nonisothermal prediction. Theoretical value of t_{90}, 156 weeks; E_a, 25 kcal/mol. (Reproduced from Ref. 348 with permission.)

estimate with a smaller bias than does isothermal prediction. Monte Carlo simulation studies showed that a t_{90} estimate close to the theoretical value was obtained for degradation exhibiting the slightly curved Arrhenius plots shown in Fig. 45 when nonisothermal prediction was used [temperature program: $T(°C) = 25 + 4t$ (week)] rather than isothermal prediction using four levels of constant temperature (50, 60, 70, and 80°C); see Table 6.[348]

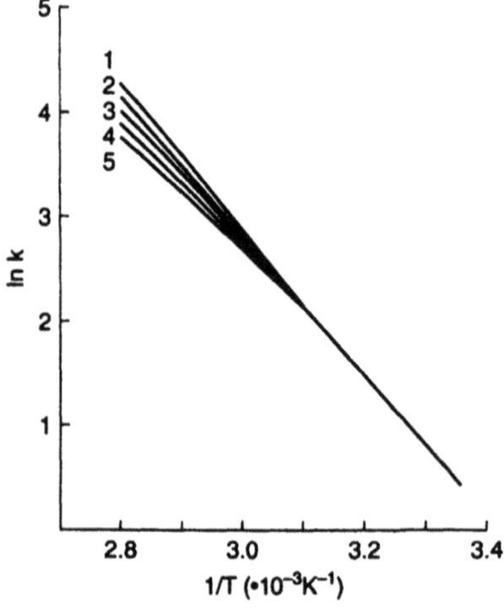

Figure 45. Slightly curved Arrhenius plots used for Monte Carlo simulation studies. (1, 2) Concave plots; (3) linear plots; (4, 5) convex plots. (Reproduced from Ref. 348 with permission.)

Table 6. Estimated t_{90} Values from Slightly Curved Arrhenius Plots[a,b]

Number in Fig. 45	Duration of experiment (weeks)	$t_{90(25)}$ (weeks) Nonisothermal	Isothermal
1	9.6	129.9 ± 21.2	176.2 ± 15.0
2	9.4	115.3 ± 20.0	132.3 ± 13.5
3	9.2	101.3 ± 16.6	100.7 ± 11.1
4	8.8	90.9 ± 15.9	77.1 ± 9.1
5	8.6	79.5 ± 15.9	60.6 ± 8.5

[a]Reference 348.
[b]Theoretical $t_{90(25)}$, 100 (weeks); E_a, 25 kcal/mol; assay error, 2%; temperature error, 0.5°C; temperature program, T(°C) = 25 + 4t.

Nonisothermal prediction using a temperature program including temperatures near room temperature reduces the effect of the deviation from the Arrhenius equation. Thus, nonisothermal prediction appears to be more suitable for t_{90} estimation when E_a might vary with temperature owing to possible changes in degradation mechanisms, an observation quite common in some drug degradation studies.

The advantage of nonisothermal prediction was shown in the estimation of t_{90} values of a vitamin A syrup.[349] Nonisothermal prediction from data observed at temperatures ranging from 25 to 54.7°C using the program $T(°C) = 25 + 0.004t^4$ (week) (Fig. 46) provided a t_{90} estimate of 83.0 days, which was close to the value observed at 25°C (82.7 days), whereas isothermal prediction from data observed at 40, 50, 60, and 70°C provided an estimate of 60.9 days. Relatively accurate t_{90} estimates can be obtained by nonisothermal prediction using a program including temperatures near room temperature even if the Arrhenius plots are not strictly linear.

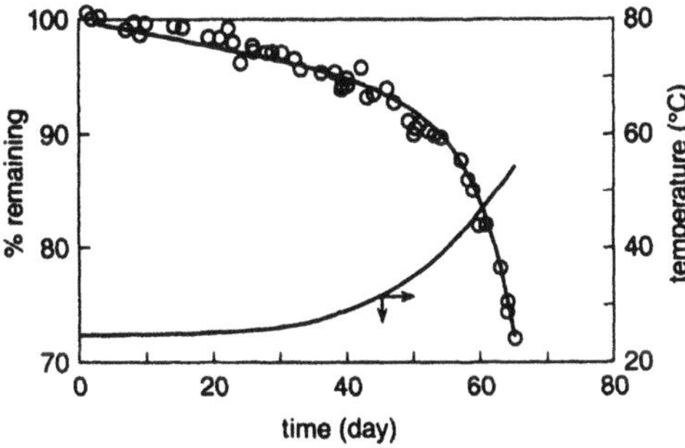

Figure 46. Time courses of degradation of vitamin A under nonisothermal conditions. Temperature program: T (°C) = 25 + 0.004t^4 (week). (Reproduced from Ref. 349 with permission.)

2.2.4.3. Stability in Frozen Solutions

Freezing is often assumed to slow chemical degradation. The assumption is correct in the majority of cases; however, freezing of aqueous solutions of drugs may increase drug degradation when the degradation occurs via bimolecular or higher orders of reaction. At temperatures just below freezing, solutes, including the drug and its reactants, become concentrated in the space between the ice crystals, thus effectively concentrating the drug and reactants. Freezing has also been proposed to enhance drug degradation by forming ice structures, resulting in an arrangement of the drug molucules that is suitable for degradation.

The epimerization rate of moxalactam in frozen solutions, which occurs in the unfrozen aqueous phase after initial ice formation, exhibits linear Arrhenius plots as shown in Fig. 47, indicating that the effect of concentration change caused by freezing is negligible in this case.[350] However, the bimolecular hydrolysis rate of *n*-propyl 4-hydroxybenzoate (propylparaben) in alkaline medium increased at temperatures between –4 and –14°C (Fig. 48). This cannot be fully explained by the increase in drug concentration in the unfrozen pockets between ice crystals.[351] Degradation of amoxicillin sodium, which undergoes bimolecular polymerization in aqueous solutions, exhibited a maximum around –7°C, as shown in Fig. 49.[352]

Freezing of phosphate buffer solutions may affect drug stability by lowering the micro-pH of the condensed aqueous phase. The degradation rate of mitomycin C in frozen phosphate buffer solutions was faster than in solutions.[353]

Care should be taken when storing, under freezer conditions, drug substances whose degradation is dependent on concentration and pH. Decreased degradation due to the decreased temperature cannot always be assumed.

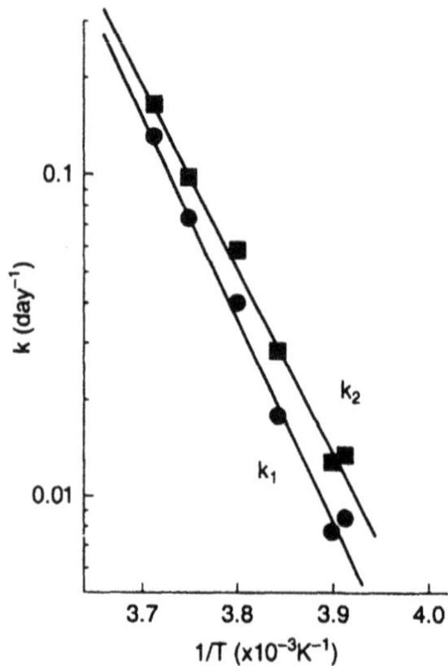

Figure 47. Arrhenius plots of rate constants (k_1 and k_2) for the epimerization of moxalactam in frozen solution, calculated according to a reversible reaction model. (Reproduced from Ref. 350 with permission.)

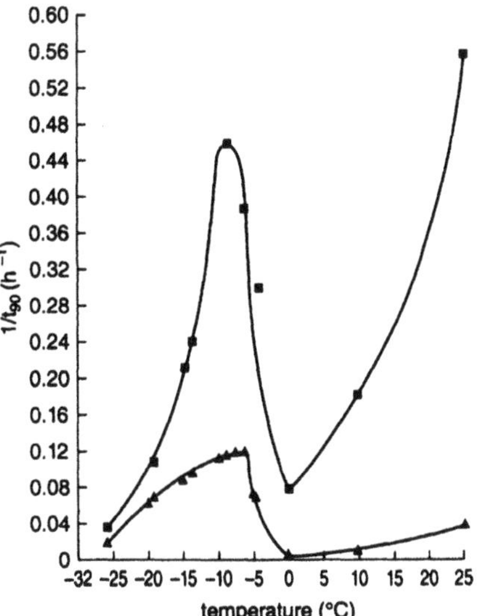

Figure 48. Degradation of *n*-propyl 4-hydroxybenzoate (propylparaben) in frozen solutions ($[OH^-] = 0.05M$). (Reproduced from Ref. 351 with permission.)

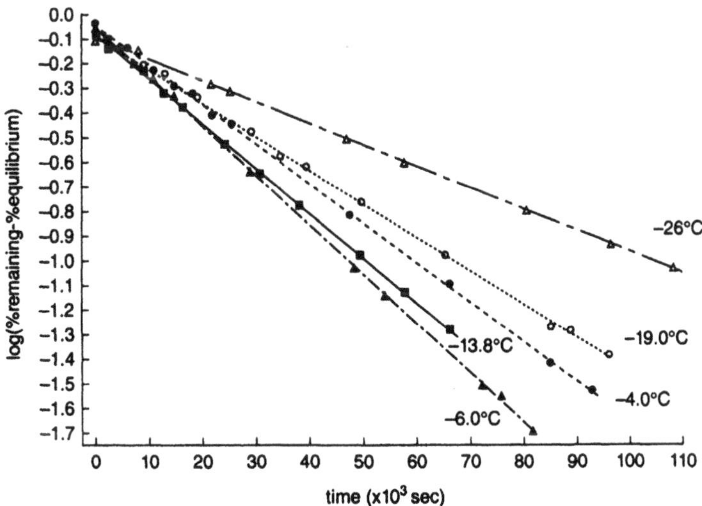

Figure 49. Temperature dependence of degradation rate of amoxicillin sodium in saline (▲) and 5% glucose solution (■). (Reproduced from Ref. 352 with permission.)

2.2.5. pH and pH–Rate Profiles

After temperature, the second most important variable affecting drug degradation is pH. The effect of pH on degradation rates of drug substances in aqueous solutions has been studied extensively, and the pH dependency of the degradation rate of benzylpenicillin was reported in the 1940s.[354] The effect of pH on degradation rate can be explained by the catalytic effects that hydronium or hydroxide ions can have on various chemical reactions. Effectively, a catalyst is a species that does not change the free energy of the reactants and products (this definition is not always followed) but acts to lower the $\Delta G^{\ddagger}$ term; that is, it lowers the energy barrier to reaction. By definition, a true catalyst is not consumed as a result of the reaction.

Degradation rates of drug substances are generally affected by pH because most degradation pathways are catalyzed by hydronium and/or hydroxide ions. Water itself is also a critical reactant. If the critical path in a reaction involves a proton transfer or abstraction step, other acids and bases present in solution (usually buffer species) can affect the reaction rate. These reactions will also be pH-dependent because the fraction of any species present in its acid or base form will be dependent on its dissociation constant and the solution pH. Also, for ionizable drugs, the fraction of drug present in any particular form will depend on the solution pH. Therefore, if the reactivity of the drug depends on its form, its reactivity will be pH-dependent.

When a reaction dependent on hydronium and hydroxide ion activity is performed at constant pH, it usually follows pseudo-first-order kinetics, which can be described by a first-order rate constant k_{obs}. A reaction in which hydronium ion, hydroxide ion, and water catalysis are observed can be described by

$$k_{obs} = k_{H^+}a_{H^+} + k_{H_2O} + k_{OH^-}a_{OH^-} \tag{2.93}$$

where k_{obs} is the sum of the specific rate constants and activities for each parallel pathway, and a_{H^+} and a_{OH^-} are the activities of hydronium and hydroxide ion, respectively. This equation is for the case when the drug itself is neutral in the pH range of study, that is, where ionization of the drug does not have to be taken into account. The pH–rate profiles for drugs meeting this criterion are relatively simple, as shown in Fig. 50.

If the contributions of the first and second terms in Eq. (2.93) are larger than that of the third term, the pH–rate profile shown in panel 1 of Fig. 50 is seen. If the second and third terms are dominant, then the profile illustrated in panel 2 is observed. If the first and third

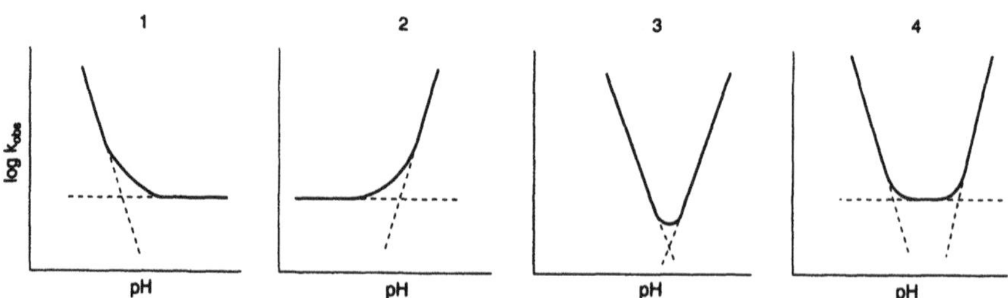

Figure 50. Simple pH–rate profiles for drug degradation.

terms are dominant, then a V-type pH–rate profile (panel 3) occurs. If all terms contribute significantly, the U-shaped pH–rate profile shown in panel 4 is observed.

Generally, drug substances capable of undergoing ionization yield more complex pH–rate profiles. For example, each ionic and nonionic form of the drug could be subject to hydronium ion, hydroxide ion, and water catalysis. When this happens, the expression for k_{obs} may contain more than three terms. For example, apparent degradation rate constants for a drug substance that is a weak base will depend on the ionization constant, K_a, of the conjugate acid of the weak base and the concentration of hydronium ion and other species, as described by

$$k_{obs} = k_{H^+}a_{H^+}\frac{a_{H^+}}{K_a + a_{H^+}} + k_{H_2O}\frac{a_{H^+}}{K_a + a_{H^+}} + k'_{H_2O}\frac{K_a}{K_a + a_{H^+}} + k_{OH^-}a_{OH^-}\frac{K_a}{K_a + a_{H^+}} \quad (2.94)$$

where k_{H^+} and k_{OH^-} are the hydronium ion- and hydroxide ion-catalyzed rate constants for ionized and nonionized drug, respectively, and k_{H_2O} and k'_{H_2O} are the H_2O-catalyzed rate constants for ionized and nonionized drug, respectively. An example of a profile for such a drug is shown in Fig. 51. The shape of the profile, especially the shoulder at higher pH values, is determined by the relative magnitudes of each of the terms in Eq. (2.94). The dashed lines, a–d, represent the contributions of each of the four terms in Eq. (2.94) A similar profile can be generated for drug substances that are weak acids.

Weak polybasic and polyacidic drugs exhibit even more complex pH–rate profiles. For example, a weak base with three ionizable groups has three ionized forms in addition to the un-ionized form, the degradation of each species being potentially catalyzed by hydronium ion, hydroxide ion, and water. Therefore, the apparent rate constant, k_{obs}, could theoretically contain up to 12 terms, and the drug would exhibit a complex pH–rate profile, as illustrated in Fig. 52.[355,356]

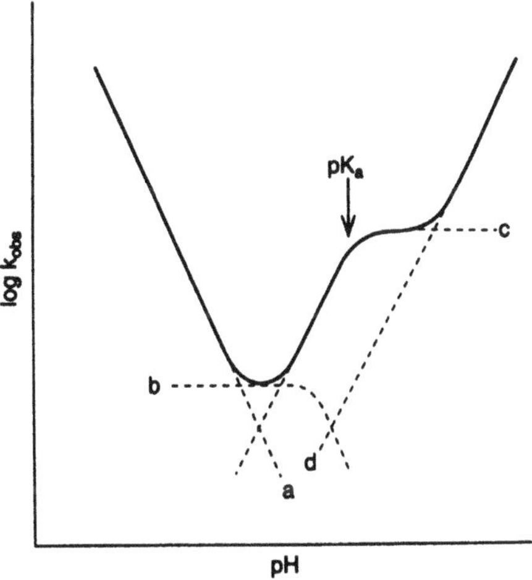

Figure 51. A typical pH-rate profile for degradation of drug substances with a single ionizable group. Curves a–d represent, respectively, the first, second, third, and fourth terms in Eq. (2.94).

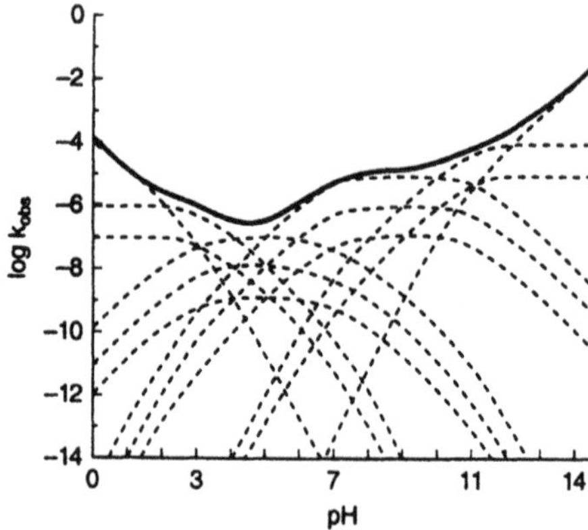

Figure 52. pH–rate profile for the degradation of a weak base with three ionizable groups. Dashed curves represent the contributions of all reactions of the different species. (Reproduced from Ref. 355 with permission.)

Although complex pH–rate profiles are often observed, some drug substances exhibit apparent degradation rate constants that are relatively independent of pH, as exemplified by the pH–rate profile of estramustine from pH 1 to 10 shown in Fig. 53.[357] The explanation for this is that estramustine undergoes a spontaneous unimolecular reaction that is catalyzed by neither acid nor base in this pH range. Other examples of pH–rate profiles for specific drug substances are presented in the following sections.

2.2.5.1. V-Type and U-Type pH–Rate Profiles

The pH–rate profiles for the hydrolysis of diltiazem,[358] fenprostalene,[359] and EO9 (a derivative of aziridinylquinone)[360] are all V-shaped (Fig. 54), indicating only apparent hydronium ion and hydroxide ion catalysis. The dehydration reaction of streptovitacin A shown in Scheme 32 also exhibits a V-type pH–rate profile.[147]

As stated earlier, if the degradation of a drug follows Eq. (2.93), its pH–rate profile will be V- or U-shaped—why? If we first consider the case where $k_{H_2O} = 0$ and $k_{H^+} = k_{OH^-}$, then when $a_{H^+} >> a_{OH^-}$, Eq. (2.93) reduces to Eq. (2.95) under these pH conditions.

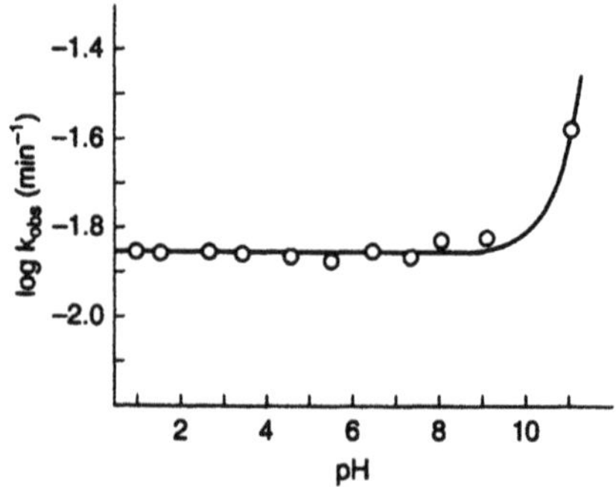

Figure 53. pH–rate profile for hydrolysis of estramustine at 80°C. (Reproduced from Ref. 357 with permission.)

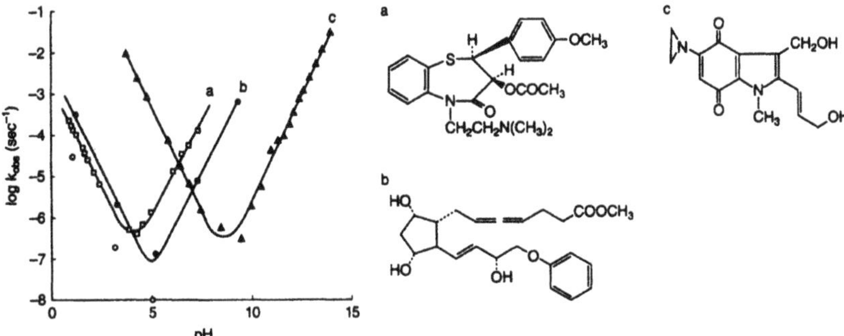

Figure 54. Representative drug substances that exhibit V-type pH–rate profiles: (a) diltiazem (80°C); (b) fenprostalene (80°C); (c) EO9 (25°C). (Reproduced from Refs. 358, 359, and 360 with permission.)

$$k_{obs} = k_{H^+}a_{H^+} \tag{2.95}$$

Taking the logarithm of both sides yields

$$\log k_{obs} = \log k_{H^+} - \text{pH} \tag{2.96}$$

Therefore, a plot of log k_{obs} versus pH should have a slope of −1. Similarly, at $a_{H^+} << a_{OH^-}$

$$\log k_{obs} = \log (k_{OH^-}Kw) + \text{pH} \tag{2.97}$$

where Kw is the ion product of water. Thus, a plot of log k_{obs} versus pH will have a slope of +1.

When $k_{H^+} \approx k_{OH^-}$, the $k_{H^+}a_{H^+}$ contributions and the $k_{OH^-}a_{OH^-}$ contributions to Eq. (2.97) will be equal at pH 7 such that the pH of maximum stability (lowest k_{obs}) will occur at around neutrality. If $k_{H^+} << k_{OH^-}$, the V shape will be shifted to lower pH values, and maximum stability (lowest k_{obs}) will occur at pH values below 7. If $k_{H^+} >> k_{OH^-}$, maximum stability will occur at pH values above 7.

If $k_{H_2O} >> (k_{H^+}a_{H^+} + k_{OH^-}a_{OH^-})$, then Eq. (2.97) leads to

$$k_{obs} = k_{H_2O} \tag{2.98}$$

As can be seen from this equation, k_{obs} should be independent of pH. Thus, when this term is numerically dominant, a pH-independent region is often seen as part of the pH–rate profile (see profile 4, Fig. 50). That is, U-shaped pH–rate profiles occur when water catalysis can compete with hydronium ion and hydroxide ion catalysis. The hydrolyses of cephalothin, cephaloridine, and cefotaxime[70,73–75] (Fig. 55) are three examples. Although these drug substances have an ionizable carboxylic group at the 4-position, apparent pH–rate profiles are U-shaped because there is no difference in degradation rate between the ionized and un-ionized forms of these substances. A U-shaped pH–rate profile has also been reported for hydrolysis of 4′-azidothymidine.[133]

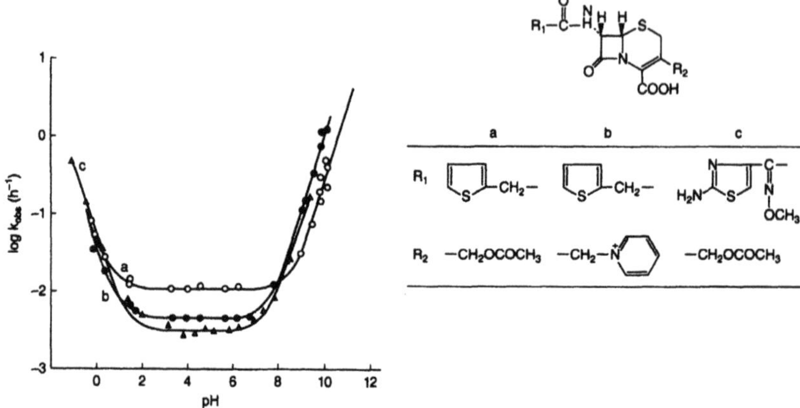

Figure 55. Representative drug substances that exhibit U-type pH–rate profiles: (a) cephalothin (35°C); (b) cephaloridine (35°C); (c) cefotaxime (25°C). (Reproduced from Refs. 70 and 73 with permission.)

2.2.5.2. *pH–Rate Profiles with Inflection Points Due to the Presence of One or More Ionized Groups*

The pH–rate profile for the ring-closure reaction of nimustine[361] reflects the effect of ionization of the amino group on the pyrimidine ring (Fig. 56). Similarly, hydrolysis of the cephalosporins cephaloglycin, cephalexin, and cephradine[70] and cefadroxil,[71] as well as loracarbef,[362] yields pH–rate profiles with an inflection point around pH 7 due to ionization of the side-chain amino group, as shown in Figs. 57 and 58.

The severe inflections seen in the profiles in Fig. 57 reflect changes in the mechanisms of degradation with changes in pH. At pH values below 6, the major reaction is cleavage of the β-lactam ring. At pH values above 9, the major reaction also involves cleavage of the β-lactam ring due to hydroxide ion attack. However, between pH 6 and 9, the major reaction

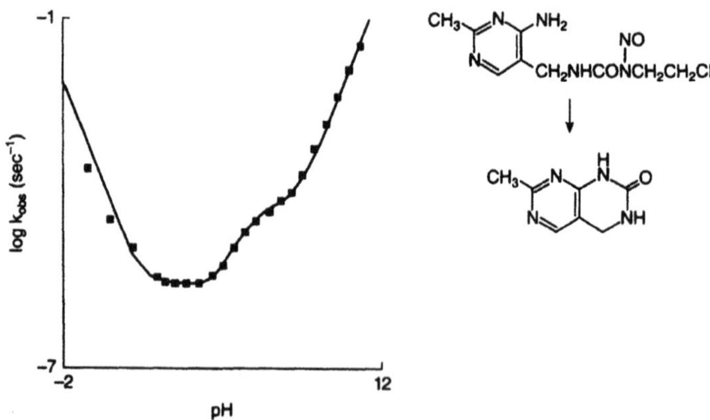

Figure 56. pH–rate profile for ring-closure reaction of nimustine at 25°C. (Reproduced from Ref. 361 with permission.)

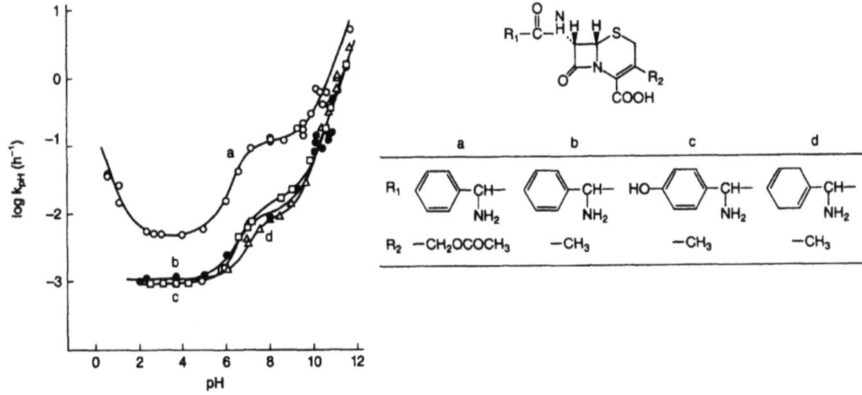

Figure 57. pH–rate profiles for hydrolysis of various cephalosporins: (a) cephaloglycin; (b) cephalexin; (c) cefadroxil; (d) cephradine. (k_{pH}): Rate constants obtained by extrapolating the buffer concentration to zero at 35°C. (Reproduced from Refs. 70 and 71 with permission.)

is an intramolecular attack of the side-chain amino group on the β-lactam ring, resulting in the formation of a diketopiperazine product (Scheme 70). The inflection in the pH–rate profile follows the change in the state of ionization of the side-chain amino group.

Although ampicillin also has a side-chain amino group in a position similar to that in the cephalosporins, the hydrolysis rate is not significantly affected by the state of ionization of the amino group. Ampicillin does exhibit an inflection point around pH 2–3 in its pH–rate profile due to ionization of the carboxyl group, as shown in Fig. 59.[65] The side-chain amino group in penicillins cannot attack the β-lactam ring to form the diketopiperazine, presumably due to conformational restrictions. A pH–rate profile similar to that of ampicillin, with an inflection point due to ionization of the carboxyl group, has been reported for the hydrolysis of carbenicillin[363] as well as other penicillins.

A classical pH–rate profile demonstrating the effect of ionization of a neighboring group on a chemical reaction can be seen in the pH–rate profile for the hydrolysis of aspirin[11,364]

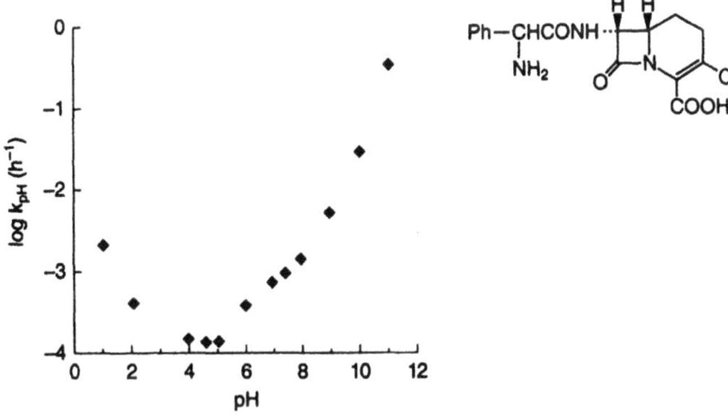

Figure 58. pH–rate profile for hydrolysis of loracarbef at 35°C. (Reproduced from Ref. 362 with permission.)

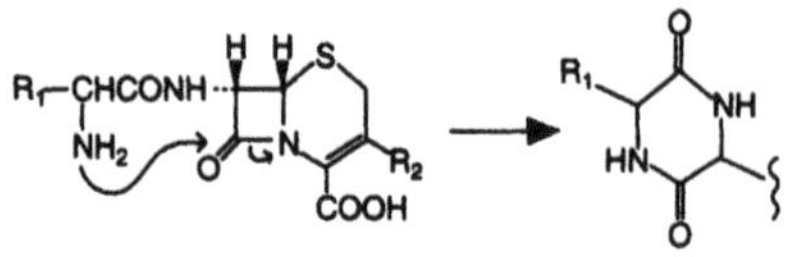

Scheme 70. Intramolecular attack by the side-chain amino group in cephalosporins in the pH range of 6–9, leading to formation of diketopiperazine degradants.

(Fig. 60). The degradation of aspirin in the pH-range 4–8 is independent of pH. The major reaction mechanism in this pH range is intramolecular general base catalysis of the attack by water on the acetyl ester functional group by the neighboring carboxylate group.[365] The pH–rate profiles for aspirin can be interpreted by considering the reaction possibilities (shown in Scheme 71a). Thus, the degradation kinetics can be defined as follows:

$$-\frac{d([\mathrm{HA}]+[\mathrm{A}^-])}{dt} = -\frac{d[\mathrm{aspirin}]_T}{dt}$$

$$= (k_{\mathrm{H}^+}a_{\mathrm{H}^+}f_{\mathrm{HA}} + k_{\mathrm{H_2O}}f_{\mathrm{HA}} + k_{\mathrm{OH}^-}a_{\mathrm{OH}^-}f_{\mathrm{HA}} + k'_{\mathrm{H}^+}a_{\mathrm{H}^+}f_{\mathrm{A}^-}$$

$$+ k'_{\mathrm{H_2O}}f_{\mathrm{A}^-} + k'_{\mathrm{OH}^-}a_{\mathrm{OH}^-}f_{\mathrm{A}^-})[\mathrm{aspirin}]_T$$

$$= k_{\mathrm{obs}}[\mathrm{aspirin}]_T \tag{2.99}$$

The various rate constants are defined in Scheme 71; f_{HA} and f_{A^-} are the fractions of aspirin present in its neutral and anionic forms, respectively. For a weak acid with one ionizable acid group, f_{HA} and f_{A^-} can be defined by

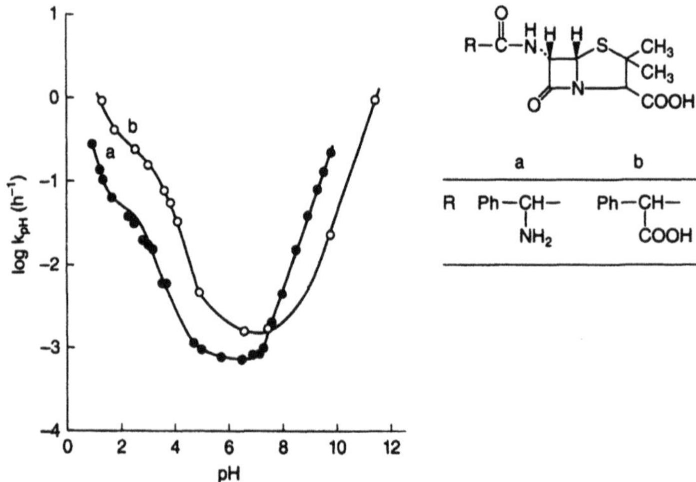

Figure 59. pH–rate profiles for hydrolysis of ampicillin (a) and carbenicillin (b) at 35°C. (Reproduced from Refs. 65 and 363 with permission.)

COOH
OCOCH3
log k_obs (day^-1)
2
1
0
-1
1 2 3 4 5 6 7 8 9 10 11 12
pH

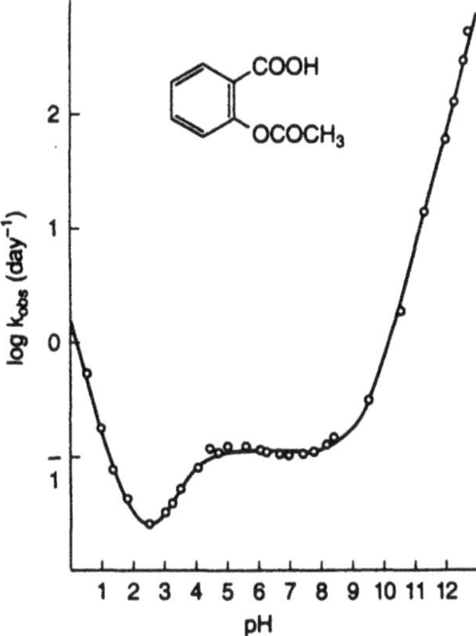

Figure 60. pH–rate profile for hydrolysis of aspirin at 17°C. (Reproduced from Ref. 11 with permission.)

$$f_{HA} = \frac{a_{H^+}}{a_{H^+} + K_a} \tag{2.100}$$

$$f_{A^-} = \frac{K_a}{a_{H^+} + K_a} \tag{2.101}$$

where K_a is defined as the dissociation constant of the acid HA, aspirin in this case.

Therefore, theoretically, the hydrolysis of aspirin may proceed, depending on solution pH, by six possible pathways. However, some of the terms are mathematically equivalent. Consider the second and fourth terms in Eq. (2.99):

$$k_{H_2O} f_{HA} = k_{H_2O} \frac{a_{H^+}}{a_{H^+} + K_a} = A \frac{a_{H^+}}{a_{H^+} + K_a} \tag{2.102}$$

where

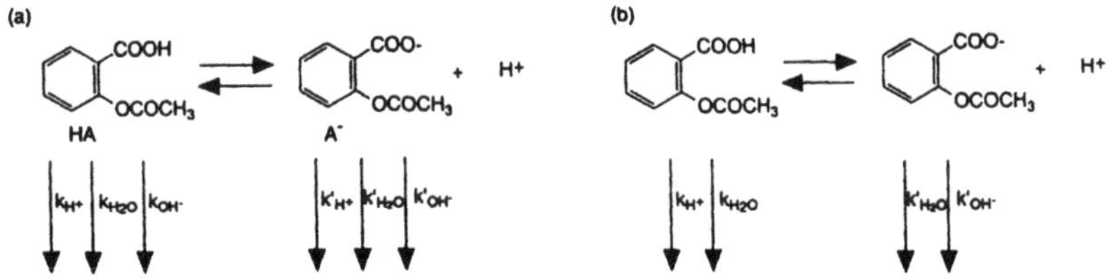

(a)
COOH
OCOCH3
HA
COO-
OCOCH3
A-
+ H+
kH+ kH2O kOH-
k'H+ k'H2O k'OH-
(b)
COOH
OCOCH3
COO-
OCOCH3
+ H+
kH+ kH2O
k'H2O k'OH-

Scheme 71. Parallel reaction pathways for the degradation of aspirin and its carboxylate. The scheme in panel a can be simplified to the scheme in panel b because of the kinetic equivalence of some of the rate pathways.

$$A = k_{H_2O}$$

$$k'_{H^+} f_{A^-} = k'_{H^+} \frac{K_a}{a_{H^+} + K_a} = B \frac{a_{H^+}}{a_{H^+} + K_a} \qquad (2.103)$$

where

$$B = k_{H^+} K_a$$

It is obvious that both terms have the same mathematical "shape" or form. Similarly, the third and fifth terms in Eq. (2.99) are indistinguishable. Therefore, Scheme 71a can be simplified to Scheme 71b. Thus, k_{obs} for the degradation of aspirin can be described by Eq. (2.104), which is identical in form to Eq. (2.94).

$$k_{obs} = k_{H^+} a_{H^+} \frac{a_{H^+}}{a_{H^+} + K_a} + k_{H_2O} \frac{a_{H^+}}{a_{H^+} + K_a} + k'_{H_2O} \frac{K_a}{a_{H^+} + K_a} + k_{OH^-} a_{OH^-} \frac{K_a}{a_{H^+} + K_a} \qquad (2.104)$$

This simplification involved choosing the k_{H_2O} term over the k'_{H^+} term and the k'_{H_2O} term over the k_{OH^-} term in Scheme 71. In such cases, one usually either selects a term based on intuition or performs a relatively sophisticated set of experiments to distinguish between the two possible pathways.

In the case of aspirin, it is interesting to compare the values of k'_{H_2O} and k_{H_2O} from a fit of the pH–rate profile. The larger value of k'_{H_2O} results from the intramolecular participation by the carboxylate group, as stated earlier. The mechanism of this participation is shown in Scheme 72.

Fenclorac also has an ionizable carboxylic acid group and exhibits a pH-independent region in its pH–rate profile above pH 4 due to intramolecular nucleophilic catalysis, as shown in Fig. 61.[366] A similar pH–rate profile has been reported for the decarboxylation, accompanied by ring opening, of a derivative of isoxazolecarboxamide (Fig. 62).[367]

In the case of aspirin, the carboxylic acid group participates in the reaction because of its proximity to the reactive ester group, whereas in the case of fenclorac the carboxylic acid group probably acts in a nucleophilic capacity. The pH–rate profiles for the amino-side-chain cephalosporins illustrated earlier (Fig. 57) also demonstrate the influence of a neighboring group that actively participates in degradation. In many other cases, neighboring groups may participate indirectly through electronic effects. Many of the examples discussed below reflect this indirect effect.

Ionizable groups other than simple amino and carboxyl groups may give rise to inflection points in pH–rate profiles. As shown in Fig. 63, the pH–rate profiles for hydrolysis of cefazolin[70] and flomoxef[368] exhibit inflection points below pH 2 owing to weakly basic

COO⁻ H
O H
OCOCH₃

Scheme 72. Participation of the *ortho*-carboxylate group in the hydrolysis of aspirin in the pH range 4–8.

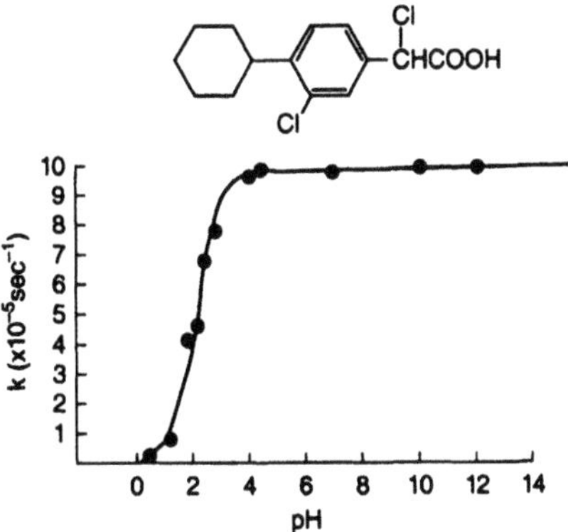

Figure 61. pH–rate profile for hydrolysis of fenclorac at 37°C by a reaction involving intramolecular nucleophilic catalysis. (Reproduced from Ref. 366 with permission.)

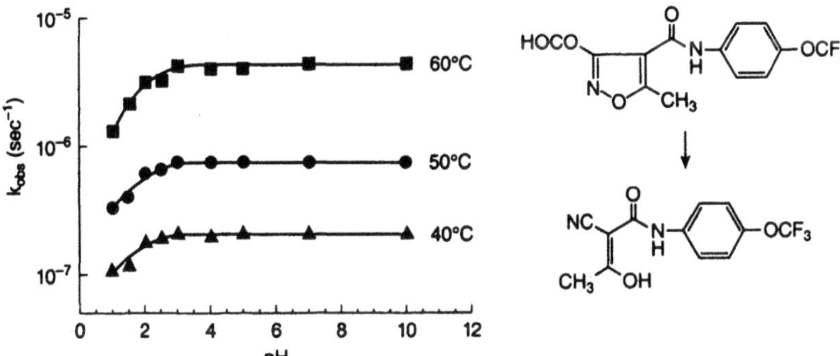

Figure 62. pH–rate profile for decarboxylation of a derivative of isoxazolecarboxamide. (Reproduced from Ref. 367 with permission.)

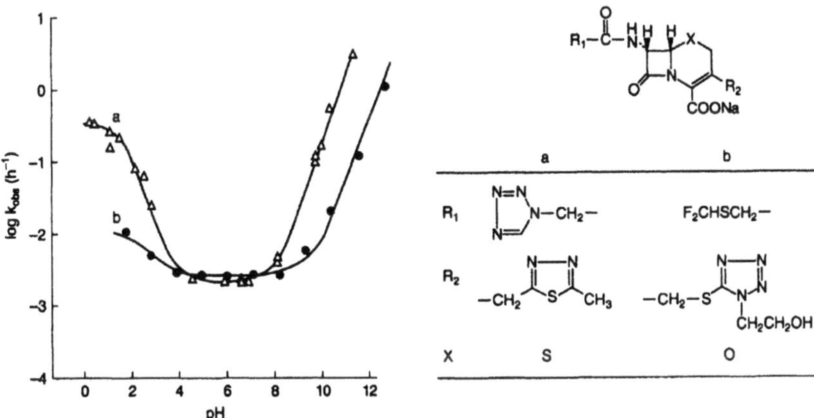

Figure 63. pH–rate profiles for degradation of cephalosporins having a basic site: (a) cefazolin (35°C); (b) flomoxef (25°C). (Reproduced from Refs. 70 and 368 with permission.)

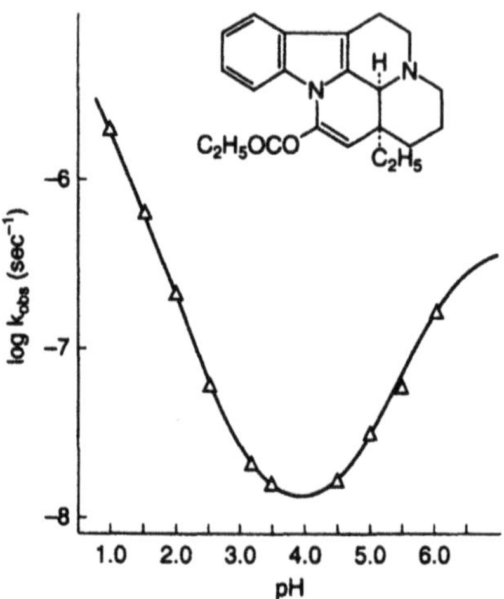

Figure 64. pH–rate profile for hydrolysis of vinpocetine at 70°C. (Reproduced from Ref. 369 with permission.)

sites in their structure. The effect of ionization of a basic site on pH–rate profiles has also been observed in the profiles for the hydrolysis of vinpocetine[369] and ciclosidomine,[370] which exhibit inflection points around pH 6 and 4–5, respectively (Figs. 64 and 65). Hydrolysis of cifenline in the pH range 8–13 mainly involves the reaction between the protonated form of cifenline and hydroxide ion. An inflection point can be seen around pH 9–10, corresponding to the change in the ionic form of cifenline (Fig. 66).[107] The long-range effect of ionization of an acidic group on a pH–rate profile can be seen for the epimerization and hydrolysis of etoposide. The inflection in the pH–rate profile is due to ionization of the phenolic hydroxyl

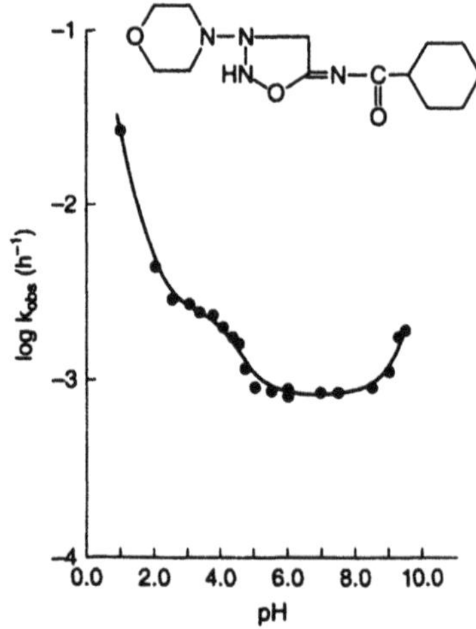

Figure 65. pH–rate profile for hydrolysis of ciclosidomine at 60°C. (Reproduced from Ref. 370 with permission.)

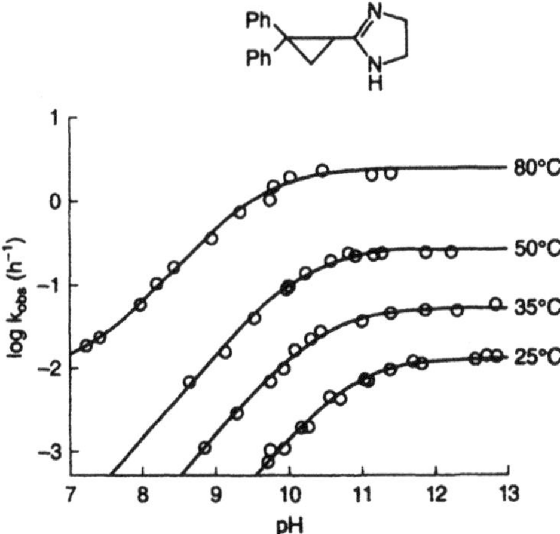

Figure 66. pH–rate profile for hydrolysis of cifenline. (Reproduced from Ref. 107 with permission.)

group, as shown in Fig. 67.[156] Similarly, ionization of the purine ring (pK_a 9.34) affects the pH–rate profile for hydrolysis of azathioprine (Fig. 68).[371]

Ionization accompanied by ring opening also affects pH–rate profiles of drug degradation. As shown in Fig. 69, the pH–rate profile for hydrolysis of the diazepinone ring of oxazolam (see Scheme 20) is affected by ionization of the oxazolidine ring.[93] Similarly, the

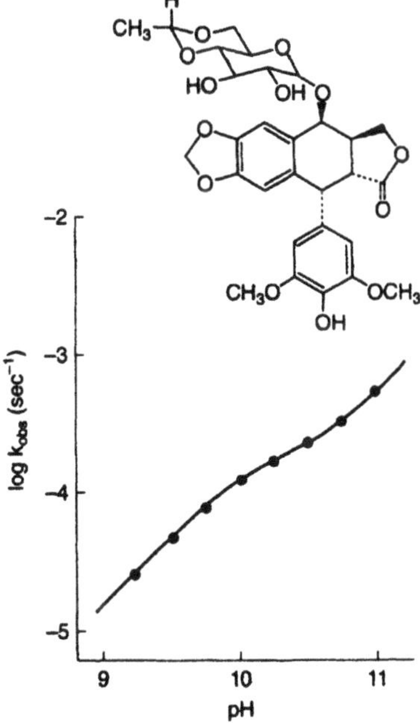

Figure 67. pH–rate profile for degradation of etoposide at 25°C. (Reproduced from Ref. 156 with permission.)

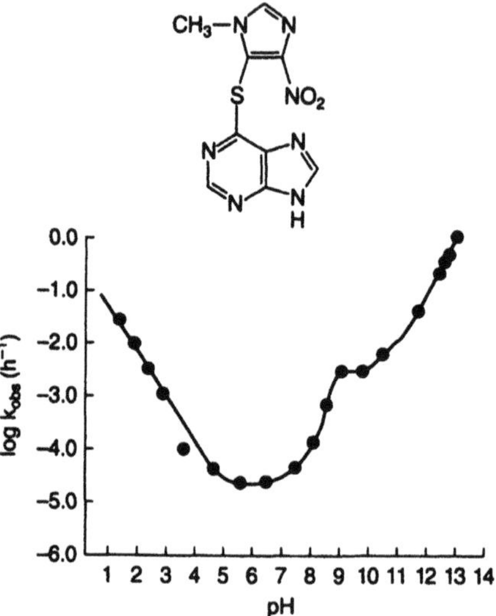

Figure 68. pH–rate profile for hydrolysis of azathioprine at 73°C. (Reproduced from Ref. 371 with permission.)

hydrolysis of benzodiazepines such as flutazolam and haloxazolam also exhibits complicated pH–rate profiles owing to ionization accompanied by ring opening (Fig. 70).[94]

Drug substances having three ionizable groups exhibit even more complex pH–rate profiles. As shown in Fig. 71, hydrolysis of cefotiam exhibits a complicated pH–rate profile

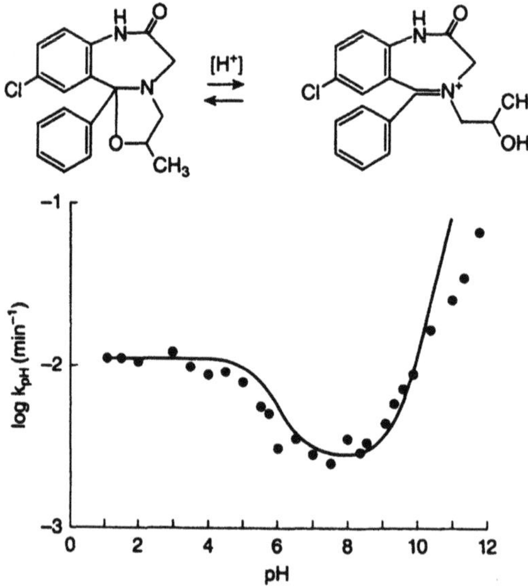

Figure 69. pH–rate profile for hydrolysis of oxazolam at 37°C. (Reproduced from Ref. 93 with permission.)

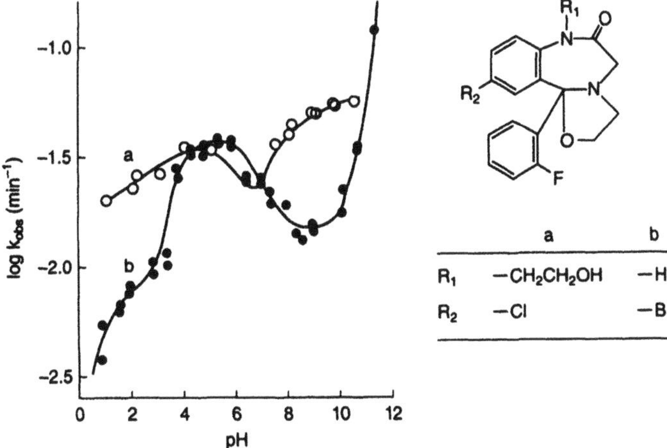

Figure 70. pH–rate profiles for hydrolysis of flutazolam (a) and haloxazolam (b) at 25°C. (Reproduced from Ref. 94 with permission.)

owing to ionization of a carboxyl group and two amino groups (the pK_a values for these groups are 2.1, 4.6, and 7.1, respectively).[372] Moxalactam has two carboxyl groups (pK_a 2.2 and 3.4) and a phenol group (pK_a 9.6); the pH–rate profile for epimerization is affected by ionization of the carboxyl group in the side chain in the acidic pH region and by ionization of the phenol group in the alkaline pH region (Fig. 72).[160–162]

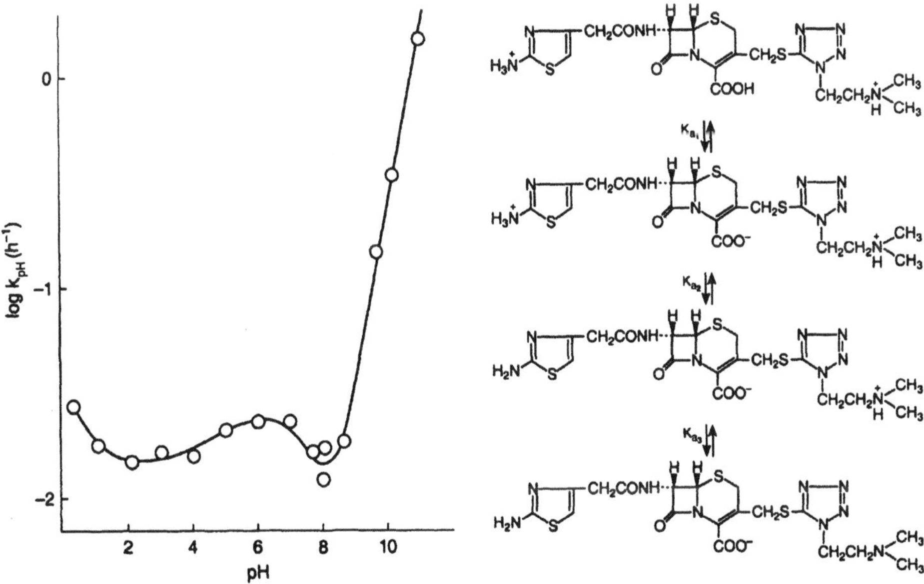

Figure 71. pH–rate profile for hydrolysis of cefotiam at 35°C. (Reproduced from Ref. 372 with permission.)

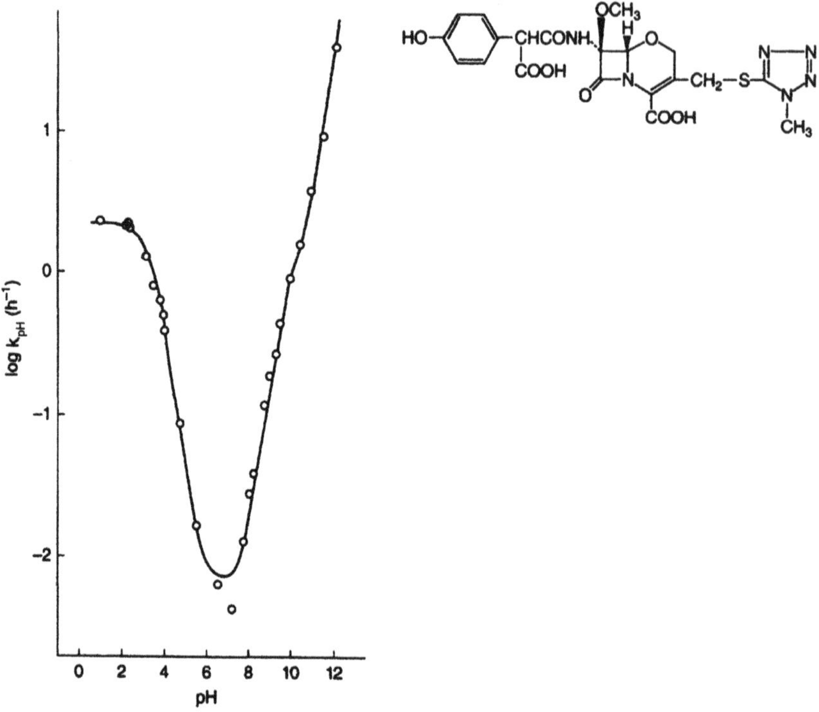

Figure 72. pH–rate profile for epimerization of moxalactam at 37°C. (Reproduced from Refs. 160 and 161 with permission.)

2.2.5.3. *Bell-Shaped pH–Rate Profiles Due to Ionization of Multiple Groups or Change in Rate-Determining Steps*

When a drug has at least two ionizable groups and the most reactive species is the intermediate species (e.g., HA^- for dibasic acid H_2A), bell-shaped pH–rate profiles can be seen with the pH of maximum degradation occurring at the pH corresponding to the isoelectric point, $(pK_{a1} + pK_{a2})/2$. At this pH, the concentration of the intermediate species is maximum. As shown in Fig. 73,[373] the decarboxylation of 4-aminosalicylic acid, an amphoteric electrolyte having a carboxylic group and an amino group, shows a bell-shaped pH–rate profile with a maximum at the isoelectric point. This is due to either the higher apparent reactivity of the amphoteric ion or the kinetically equivalent reaction of acid attack on the anionic form of 4-aminosalicylic acid. Hydrolysis of estrone phosphate exhibits a bell-shaped pH–rate profile with maximum degradation occurring in the pH range 3–5 owing to the greater reactivity of its monoanion form (Fig. 74).[374] A similar bell-shaped pH–rate profile has been reported for hydrolysis of dacarbazine.[375]

Another type of degradation leading to a bell-shaped pH–rate profile is a successive reaction in which a pH-dependent change in the rate-determining step occurs. Hydrolysis of benzothiadiazines such as hydrochlorothiazide shows a bell-shaped pH–rate profile, as seen in Fig. 75.[253] An inflection point observed at pH 8–10 is attributed to ionization of the sulfamide group, whereas one occurring around pH 4.5 results from a change in the

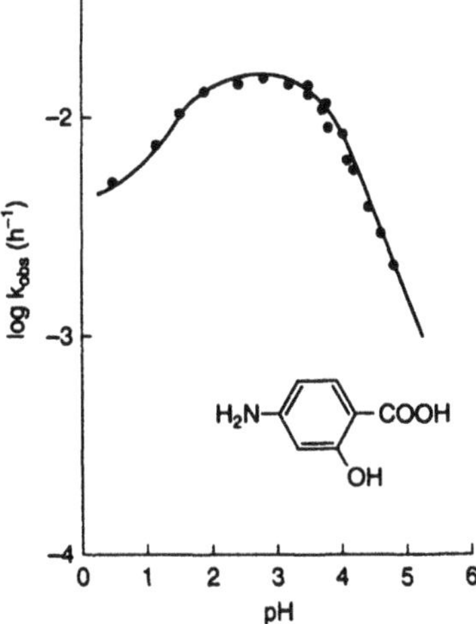

Figure 73. pH–rate profile for decarboxylation of 4-aminosalicylic acid at 25°C. (Reproduced from Ref. 373 with permission.)

rate-determining step (the drug has no ionizable group with a pK_a corresponding to this pH value). The rate-determining step for the degradation, which proceeds through an imine intermediate, may change from formation to decomposition of the carbinolamine intermediate around pH 4.5.[253] Hydrolysis of 4-methylpiperazine-2,6-dione through a tetrahedral intermediate exhibits a bell-shaped pH–rate profile in the pH range 2–4 because of a change in the rate-determining step from formation to decomposition of the intermediate (Fig. 76).[376] Similar bell-shaped pH–rate profiles due to changes in rate-determining steps have been reported for hydrolysis of tyrphostins.[377]

Bell-shaped pH–rate profiles are also observed for reactions between acidic and basic species. For example, hydrolysis of penicillins in the presence of 3,6-bis(dimethylaminomethyl)catechol is catalyzed by the monoanionic form of the catechol and exhibits a bell-shaped pH–rate profile with a maximum at pH 8 (Fig. 77).[378]

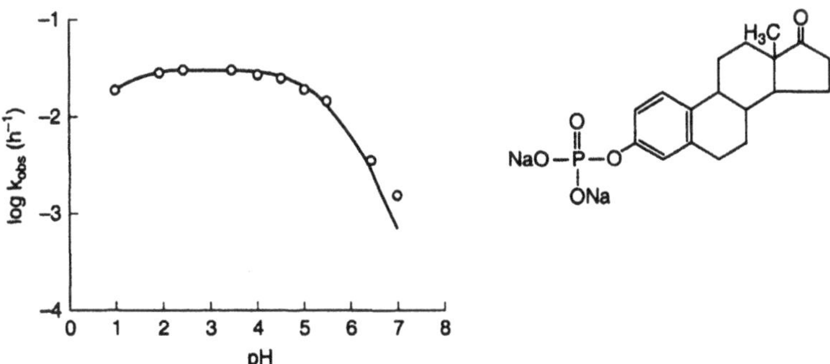

Figure 74. pH–rate profile for hydrolysis of estrone phosphate at 70°C. (Reproduced from Ref. 374 with permission.)

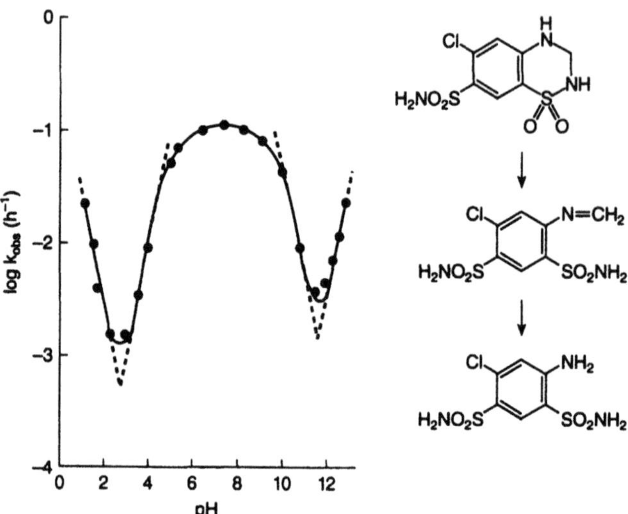

Figure 75. pH–rate profile for hydrolysis of hydrochlorothiazide at 60°C. (Reproduced from Ref. 253 with permission.)

2.2.5.4. *Miscellaneous pH–Rate Profiles*

Rifampicin undergoes hydrolysis of the azomethine double bonds (see Scheme 23), which involves reversible hydrolysis followed by multiple secondary degradation pathways with complex pH dependencies (Fig. 78).[379]

The effect of pH on the stability of solid drug substances is generally more complex than that on the stability of drugs in solution. In the case of freeze-dried moexipril,[380] degradation rates plotted against the pH of the solutions prior to freeze-drying generally provide a pH dependency different from that obtained for solution stability of moexipril.

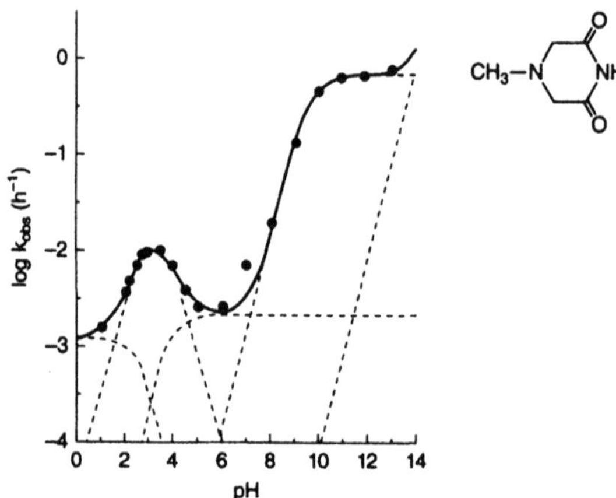

Figure 76. pH–rate profile for hydrolysis of 4-methylpiperazine-2,6-dione at 25°C. (Reproduced from Ref. 376 with permission.)

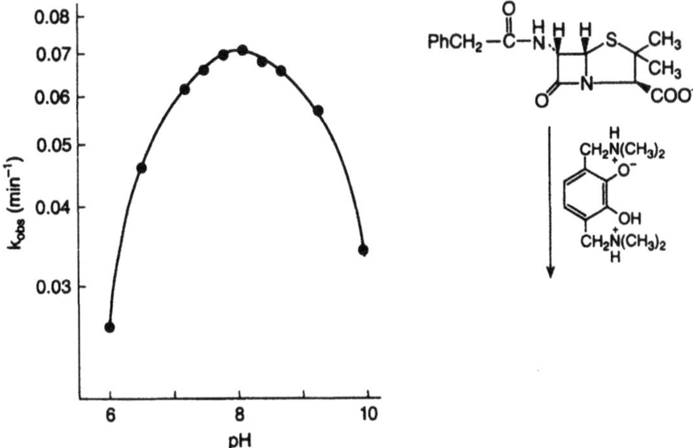

Figure 77. pH–rate profile for hydrolysis of benzylpenicillin catalyzed by 3,6-bis(dimethylaminomethyl)catechol at 31.5°C. (Reproduced from Ref. 378 with permission.)

2.2.6. Buffer, General Acid–Base, and Nucleophilic–Electrophilic Catalysis

The effect of buffer species on the stability of drug substances has been well recognized in the chemical and pharmaceutical literature. For example, the catalysis of chloramphenicol hydrolysis by phosphate and acetate buffer was reported in the 1950s (Fig. 79).[14] These buffer species, like hydronium ion and hydroxide ion, participate in formation or breakdown of activated complexes of various reactions and determine their reaction rate according to Eq. (2.4). Equation (2.105) describes the degradation rate constant, assuming that the monoanion or the dianion of phosphoric acid, or both, participates in degradation of the drug.

$$k_{obs} = k_{H^+}a_{H^+} + k_{H_2O} + k_{OH^-}a_{OH^-} + k_{H_2PO_4^-}a_{H_2PO_4^-} + k_{HPO_4^{2-}}a_{HPO_4^{2-}} \tag{2.105}$$

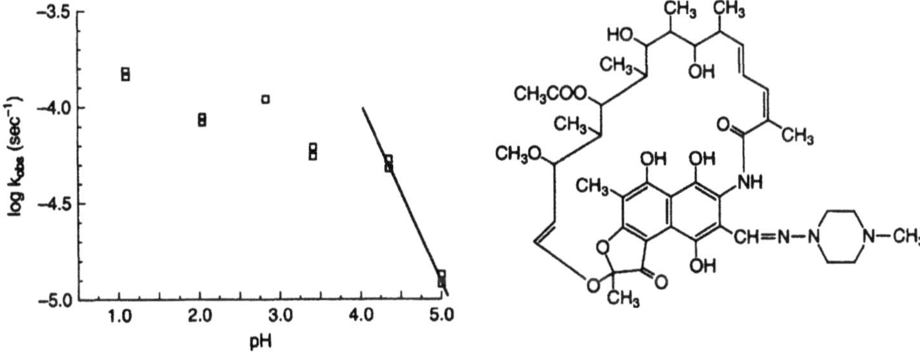

Figure 78. pH–rate profile (initial hydrolysis) for rifampicin at 37°C. (Reproduced from Ref. 379 with permission.)

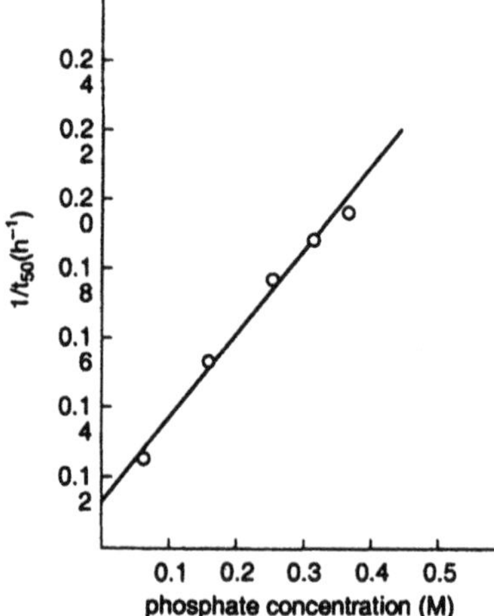

Figure 79. Effect of phosphate concentration on hydrolysis rate (reciprocal of the half-life, t_{50}) of chloramphenicol (pH 7.00, 97.3°C) (Reproduced from Ref. 14 with permission of the American Pharmaceutical Association.)

These catalytic species are often referred to as general acid–base catalysts, in contrast to specific acid–base catalysts. The effects of these species are also sometimes referred to as secondary salt effects in contrast to the primary salt effect of ionic strength described in Section 2.2.7.

Many studies on general acid–base catalysis have been conducted with phosphate as the buffer species. It has been reported that various phosphate species (there are four possible phosphate species) enhance degradation of various drug substances such as benzylpenicillin,[381] cefadroxil,[71] and carbenicillin[382] (Fig. 80). Degradation enhanced by phosphate has also been reported for codeine,[383] spironolactone,[384] and heroin[385] as well as many other drug substances.

In addition to possible general acid–base catalysis where a buffer can act as either a proton donor or acceptor (Brønsted acid or base), buffer species can also act as a Lewis acid or base through nucleophilic or electrophilic mechanisms.

As shown in Scheme 73 and discussed earlier, aspirin anion undergoes intramolecular general base catalysis in the neutral pH region. In contrast, intramolecular nucleophilic catalysis to form a tetrahedral intermediate that goes on to form the mixed anhydride has been demonstrated for the hydrolysis of 3,5-dinitroaspirin.[365,386] In the hydrolysis of *p*-nitrophenyl esters, polyalcohol anions such as the glucose anion act by a nucleophilic mechanism.[387] A plot of the log of the rate constant against the pK_a of the anion (Fig. 81) exhibits a deviation from the linear Brønsted relationship observed for catalysis by various phenoxide anions when the carbohydrate species are included on the plot. These deviations may be due to the very high basicity of the polyalcohol anions, which leads to very large solvation energy requirements.[387]

Other examples of general base catalysis and nucleophilic catalysis that have been reported include the hydrolysis of cefotiam[388] and cefsulodin[166] catalyzed by amikacin and kanamycin and the hydrolysis of moxalactam[389] catalyzed by various amines.

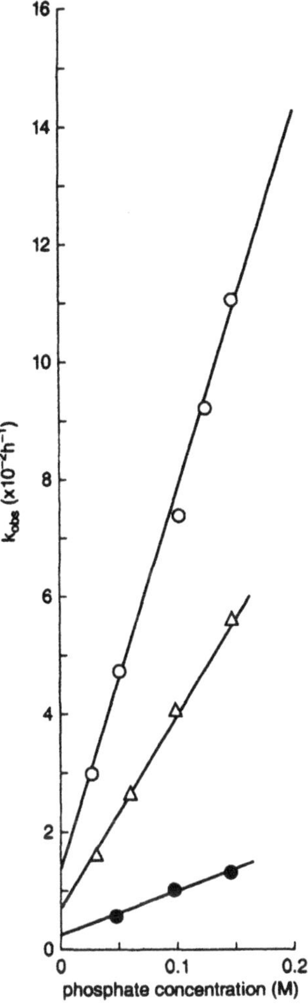

Figure 80. Effect of phosphate concentration on hydrolysis rate of representative drug substances for which phosphate buffer catalysis is observed. ○, Benzylpenicillin, 60°C, pH 7.05; △, cefadroxil, 35°C, pH 7.20; ●, carbenicillin, 35°C, pH 7.48. (Reproduced from Refs. 71, 381, and 382 with permission.)

Because significant catalysis by buffer species can occur, it is important, especially in mechanistic studies, to separate the buffer catalytic component of any observed rate constant from the contribution by hydronium and hydroxide ions or by water itself.

2.2.7. Ionic Strength (Primary Salt Effects)

For drug degradation involving reactions with or between ionic species, the rate is affected by the presence of other ionic species such as salts like sodium chloride. Ionic strength affects the observed degradation rate constant, k, by its effect on the activity coefficients, f, in Eq. (2.6). Ionic strength, μ, is described by

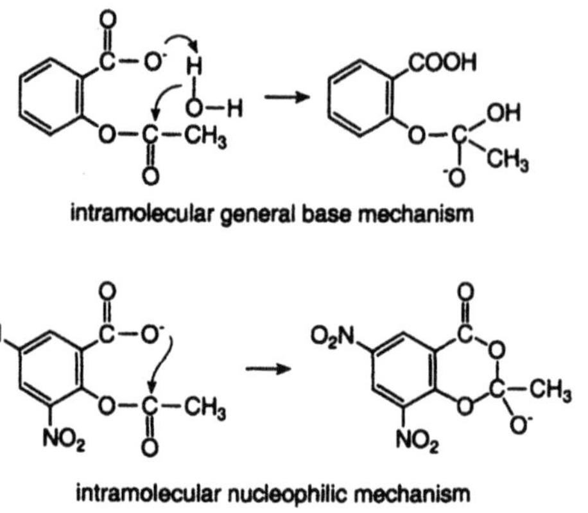

intramolecular general base mechanism

intramolecular nucleophilic mechanism

Scheme 73. Intramolecular general base catalysis and intramolecular nucleophilic catalysis: Hydrolysis of aspirin and 3,5-dinitroaspirin.

$$\mu = \frac{1}{2} \sum_i C_i Z_i^2 \tag{2.106}$$

where C_i is the concentration of ionic species i and Z_i is its electric charge. When an ionic species participates in a reaction as a reactant, the activity coefficient f for that species is generally described by γ, which is related to μ through Eq. (2.107) according to the Debye–Hückel theory.

$$(f = \gamma \text{ in the Debye–Hückel Eq.})$$

$$\log \gamma = -QZ^2 \sqrt{\mu} \tag{2.107}$$

Consequently, a rate constant for a reaction between ionic species depends on ionic strength according to Eq. (2.108):

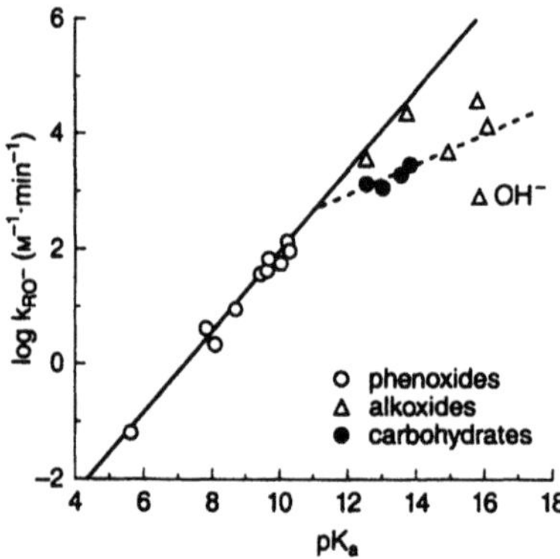

Figure 81. Brønsted plots of hydrolysis rate of *p*-nitrophenyl esters susceptible to nucleophilic catalysis at 25°C. (Reproduced from Ref. 387 with permission.)

$$\log k = \log k_0 + 2QZ_AZ_B\sqrt{\mu} \quad (2.108)$$

where Z_A and Z_B are the charges of A and B, and k_0 is the rate constant when $\mu = 0$. The term $2Q$ is a function of the dielectric constant, density, and temperature and is 1.018 for aqueous solutions at 25°C. Equation (2.108) was established by Brønsted and Bjerrum in the 1920s and is referred to as the Brønsted–Bjerrum equation.[390–393]

Equation (2.108) is dependent on Eq. (2.107) and is applicable to reactions at ionic strength less than 0.01. Therefore, Eq. (2.108) cannot be applied to most drug degradation studies because the ionic strength is usually much higher than the limiting value of 0.01. The following modified equation is generally applicable to drug degradation studies performed at higher ionic strength:

$$\log k = \log k_0 + 2QZ_AZ_B\frac{\sqrt{\mu}}{1+\sqrt{\mu}} \quad (2.109)$$

As shown in Fig. 82, the effect of ionic strength on the degradation of thiamine hydrochloride is described by Eq. (2.109) rather than by Eq. (2.108).[394]

Equation (2.109) has also been used to describe the degradation rates of barbiturates. However, Eq. (2.108) described better the degradation of benzylpenicillin[394] and carbenicillin[382] (Fig. 83). Equations (2.108) and (2.109) indicate that rate constants are independent of ionic strength when at least one of the reactants is un-ionized (when Z_A or Z_B is zero). As ionic strength increases, the rate of reaction between ions of opposite charge decreases and the rate of reaction between ions of similar charge increases. Therefore, studying the effect of ionic strength can help our understanding of the possible charges of the species involved in the degradation. For example, the degradation of barbituric acids in the alkaline pH region is proposed to be due to the attack of hydroxide ion on the monoanion of the barbituric acid, as supported by the increase in the degradation rate with increasing ionic strength.[80] Norfloxacin is most stable at an ionic strength of 0.2, suggesting a change in degradation mechanism around this ionic strength.[395]

The use of ionic strength changes to probe mechanisms has its limitations. For example, it is necessary to consider changes in the pK_a values of ionic species with changes in ionic

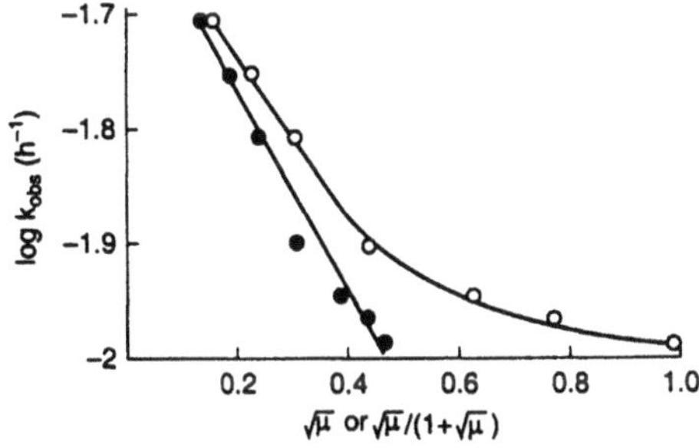

Figure 82. Effect of ionic strength on degradation rate of thiamine hydrochloride (pH 6.40, 96.4°C). The logarithm of the degradation rate is plotted versus $\sqrt{u}$ (○), in accordance with Eq. (2.108), and versus $\sqrt{u}/(1+\sqrt{u})$ (●), in accordance with Eq. (2.109). (Reproduced from Ref. 394 with permission.)

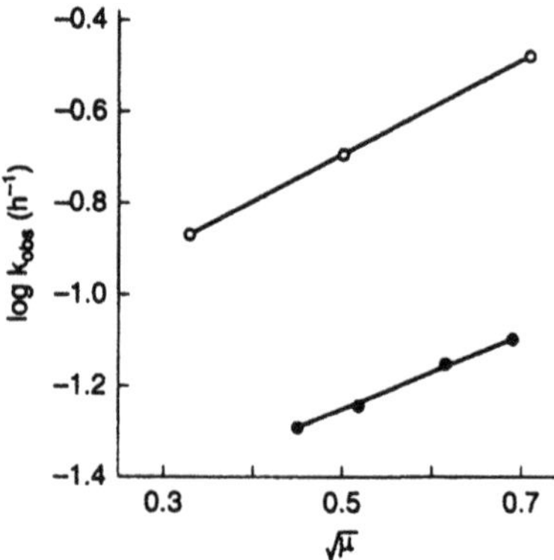

Figure 83. Effect of ionic strength on degradation rates described by Eq. (2.108). ●, Benzylpenicillin, pH 8.75, 60°C; ○, carbenicillin, pH 10.47, 35°C. (Reproduced from Refs. 382 and 394 with permission.)

strength in assessing degradation mechanisms. Often, this is not taken into account in the interpretation of data on the effect of ionic strength.

2.2.8. Dielectric Constant of Solvents

Rates of degradation between ions and dipoles in solutions depend on the bulk properties of the solvent, such as the dielectric constant. Variation in the dielectric constant of a solvent can cause $\Delta G^{\ddagger}$ in Eq. (2.7) to vary, leading to a variation in rate constants with changes in dielectric constant. For example, ion–dipole reaction rate constants have been related to the dielectric constant D of the solvent through Eq. (2.110), which was developed by Amis.[396]

$$\ln k = \ln k_{D=\infty} + \frac{Z_A \mu \cos \theta}{D \kappa T r^2} \tag{2.110}$$

In Eq. (2.110), $k_{D=\infty}$ is the rate constant at infinite dielectric constant, Z_A, μ, and r are ion charge, dipole moment and the shortest ion–dipole distance, respectively, and κ is the Boltzmann constant. The term θ represents the alignment of reactants, and $\cos \theta$ is unity in the case of head-on alignment. Thus, as the dielectric constant decreases, the rates of anion–dipole reactions decrease and the rates of cation–dipole reactions increase. As indicated by a linear relationship with a positive slope in log k versus $1/D$ plots (Fig. 84), the hydrolysis rate constant for chloramphenicol in water–propylene glycol mixtures increases with decreasing dielectric constant, suggesting a hydronium ion–dipole reaction.[397]

In acid, un-ionized azathioprine undergoes degradation catalysis by hydronium ion, whereas in alkali its anionic form is degraded by hydroxide ions.[398] As shown in Fig. 85, the dipole–cation reaction at pH 1 exhibits log k versus $1/D$ plots with a negative slope, suggesting that reactant alignment is opposite to the head-on alignment. The observation that the rate of anion–anion reaction at pH 11 is independent of dielectric constant has been explained by assuming that a change in the bulk dielectric constant is not reflected in the microscopic dielectric constant and has no effect on the reaction rate.[398]

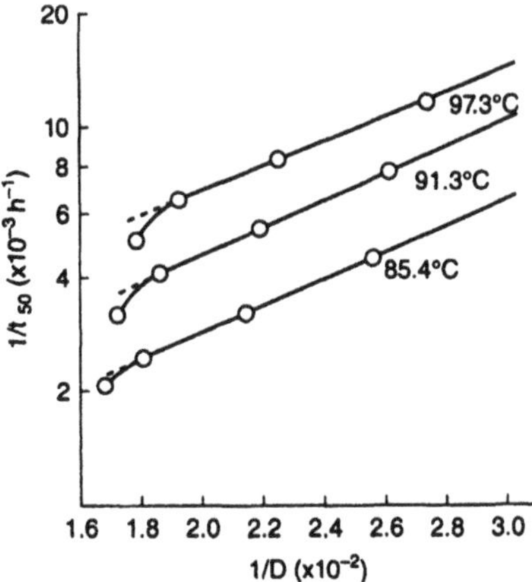

Figure 84. Effect of changes in solvent dielectric constant on the degradation rate of chloramphenicol in the presence of perchloric acid. (Reproduced from Ref. 397 with permission of the American Pharmaceutical Association.)

The dependence of ion–ion reaction rate constants on the dielectric constant of the solvent is given by

$$\ln k = \ln k_{D=\infty} - \frac{Z_A Z_B}{D \kappa T r} \tag{2.111}$$

where Z_A and Z_B are the charges of the ions, and r is ion–ion distance. This equation indicates that as the dielectric constant decreases, the rate of reaction between ions of similar charge decreases, and the reaction rate between ions of opposite charge increases. For example, Eq. (2.111) can be used to describe the degradation rate of barbiturates in alcoholic solvents. Plots of log k versus $1/D$ have a negative slope, as shown in Fig. 86.[399,400] Similar log k versus $1/D$ plots have been reported for the degradation of phenoxybenzamine.[401]

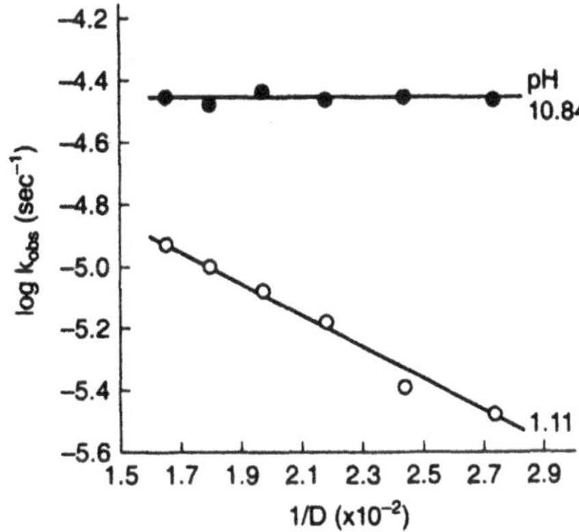

Figure 85. Effect of dielectric constant changes on the degradation rate of azathioprine at 80°C. (Reproduced from Ref. 398 with permission.)

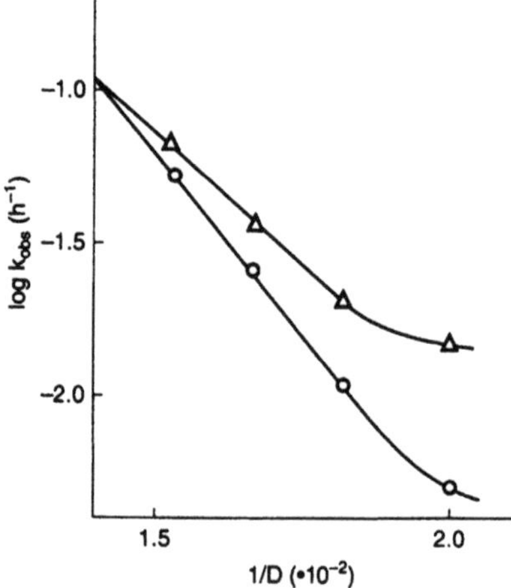

Figure 86. Effect of dielectric constant changes on the degradation rate of hexobarbiturate at 50°C. The plotted values are the apparent rate constants obtained by extrapolating to zero ionic strength. △, C_2H_5OH/H_2O system; ○, CH_3OH/H_2O system. (Reproduced from Ref. 399 with permission.)

A complication in interpreting the effect of solvent dielectric constants on kinetic data is that the dissociation constants of various species can change with changing solvent properties. Also, the possible role that a solvent modifier plays in the reaction must be considered. For example, an alcohol or glycol or some other co-solvent might be used to modify the dielectric constant of water. The co-solvent molecule may react with the drug in question, thus complicating the interpretation of the solvent dielectric effect.

An alternative qualitative approach to the use of Eq. (2.111) is to consider the differences in solvation between the ground state and the transition state of a particular reaction. For example, in an S_N1 reaction, a molecule may go from a neutral ground state to a transition state with a great deal of charge separation. The stability of the highly polar transition state would be enhanced by a solvent that has a high dielectric constant whereas the free energy of the ground-state reactant may be only slightly affected. As the dielectic constant of the solvent is decreased, for example, by addition of a co-solvent, the transition state would be relatively destabilized, resulting in a decrease in the reaction rate. By considering the expected polarity of the ground state of a molecule and possible transition states, the effect of changes in solvent dielectic constant can often be rationalized.

2.2.9. Oxygen

The kinetics of the oxidation of drug substances can be affected by the availability of oxygen. Also, some photodegradation reactions involve photooxidative mechanisms that are dependent on oxygen concentration. An example is the increased photodegradation of cianidanol with increasing oxygen concentration.[402] Oxygen participates in Eq. (2.4) as a reactant and alters the degradation rate. Although the concentration of oxygen in the atmosphere and in various solutions is known to often affect drug degradation significantly, only a few studies relating drug degradation kinetics and oxygen concentration are available. The rate of oxidation of ascorbic acid depends on oxygen concentration, as shown in Fig. 87.[177]

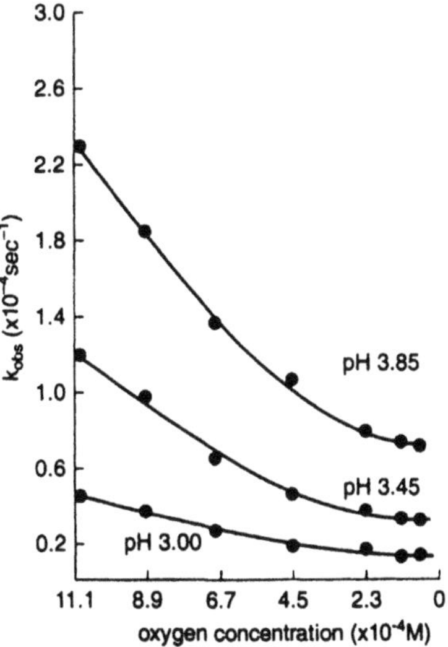

Figure 87. Effect of oxygen concentration on the apparent first-order rate constants for ascorbic acid oxidation at 25°C. (Reproduced from Ref. 177 with permission.)

Oxygen exists in various states. In its ground state, oxygen exists as a diradical or what is called triplet oxygen, but it can be excited by light to singlet oxygen, as shown in Scheme 74. Singlet oxygen is highly oxidizing and capable of attacking olefinic bonds. Oxygen can also form other oxidizing species, as shown in Scheme 75. The superoxide species is a mild reductant whereas hydrogen peroxide is a fairly specific oxidant. The hydroxyl radical is highly reactive but has low selectivity. Therefore, the general term oxidation does not refer just to exposure of oxidatively labile drugs to oxygen but also to conditions that favor oxidation, such as photolysis; the presence of various oxygen species, other radicals, metal ions and pro-oxidants, and high oxygen pressure.

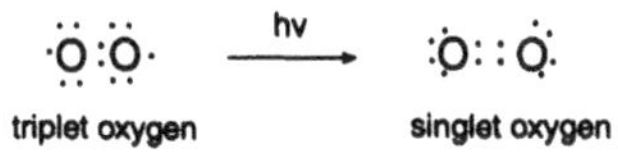

$$\cdot\ddot{O}:\ddot{O}\cdot \xrightarrow{h\nu} :\ddot{O}::\ddot{O}:$$

triplet oxygen → singlet oxygen

Scheme 74. Activation of triplet oxygen to singlet oxygen by light.

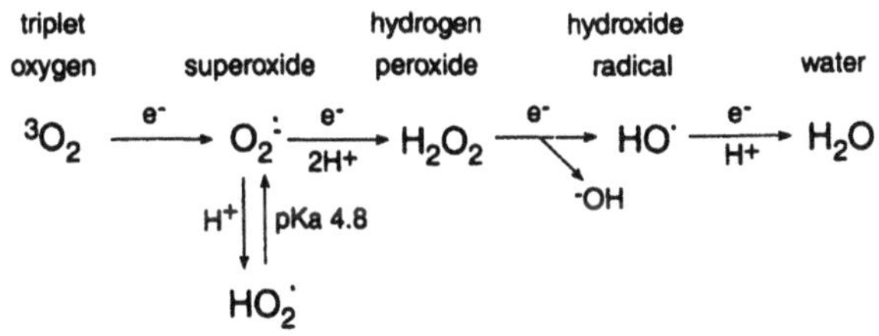

$$^3O_2 \xrightarrow{e^-} O_2^{\cdot -} \xrightarrow[2H^+]{e^-} H_2O_2 \xrightarrow{e^-} HO^{\cdot} + {}^-OH;\quad HO^{\cdot} \xrightarrow[H^+]{e^-} H_2O$$

$$O_2^{\cdot -} \underset{pKa\ 4.8}{\overset{H^+}{\rightleftharpoons}} HO_2^{\cdot}$$

triplet oxygen; superoxide; hydrogen peroxide; hydroxide radical; water

Scheme 75. Various oxygen species capable of catalyzing reactions.

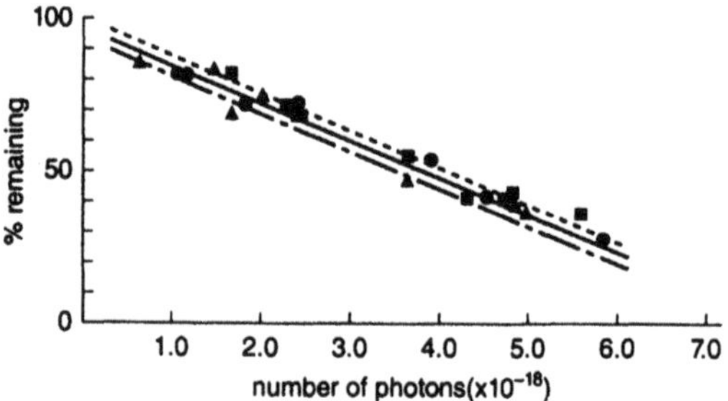

Figure 88. Effect of light intensity on the photodegradation of nifedipine at room temperature. Light source: ●, high-pressure mercury lamp; ■, sunlight; ▲, fluorescent lamp. (Reproduced from Ref. 403 with permission.)

2.2.10. Light

The number and the wavelength of incident photons affect the photodegradation rate of drugs. It is not easy to study the effect of light quantitatively because the wavelength dependence of degradation varies among drug substances and because light sources have different spectral distributions. In many instances, only qualitative data on the photodegradation of drugs have been reported.

As shown in Fig. 88, the amount of photodegraded nifedipine was proportional to the number of incident photons.[403] Maximum photodegradation of nifedipine in tablets occurred at 420 nm (Fig. 89).[404] On the other hand, the relationship between the discoloration rate of sulfisomidine in tablets irradiated by a mercury lamp versus ultraviolet light intensity was complex.[405] The values of L, a, and b determined for the discoloration of the tablet depended on the energy of the mercury lamp.[406] Photodegradation of menatetrenon yielded linear plots of log k versus the reciprocal of the illumination intensity, as shown in Fig. 90.[407]

Photodegradation of drug substances strongly depends on the spectral properties of the drug substance and the spectral distribution of the light source. Discoloration of sulpyrine is significant in the presence of a mercury lamp, which is a good source of UV energy; however, little discoloration occurs in the presence of a fluorescent lamp, which radiates mainly visible light.[408]

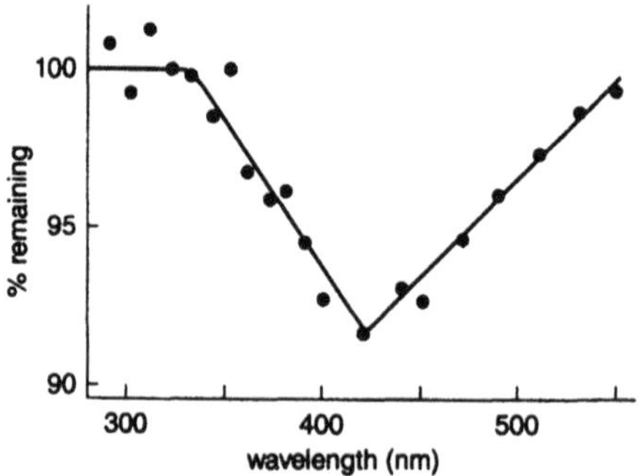

Figure 89. Wavelength dependence of photodegradation of nifedipine tablets at 15°C. Light intensity: 1.23×10^8 erg/cm^2. (Reproduced from Ref. 404 with permission.)

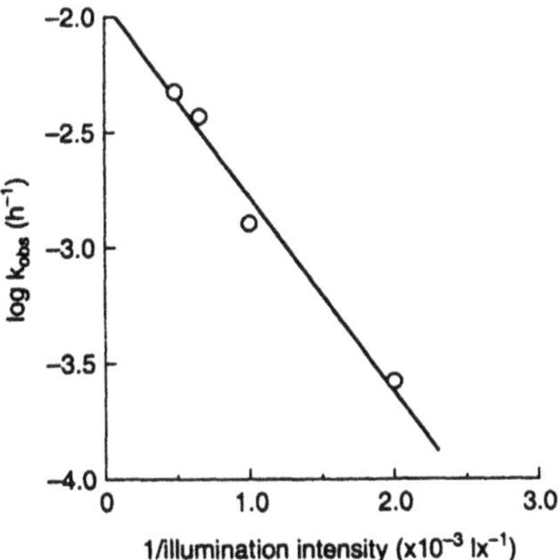

Figure 90. Effect of illumination intensity on the degradation of menatetrenon (25°C). Light source: White fluorescent lamp. (Reproduced from Ref. 407 with permission.)

2.2.11. Crystalline State and Polymorphism in Solid Drugs

The chemical stability of solid drugs is affected by the crystalline state of the drug. Drugs in the crystalline state have lower ground-state free energy and exhibit higher $\Delta G^{\ddagger}$ (Eq. 2.7) and, therefore, slower reactivity. This is exemplified by the solid-state chemical degradation of various vitamin A derivatives as shown in Fig. 91, where a linear relationship between the observed degradation rate constant and the inverse of the melting point of the derivative can be seen.[190]

Many drug substances exhibit polymorphism. Each crystalline state has a different ground-state free-energy level and a different chemical reactivity. For example, ramified crystals of 5-nitroacetylsalicylic acid are more susceptible to hydrolysis than are column-

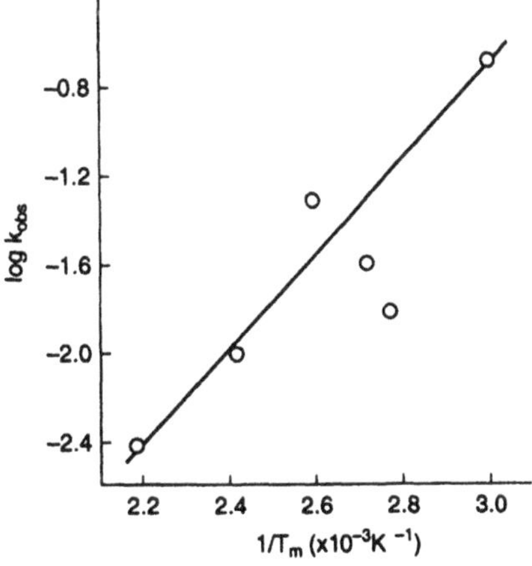

Figure 91. Relationship between the degradation rates of various vitamin A derivatives at 50°C and their melting points (T_m). (Reproduced from Ref. 190 with permission.)

shaped crystals.[409] Solid-state hydrolysis of carbamazepine from needle-shaped crystals with a higher crystalline order is faster than that of beam-shaped and prismatic forms.[410] Reactivity of carbamazepine to light also depends on the crystalline form of the drug.[411] Differences in reactivity among different crystalline forms have also been reported for photodegradation of furosemide.[412]

The stability of drugs in their amorphous form is generally lower than that of drugs in their crystalline form, because of the higher free-energy level of the amorphous state. The relationship between degradation rate and crystallinity determined from heats of dissolution for β-lactam antibiotics such as sodium cefazolin indicates that a drug with low crystallinity tends to have decreased chemical stability.[413]

Decreased chemical stability of solid drugs brought about by mechanical stresses such as grinding is said to be due to a change in crystalline state. For example, grinding of aspirin increased its degradation rate in suspension form. A relationship between the stability and grinding time was reported[414] (Fig. 92) and was attributed to the increased solubility of aspirin due to a change in its crystalline state rather than to any increase in surface area.

The chemical stability of solid drugs is also affected by the crystalline state of the drug through differences in surface area. For reactions that proceed on the solid surface of the drug, an increase in the surface area can increase the amount of drug participating in the reaction. For example, in a study of the reaction between solid-state sulfacetamide and phthalic anhydride, the percent of the drug that reacted within 3 h increased with increasing surface area, as shown in Fig. 93.[415]

2.2.12. Effect of Moisture and Humidity on Solid and Semisolid Drugs

Drug degradation in heterogeneous systems such as solids and semisolid states is affected by moisture. The effect of moisture and humidity on the degradation kinetics of various drug substances including ascorbic acid,[416,417] thiamine salts,[418,419] aspirin,[420] vitamin A[421] (Fig. 94), and ranitidine hydrochloride,[422] to name a few examples, has been reported.

Moisture plays two primary roles in catalyzing chemical degradation. First, water participates in the drug degradation process itself as a reactant, leading to hydrolysis,

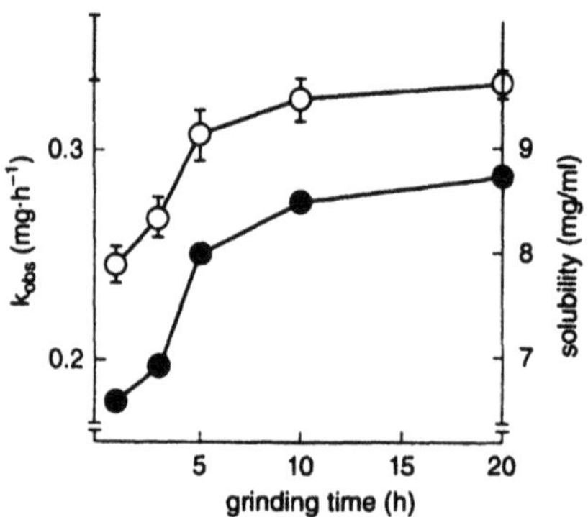

Figure 92. Effect of grinding time on the degradation of aspirin in suspension at 40°C. ○, Zero-order rate constant; ●, solubility. (Reproduced from Ref. 414 with permission.)

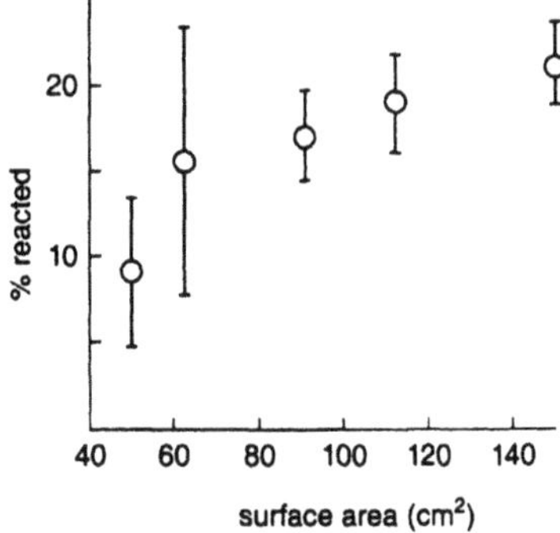

Figure 93. Effect of surface area on the reaction between sulfacetamide and phthalic anhydride. The plotted values are the percentages of the drug that reacted during 3 h at 95°C. (Reproduced from Ref. 415 with permission.)

hydration, isomerization, or other bimolecular chemical reactions. In these cases, the degradation rate is directly affected by the concentrations of water, hydronium ion, or hydroxide ion according to Eq. (2.4). Second, water adsorbs onto the drug surface and forms a moisture-sorbed layer in which the drug is dissolved and degraded. Water adsorption may also change the physical state of drugs, thereby affecting their reactivity. Thus, water affects drug degradation indirectly by providing a favorable environment for degradation.

The mechanisms for these effects of water are complicated and are determined by the physical state of water molecules. For example, for drugs that form hydrates, water of crystallization is trapped in the crystals and, generally, cannot participate in chemical reactions. This is exemplified by the discoloration of glucuronic acid derivatives.[423] As shown in Fig. 95, addition of water in amounts that form water of crystallization does not cause discoloration; however, addition of excess water does accelerate the discoloration at a rate proportional to the amount of water not involved in hydrate formation.

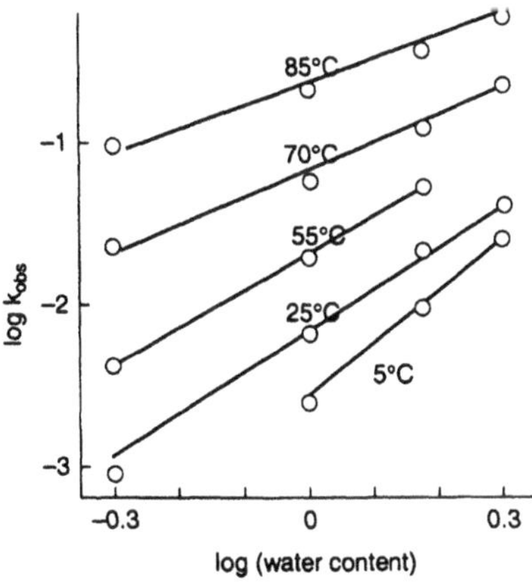

Figure 94. Effect of water content on the degradation of vitamin A tablets. (Reproduced from Ref. 421 with permission.)

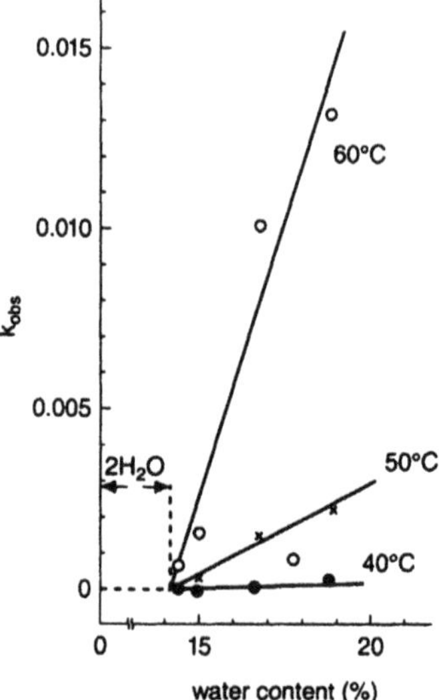

Figure 95. Effect of water content on the discoloration rate of potassium glucuronate. (Reproduced from Ref. 423 with permission.)

Water of crystallization can participate in drug degradation when it is released from the crystalline state by actions such as grinding. This has been reported for the hydrolysis of sodium prasterone sulfate[424] and ampicillin trihydrate.[425] The degradation rate of ampicillin trihydrate increased with increasing grinding time as shown in Fig. 96. Similarly, cefixime trihydrate exhibited decreased stability as a result of grinding,[426] and rapid degradation was seen when its crystalline structure was disordered by dehydration upon storage under humidity conditions below the critical relative humidity.[427]

The effect of moisture on drug degradation has also been studied for various pharmaceuticals in the presence of excipients. This will be described in the next section.

A general equation analogous to the Arrhenius and Eyring equations for the effect of temperature on chemical degradation has not been developed to describe the effect of humidity. A linear relationship between the logarithm of the rate constant (or a constant

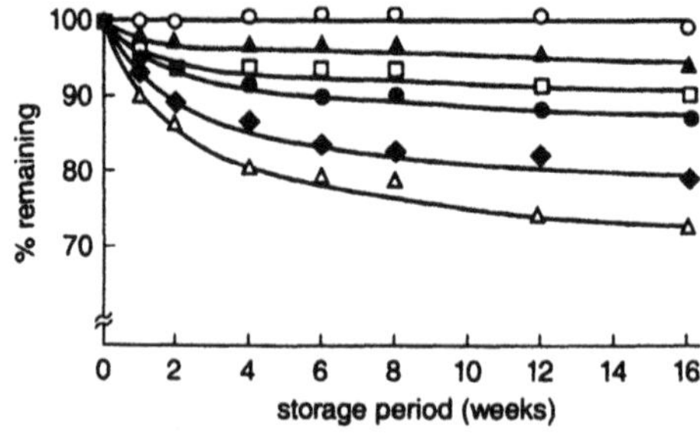

Figure 96. Effect of grinding on the degradation of ampicillin trihydrate during storage at 40°C. Grinding time: ○, 0, ▲, 15 min; □, 30 min; ●, 60 min; ♦ 120 min; △, 180 min. (Reproduced from Ref. 425 with permission.)

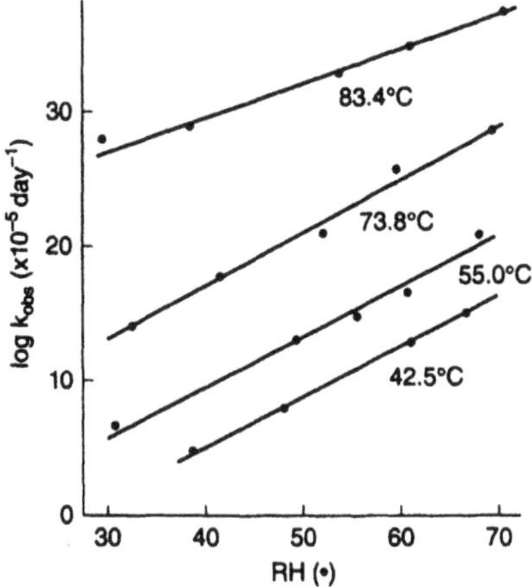

Figure 97. Effect of humidity on the degradation of nitrazepam. (Reproduced from Ref. 428 with permission.)

related to rate such as $1/t_{90}$) and relative humidity has been used empirically to describe the effect of humidity on degradation rate. For example, the effect of humidity on the degradation rate constants of nitrazepam (Fig. 97),[428] mecillinam (Fig. 98),[429] and penicillins[430] has been described by

$$\ln k = \frac{-E_a}{RT} + \ln A' + C \cdot \text{RH} \tag{2.112}$$

where RH is relative humidity and C is a constant. However, Eq. (2.112) cannot be applied in all cases. For example, a linear relationship was not observed for the degradation rates of acetyl-5-nitrosalicylic acid and glutathione with changing humidity, as shown in Fig. 99.[431]

The effect of humidity on the solid-state degradation rates of propantheline bromide and meclofenoxate hydrochloride has been described by

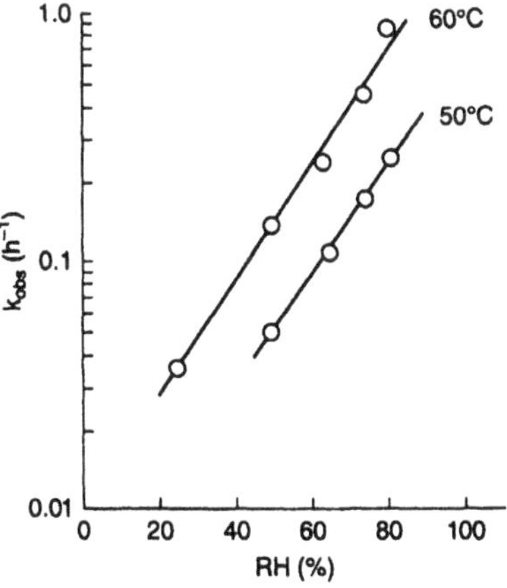

Figure 98. Effect of humidity on the degradation of mecillinam. (Reproduced from Ref. 429 with permission.)

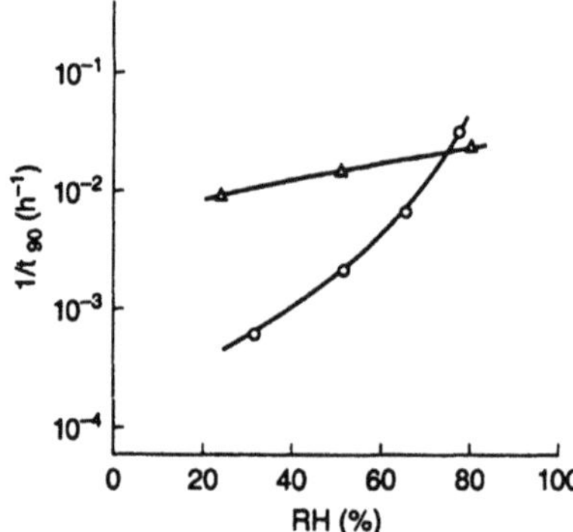

Figure 99. Effect of humidity on the degradation of acetyl-5-nitrosalicylic acid (△) and glutathione (○) at 80°C. (Reproduced from Ref. 431 with permission.)

$$k = k' \cdot \exp\left(\frac{-E_a}{RT}\right) \cdot P^s \tag{2.113}$$

where P is the vapor pressure of water and s is a constant. The constant k, calculated according to Eq. (2.58), was related to temperature and the vapor pressure of water using Eq. (2.113).[432,433] As shown in Fig. 100, the time course of degradation was well represented by Eq. (2.114), which was obtained by combining Eqs. (2.113) and (2.58):

$$x = x_0 \cdot \exp\left[\frac{E_a}{R}\left(\frac{1}{298} - \frac{1}{T}\right)\right] \cdot \left(\frac{P}{23.756}\right)^s \cdot \left(\frac{t}{50}\right)^{n'} \tag{2.114}$$

where x is the ratio of drug degraded at time t, and x_0 is the ratio of drug degraded at 25°C and 100% RH ($P = 23.756$ mm Hg) when $t = 50$ (day).

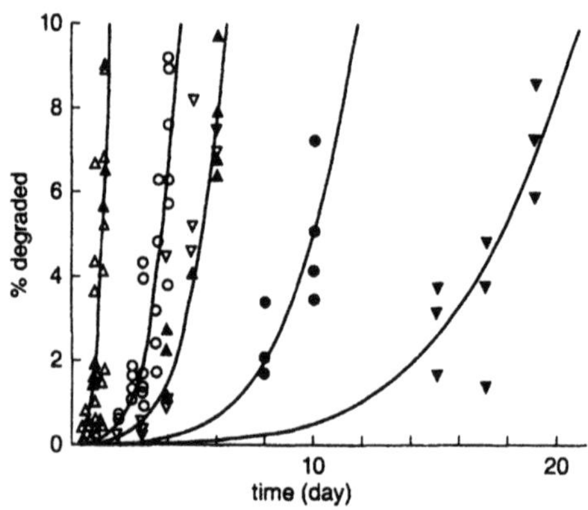

Figure 100. Time courses of the degradation ratio of meclofenoxate hydrochloride defined according to Eq. (2.114). △, 90°C, 23.4% RH; ○, 70°C, 37.5% RH; ▽, 80°C, 22.6% RH; ▲, 60°C, 49.9% RH; ●, 60°C, 43.1% RH; ▼, 70°C, 22.0% RH. (Reproduced from Ref. 433 with permission.)

Estimates of E_a and S were obtained by nonlinear regression analysis. These equations appear to be useful for predicting the solid-state hydrolysis rate of water-soluble drugs as functions of changes in both temperature and humidity.

2.2.13. Excipients

The role that excipients play in drug stability has been extensively reported. Examples include the stabilizing effect of sugars on the degradation of ascorbic acid in aqueous solutions[434]; the accelerating effect of talc on the hydrolysis of thiamine hydrochloride powders[435,436]; the accelerating effect of magnesium stearate on discoloration of tablets containing amines and lactose[437]; and the effect of talc impurities,[438] stearic acid,[439,440] and calcium succinate[441] on the degradation of aspirin tablets. Additional information includes reports on the compatibility and incompatibility of drugs[442–444]; the incompatibility of flomoxef and sugars[368]; and the incompatibility of components of intravenous hyperalimentation fluids.[445] However, detailed mechanistic studies on drug/excipient interactions and incompatibilities have not received much attention.

Excipients may affect drug stability via various mechanisms. The most obvious examples are those in which the excipients may participate directly in degradation as reactants, such as addition reactions with drugs. Excipients may also exhibit catalyzing effects toward drug degradation, as described in Section 2.2.6. This is exemplified by the nucleophilic catalysis effect of sugars (such as glucose and sucrose) and amines on the degradation of ester or amide drugs.[387,388] Other mechanisms include the effect of moisture present in excipients, the effect of pH changes caused by excipients and other such effects. In following sections, the effects of excipients on chemical degradation, excluding their role as reactants or catalysts, are described.

2.2.13.1. *Effect of the Amount of Moisture Present in Excipients*

Excipients can affect drug stability by being a source of moisture. For example, owing to the high moisture content of polyvinylpyrrolidone and urea, aspirin hydrolysis was enhanced in solid dispersions with these excipients.[446] Decreased drug stability caused by excipients having higher moisture-containing ability has been reported for tablets of aspirin and ascorbic acid,[447] a urea–linoleic acid inclusion complex,[448] powders of cysteine derivatives,[449] and dry syrups of cephalexin.[450]

The effects of moisture from excipients on drug degradation rates are often difficult to interpret. Degradation of ascorbic acid in the presence of silica gel increased with increasing water content, as shown in Fig. 101.[451] The higher degradation rate observed in the presence of silica gel compared to that for ascorbic acid alone at the same moisture content suggested an accelerating effect of silica gel itself or of one of its impurities. On the other hand, degradation decreased when the ratio of silica gel to ascorbic acid was increased. This suggested that most of the moisture was adsorbed onto the silica gel and that the entrapped water was unable to participate in the degradation. Thus, silica gel appeared to exhibit both an inhibiting effect via water entrapment and an accelerating effect toward ascorbic acid degradation. Similar inhibiting effects of excipients via water entrapment/adsorption were shown in tablets containing aspirin and colloidal silica.[452]

Degradation of thiamine hydrochloride in tablets containing magnesium stearate and microcrystalline cellulose exhibits a maximum at a water content of about 5%, as shown in

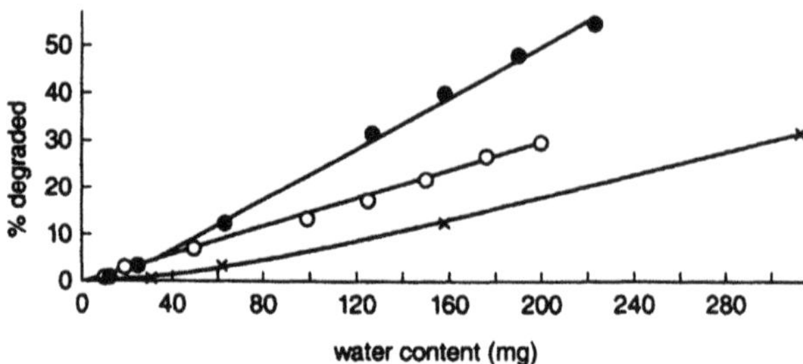

Figure 101. Effect of silica gel and moisture content on the solid-state degradation of ascorbic acid during storage at 45°C for 3 weeks. ○, 300 mg ascorbic acid; ●, 300 mg ascorbic acid and 80 mg silica gel; ×, 300 mg ascorbic acid and 640 mg silica gel. (Reproduced from Ref. 451 with permission.)

Fig. 102.[453] It was proposed that the degradation proceeds through the catalyzing effect of an alkaline component on the dissolved drug. Thus, the degradation rate increases with increasing water content, as more drug is dissolved and comes into contact with the catalytic species. At higher water content, however, where the drug is completely dissolved, the degradation rate decreases with increasing water content as the two reactive species are diluted.[453] A similar maximum degradation rate observed at a certain water content has been reported for the degradation of propantheline bromide in the presence of sodium aluminum gel (Fig. 103).[454]

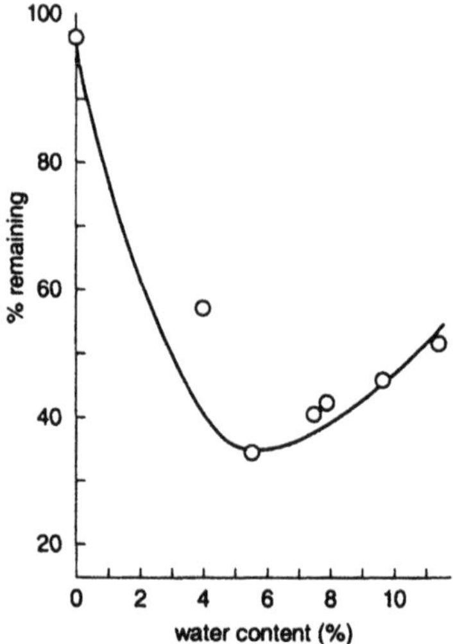

Figure 102. Effect of water content on the degradation of thiamine hydrochloride tablets composed of magnesium stearate and microcrystalline cellulose. The percent of drug remaining after reaction equilibrium at 55°C is plotted versus water content. (Reproduced from Ref. 453 with permission.)

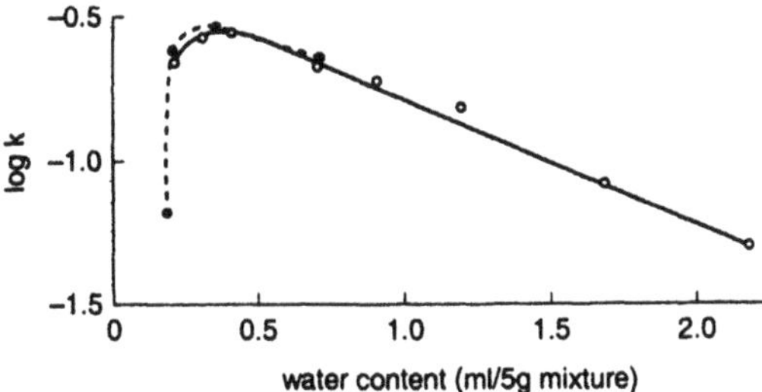

Figure 103. Effect of water content on degradation of propantheline bromide in the presence of sodium aluminum gel. The values of the rate constant (in units of $h^{-0.5}$) calculated according to the Jander equation at 37°C. ●, Moisture equilibrium; ○, water added. (Reproduced from Ref. 454 with permission.)

2.2.13.2. *Effect of the Physical State of Water Molecules in Excipients*

As described earlier, the contribution of water to the chemical degradation of drug substances is determined by the physical state of water. This is also the case for drug–excipient mixtures. Recent studies have shown that water present in excipients exists in various physical states, being either weakly or strongly adsorbed to the excipient. The physical state of water can affect drug degradation.

Excipients having strong water-entrapping abilities tend to inhibit drug degradation, as exemplified by silica gel[451] and colloidal silica.[452] The hydrolysis rate of nitrazepam in the presence of various excipients such as microcrystalline cellulose was inversely proportional to the nitrogen-adsorption energy of the excipients, as shown in Fig. 104.[455] This would suggest that excipients having higher adsorption energy decrease water reactivity and

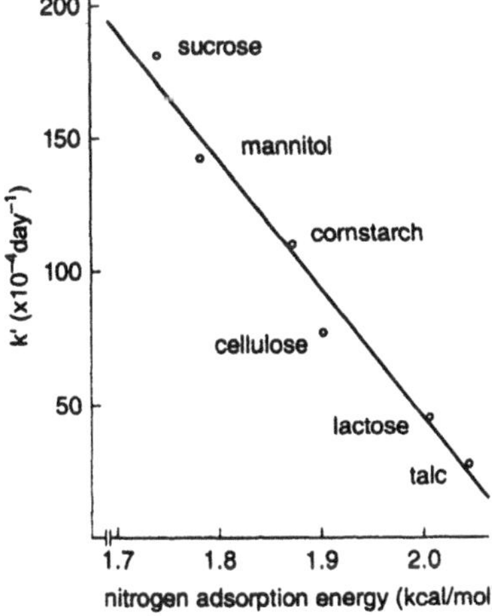

Figure 104. Plot showing the relationship between the nitrogen-adsorption energy of various excipients and the hydrolysis rate of nitrazepam. The values of the rate constant (k') have been normalized with respect to the specific surface area for the sample with 0.5% nitrazepam (70°C, 60% RH). (Reproduced from Ref. 455 with permission.)

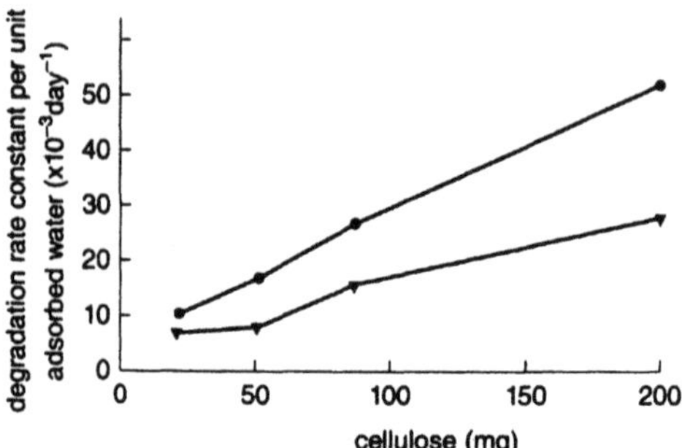

Figure 105. Effect of two cellulose forms on the degradation of aspirin (55°C, 75% RH). ●, Microcrystalline cellulose; ▼, microfine cellulose. (Reproduced from Ref. 456 with permission.)

thereby decrease the relative hydrolysis rates. However, the relationship between nitrogen- and water-adsorption energies requires further clarification.

The degradation rate of aspirin in the presence of cellulose compounds does not necessarily increase with an increase in the moisture-containing capacity of the compound.[456] Microcrystalline cellulose provided a higher degradation rate constant per unit of water content than seen with microfine cellulose, suggesting a more highly reactive water content in the case of the microcrystalline material (Fig. 105). That is, this observation may be due to a lower proportion of strongly adsorbed water in microcrystalline cellulose. It has been reported that the amount of water that is strongly adsorbed on microcrystalline cellulose is only 0.856 mol/100 g and that most water in microcrystalline cellulose is only weakly adsorbed and, therefore, evaporates readily.[457]

The degradation rate of a drug substance can also depend on the time required to reach moisture-adsorption equilibrium, rather than on the amount of water adsorbed at equilibrium.[458] As shown in Fig. 106, the degradation of tablets of a proprietary drug formulated with various excipients such as microcrystalline cellulose decreased as the amount of water adsorbed by the excipients during storage increased. An explanation for this observation is

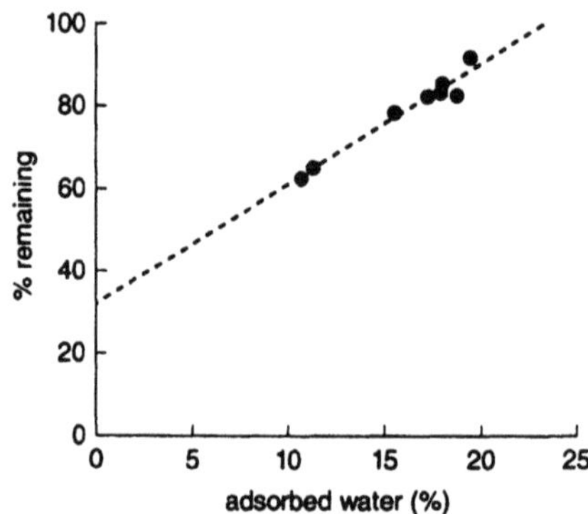

Figure 106. Effect of the amount of water adsorbed by excipients on the degradation of drug A (storage at 40°C, 80% RH for 4 weeks). (Reproduced from Ref. 458 with permission.)

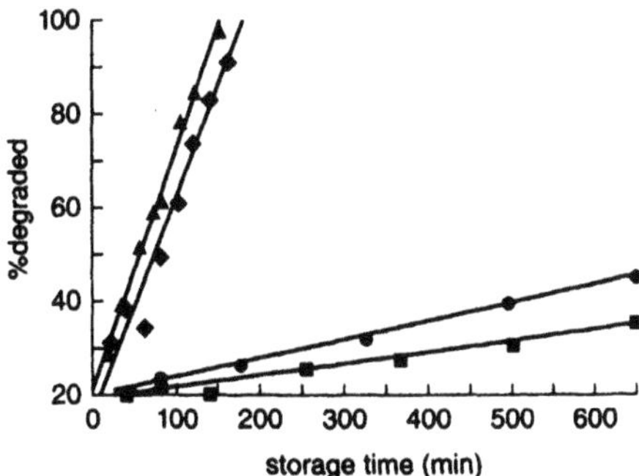

Figure 107. Effect of grinding of lactose monohydrate on the degradation rate of 4-methoxyphenylaminoacetate hydrochloride at 37°C. Mixtures of the drug and lactose monohydrate were prepared in four ways: ■, mixing at 10% RH; ●, mixing at 80% RH; ♦, mixing after grinding lactose for 10 min; ▲, mixing after grinding both drug and lactose for 10 min. (Reproduced from Ref. 459 with permission.)

that an excipient that adsorbs more moisture adsorbs it more strongly. Thus, the amount of free water is less for the strongly adsorbing excipients before moisture-adsorption equilibrium is reached. Because of a decrease in free water, the relative reactivity is decreased.

As in the case of drugs that are hydrated, excipients that can form hydrates may enhance drug degradation by giving up their water of crystallization during grinding. For example, the excipient α-lactose hydrate has been reported to enhance degradation of 4-methoxyphenylaminoacetate hydrochloride upon grinding, as shown in Fig. 107.[459]

2.2.13.3. Effect of the Mobility of Water Molecules in Excipients on Drug Degradation

In the previous section, drug stability was shown to depend on the physical state of water in excipients. Detailed information on the physical state of water can be obtained by measuring the dynamics or the mobility of water molecules. The effect of water mobility on drug stability has been studied by determining water mobility in mixtures of water and polymers used as pharmaceutical excipients. Methods used include the measurement of spin–lattice relaxation time and spin–spin relaxation time by nuclear magnetic resonance (NMR) spectroscopy as well as of dielectric relaxation time by dielectric relaxation spectroscopy.

The spin–lattice relaxation time, T_1, of water ($H_2{}^{17}O$) in aqueous solutions of water-soluble polymers such as polyvinylpyrrolidone, polyethylene glycol, and gelatin depends on polymer concentration as shown in Fig. 108.[460] T_1 is inversely related to the correlation time, τ_C, which corresponds to the time required for the rotation of a water molecule. T_1 increases with increasing rotation rate in this range of water content; thus, T_1 can be used as a measure of water mobility. T_1 can be described using Eq. (2.115) by assuming that there are two kinds of water molecules present, freely mobile water and water whose mobility is decreased by interaction with the polymer. In Eq. (2.115), $T_{1(1)}$ and f_1 represent the spin–lattice relaxation time and the fraction of freely mobile water, respectively, while $T_{1(2)}$ and f_2 represent the corresponding values for water with restricted mobility. Assuming that the value of $T_{1(1)}$ is that of pure water, $T_1{}^{\circ}$, rearrangement of Eq. (2.115) yields Eq. (2.116).

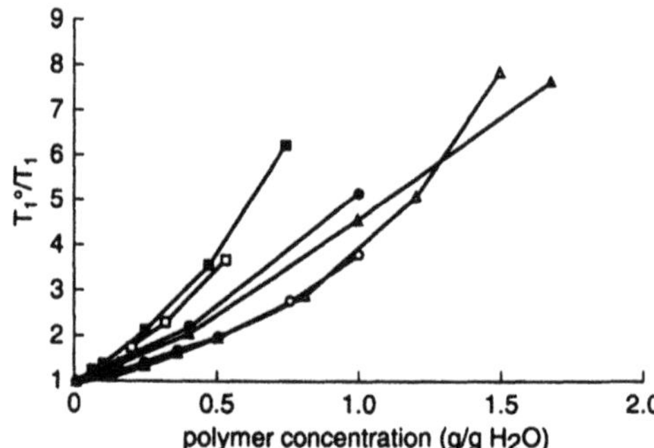

Figure 108. Spin–lattice relaxation time of water in aqueous solutions of polymer excipients. ■, Polyvinylpyrrolidone; □, gelatin; ●, polyethylene glycol (PEG) 20,000; ▲, PEG400; △, sucrose; ○, glucose. $T_1°$, Spin–lattice relaxation time of pure water. (Reproduced from Ref. 460 with permission.)

$$\frac{1}{T_1} = \frac{f_1}{T_{1(1)}} + \frac{f_2}{T_{1(2)}} \tag{2.115}$$

$$\frac{T_1°}{T_1} = 1 + \left(\frac{T_1°}{T_{1(2)}} - 1\right) f_2 \tag{2.116}$$

Since f_2 is proportional to polymer concentration (weight of polymer/total weight of water), plotting the term on the left-hand side of Eq. (2.116) ($T_1°/T_1$) against polymer concentration should yield a linear relationship. As shown in Fig. 108, the plots obtained for aqueous solutions of polyvinylpyrrolidone, polyethylene glycol, and gelatin exhibited nonlinear relationships at higher polymer concentrations, indicating that T_1 of water in these polymer solutions cannot be adequately described in terms of two groups of water having constant T_1 values (namely, $T_{1(1)}$ and $T_{1(2)}$) and that both $T_{1(1)}$ and $T_{1(2)}$ decrease with increasing polymer concentration. This has been confirmed by measurements of the dielectric relaxation time of water with these polymer solutions.[461] Polyvinylpyrrolidone exhibited the most significant increase in $T_1°/T_1$ values with increasing polymer concentration, suggesting its stronger tendency to decrease water mobility.

Water mobility in polymer solutions, as assessed by the methods discussed above, appears to be related to drug degradation rate in the presence of the polymers. The rate of kanamycin-catalyzed hydrolysis of flomoxef in gelatin gels, which depends on the diffusion rates of the drug and the catalyst, was related to the molecular mobility of water as determined by measurements of its spin–lattice relaxation time.[462] The decrease in the T_1 of water with increasing polymer concentration may reflect an increase in the microviscosity of the gelatin gel, which would result in a decrease in the diffusion-controlled reaction rate.

The rate of microviscosity as a factor affecting drug degradation rate has also been reported by Spancake *et al.*[463] For example, the degradation rate of aspirin in Tetronic gels decreased with increasing polymer concentration owing to increasing microviscosity, as shown in Fig. 109. The decrease in degradation rate was less than expected based on the changes in macroviscosity. This was explained by assuming that the microviscosity did not

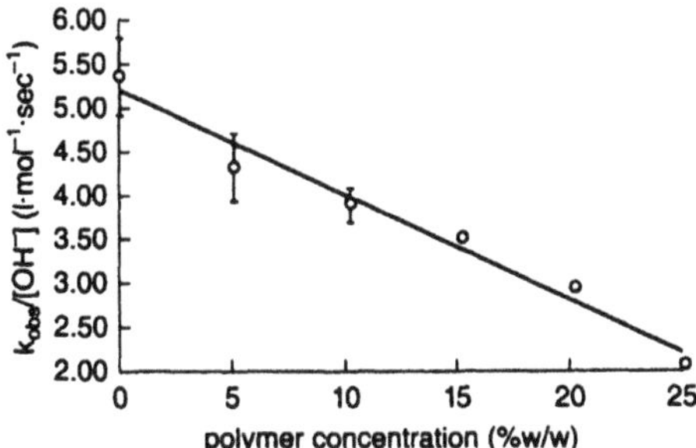

Figure 109. Effect of polymer concentration on the degradation of aspirin in alkaline media as a function of Tetronic polymer concentration (pH 10.0, 50°C). (Reproduced from Ref. 463 with permission.)

change as much as the macroviscosity, with water maintaining a high mobility within the gel structure.

Drug degradation rate in gels of lower water content was determined by the concentration of water with high mobility. The apparent hydrolysis rate of trichlormethiazide in gelatin gels increased with increasing water content at a water content of more than 15%, as shown in Fig. 110.[462] Dielectric relaxation spectroscopy showed that there are three kinds of water with different mobilities within gelatin gels: water bound to gelatin (both strongly and weakly bound) and water with a high mobility close to that of pure, or bulk, water.[461] The amount of water with high mobility as determined by dielectric relaxation spectroscopy, increased with increasing water content at a water content greater than 15%. This increase correlated well with the degradation rate.[462]

The effect of water with high mobility on drug degradation has also been demonstrated in the hydrolysis of a solid cephalothin mixture with microcrystalline cellulose.[464] The hydrolysis rate was found to be proportional to the amount of water with high mobility rather than to the total amount of water, which included water with strongly reduced mobility.

The increase in drug degradation rate in solid polymer matrices with increasing water content can also be attributed to the plasticizing effect of water. As shown in Fig. 111,[465] the

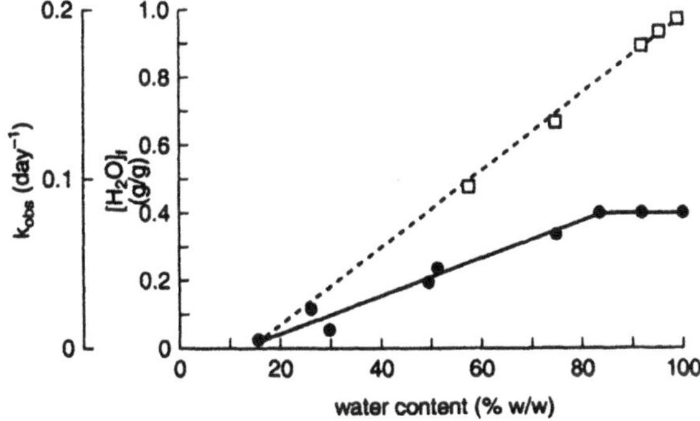

Figure 110. Effect of percent water content on the first-order degradation rate constant for trichlormethiazide in a gelatin gel at 27°C. ●, Apparent first-order rate constant; □, $[H_2O]_f$ (concentration of free water). (Reproduced from Ref. 462 with permission.)

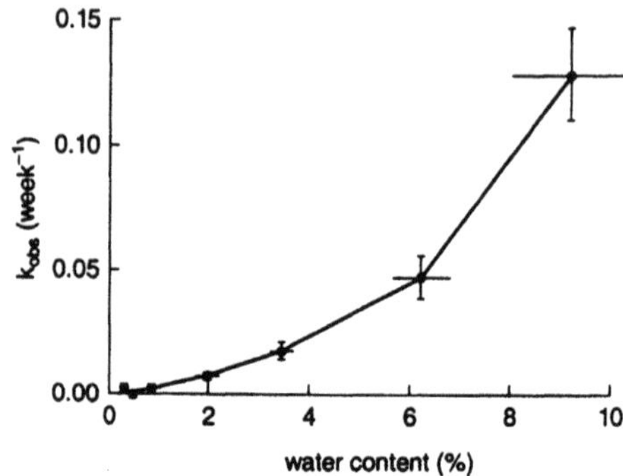

Figure 111. Effect of water content on the dehydration of misoprostol dispersed in hydroxymethyl cellulose (55°C) at a function of water content. (Reproduced from Ref. 465 with permission.)

dehydration rate of misoprostol dispersed in hydroxymethyl cellulose increased significantly as water content increased, presumably due to the plasticizing effect of water. Drugs in polymer matrices in a glassy state with low water content exhibit high stability owing to the restricted mobility, whereas those in polymer matrices plasticized by water exhibit increased degradation. Few studies have been performed on the plasticizing effect of water on the chemical degradation of pharmaceuticals. However, there are some studies on the plasticizing effect of water on physical degradation in relation to glass-transition temperature. This effect will be described in Chapter 3. It is difficult to make generalizations about the role of water in chemical degradation because of the multiple roles (reactant, plasticizer etc.) that water can play.

2.2.13.4. Other Properties of Excipients

Excipients can also affect drug stability by altering microclimate pH. The surface acidity of excipients has been reported to be a factor contributing to drug degradation, for example, in the isomerization of vitamin D_2.[466] Carboxylic acid groups on the solid surface furnish a representative example. Lomustine exhibited faster degradation in poly(*d*,*l*-lactide) microspheres than in its pure crystalline state.[467] Although this enhanced degradation has been attributed to molecular dispersion of the drug in the microspheres, the possibility that the terminal carboxylic acid groups of poly(*d*,*l*-lactide) effect micro-pH changes cannot be excluded. Degradation of etoposide entrapped in poly(*l*-lactide) microspheres increased as the number of terminal carboxylic acid groups increased during poly(*l*-lactide) decomposition, suggesting the degradation enhancing effect of these groups.[468] Enhanced degradation of solid oxazolam in the presence of microcrystalline cellulose may be attributed to carboxylic acid groups on the cellulose surface in addition to the effect of moisture.[469,470]

Excipients affect drug degradation via various mechanisms other than pH changes. The effect of stearate on the degradation of aspirin has been explained by a change in melting behavior rather than pH changes (Fig. 112).[471] Dye excipients may enhance oxidation and photodegradation of drugs by producing singlet oxygen that participates in chain reactions. Examples are enhancement of the oxidation of phenylbutazone[472] and ascorbic acid[473] by dye excipients.

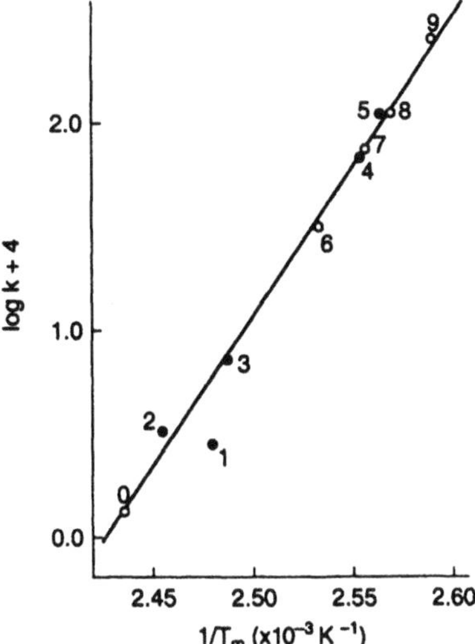

Figure 112. Relationship between degradation rate and melting point of aspirin in the presence of various stearate salts. (0) No additive; (1) 3% Zn salt; (2) Al salt; (3) Na salt; (4) Ca salt; (5) and (8) Mg salt; (6) 1% Mg salt; (7) 2% Mg salt; (9) 5% Mg salt. The values of k were calculated according to the equation $[1-(1-x)^{1/2}]^2 = kt$. (Reproduced from Ref. 471 with permission.)

Metal ions used as pharmaceutical excipients, or present as impurities, often catalyze drug degradation. Metal ions are well known to be catalysts of oxidation and photodegradation of drugs, as described in 2.1.5 and 2.1.6. Metal ions may also cause drug degradation by forming complexes with drugs. A good example is the rearrangement reaction of fosinopril caused by magnesium ion.[474]

The role that surfactants play in drug degradation has received considerable attention.[475] Mechanisms for the effect of surfactants are complex and depend on various factors. For example, the effect of cetyltrimethylammonium bromide (CTAB) on the kinetics of the hydrolysis of phenyl acetates and ethyl benzoates varied depending on the substituent on the phenyl ring of these esters (amino or nitro groups etc.) (Fig. 113).[476] The observation that alkaline hydrolysis of acetylcholine is decreased by dodecyltrimethylammonium chloride (DTAC), as shown in Fig. 114, has been explained by assuming that the drug molecule penetrates the micellar phase and is shielded from the attack of hydroxide ion.[477] However, it is hard to imagine why the polar acetylcholine should have an affinity for the hydrophobic core of the micelle. Alkaline hydrolysis of benzocaine is inhibited by cetyltrimethylammonium chloride (CTAC).[478] Alkaline hydrolysis of indomethacin is inhibited by nonionic surfactants such as ethoxylated lanolin and anionic surfactants such as sodium dodecyl sulfate but enhanced by a cationic surfactant, CTAB, as shown in Fig. 115.[479,480] The latter enhancing effect has been explained by increased concentration of the reactants upon micelle formation. This effect decreased at higher concentrations of surfactant. Similar enhancing effects of CTAB have been reported for hydrolysis of various naphthyl[481] and carbaryl esters.[482]

1,4-Benzodiazepines undergo hydrolysis of the azomethine and amide groups by acid catalysis. Anionic surfactants such as sodium dodecyl sulfate inhibited hydrolysis of the

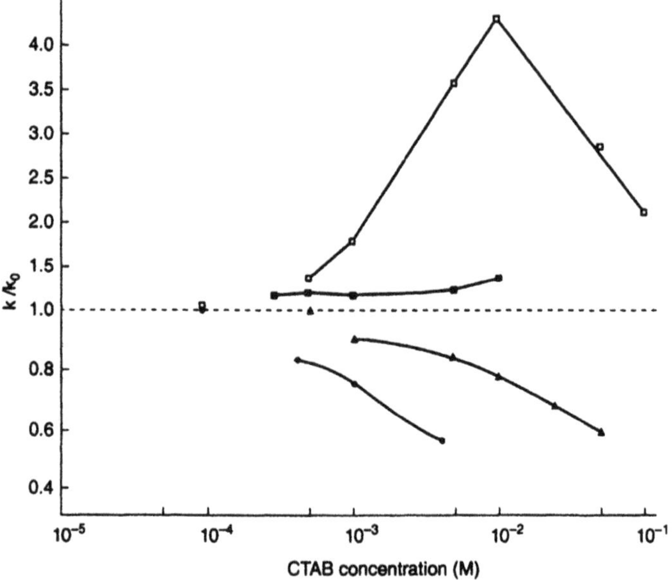

Figure 113. Effect of CTAB on the degradation rate of various acetate and benzoate esters. The ratio of the rate constant in the presence of CTAB (k) to that in its absence (k_o) is plotted as a function of CTAB concentration. □, *p*-Nitrophenyl acetate (pH 9.2, 25°C); ■, ethyl *p*-nitrobenzoate (pH 10.64, 25°C); ▲, *p*-aminophenol acetate (pH 10.64, 50°C); ●, ethyl *p*-aminobenzoate (pH 10.55, 50°C). (Reproduced from Ref. 476 with permission.)

azomethine group while enhancing amide hydrolysis.[483] The inhibiting effect increased as the hydrophobicity of the surfactants increased.[484]

The effect of surfactants on the degradation of β-lactam antibiotics is difficult to interpret. As shown in Fig. 116, acid degradation of propicillin in solutions was inhibited by polyoxyethylene-23-laurylether (a nonionic surfactant) and CTAB (a cationic surfactant) but enhanced by sodium lauryl sulfate (an anionic surfactant).[485] Degradation of cefaclor (an α-aminophenyl cepalosporin) was enhanced by CTAB. The dependence on salt concentration suggests a complex mechanism for the effect of surfactants on degradation of cefaclor.[486]

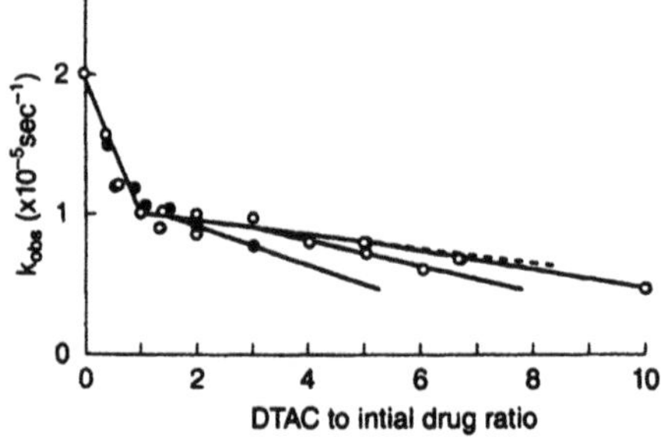

Figure 114. Effect of DTAC on the degradation of acetylcholine (pH 9.0, 25°C). Initial drug concentration: ◑, 3 m*M*; ○, 5 m*M*; ●, 10 m*M*. (Reproduced from Ref. 477 with permission.)

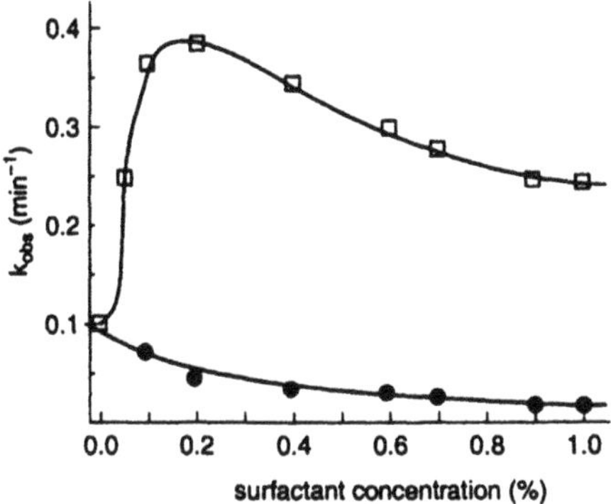

Figure 115. Effect of surfactants on the degradation of indomethacin (30.3°C, [OH$^-$] = 5 m*M*). ●, Ethoxylated lanolin; □, CTAB. (Reproduced from Ref. 479 with permission.)

Phospholipid liposomes can affect drug degradation. The hydrolysis rate of procaine free base decreased in the presence of liposomes (Fig. 117).[487,488] The inhibiting effect of liposomes has also been reported for alkaline hydrolysis of various esters such as aspirin.[489–491] An electron spin resonance (ESR) study confirmed that the decrease in rate is due to partitioning of the drug substances into the liposomes, thus altering the reactant concentration in the aqueous phase, where the hydrolysis reaction is presumed to occur.[492] Liposomes inhibited the degradation of the free base of 2-diethylaminoethyl *p*-nitrobenzoate but enhanced the degradation of the protonated form of the drug. An explanation is the differing interaction between drug and liposome, depending on the charge of the drug molecule.[493]

Excipients such as cyclodextrins, which form inclusion complexes with drug substances, can have a significant effect on drug stability. These effects will be described in

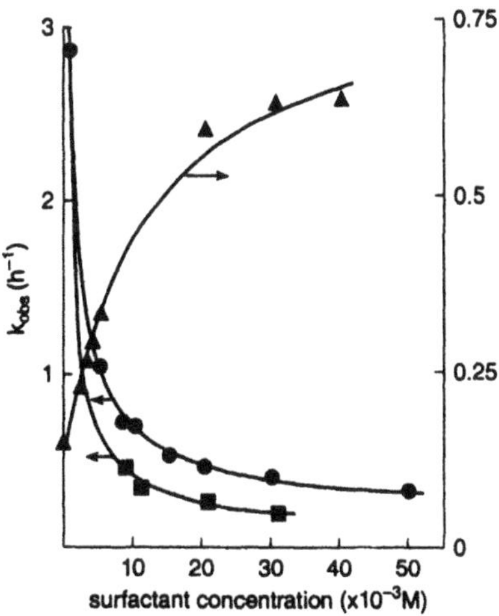

Figure 116. Effect of surfactants on the degradation of propicillin at 37°C. ●, Polyoxyethylene-23-laurylether, pH 1.10; ▲, sodium lauryl sulfate, pH 3.00; ■, CTAB, pH 1.10. (Reproduced from Ref. 485 with permission.)

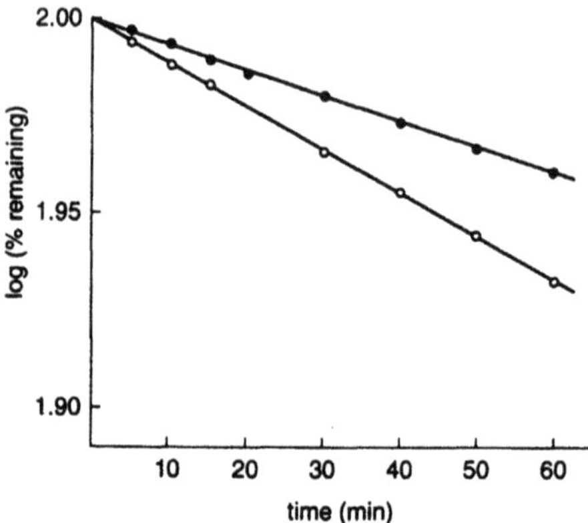

Figure 117. Effect of a liposome formulation on the degradation of procaine (pH 10.2, 40°C). Lecithin concentration: ○, 0; ●, $1.01 \times 10^{-2} M$. (Reproduced from Ref. 487 with permission.)

greater detail in Section 2.3. Some polymers affect drug stability by mechanisms different from those described earlier. For example, it has been proposed that hydroxypropyl methyl cellulose decreased the rate of dehydration of prostaglandin E_1 by protecting the drug from water through entanglement of the drug in a polymer environment.[494]

Detailed mechanisms are not clear in other reports of the effects of excipients on drug stability. For example, bropirimine was destabilized by solubilizers (such as polyethylene glycol),[495] whereas water adsorbed on porous montmorillonite exhibited a catalyzing effect on the degradation of aspirin.[496] Some excipients appear to have minimal effects on drug stability; for example, fluocinolone acetonide in a cream formulation exhibited the same degradation as in aqueous solution without the excipients present in the cream formulation.[497]

2.2.14. Miscellaneous Factors

Although the effect of γ irradiation is not a common variable considered during most drug stability studies, γ irradiation is employed for the sterilization of some pharmaceutical products, and its effect on drug stability should therefore be considered. The decrease in bioactivity of β-lactam antibiotics[498,499] and insulin[500] after γ irradiation has been reported, and an acceptable irradiation dose for sterilization has been proposed. Similar effects of γ irradiation have been reported for various other drug substances.[501,502] Altered drug degradation as a result of γ irradiation generally takes the form of changes in the degradation products; for example, γ irradiation of corticosteroids yielded the C_{17}-ketone as a principal degradation product.[503] The formation of various products upon γ irradiation of medazepam in a methanolic solution was inhibited by ascorbic acid.[504] Stable free radicals formed upon γ irradiation of ceftazidime were suggested to exist for more than 100 days at room temperature.[505]

Another factor affecting drug stability is electrical current. The stability of hydrocortisone hemisuccinate and phosphate esters as their sodium salts was affected by a change in pH that occurred when current was applied during an iontophoresis study.[506] Degradation of propranolol hydrochloride in an electrically modulated drug delivery device has been reported.[507]

2.3. Stabilization of Drug Substances against Chemical Degradation

As discussed above, the chemical degradation rate of drug substances is affected by various factors. These effects can be related to the rate equations, Eqs. (2.1), (2.3) and (2.4). For example, in a given formulation strategy, such factors as pH, ionic strength, and solvent dielectric constant could be controlled so as to attain the lowest possible degradation rate for a particular drug. Other molecules such as buffers and excipients, which can possibly interact with the drug directly or act as general acid–base catalysts, nucleophilic or electrophilic catalysts etc., should be excluded or their concentrations should be minimized. Other approaches include minimization of moisture, particularly freely mobile water present in excipients, especially in solid-dosage-form formulations.

Pharmaceuticals should be stored under conditions favorable to greater drug stability. The temperature dependence of the rate of drug degradation described in Section 2.2.4 indicates that pharmaceuticals should be stored at low temperature if they lack sufficient stability at normal temperatures. The exceptions to this rule are the occasional drugs on which freezing has an adverse effect. It is also important to avoid oxygen and light by using suitable containers and packaging in cases where these variables enhance drug degradation. In the following sections, other methods for stabilizing pharmaceuticals are described.

2.3.1. Stabilization by Modification of Molecular Structure of Drug Substances

Drug degradation rates depend on the chemical structure of the drug, as described earlier. Most often, structure modifications are performed to enhance activity or to have a positive impact on the *in vivo* properties of the drug. However, for drugs that are very chemically unstable, development of more stable analogs should be possible through appropriate structure modifications.

An example of analog development to effect stabilization is the masking of reactive hydroxyl groups. Degradation of erythromycin via 6,9-hemiketal breakdown under acidic pH conditions is inhibited by substituting a methoxy group for the C-6 hydroxyl. For example, the acid stability of clarithromycin is 340 times greater than that of erythromycin (Fig. 118).[508]

The effect of substituents on degradation rate has been reported for hydrolysis of water-soluble prodrugs of phenytoin,[509] and (aminomethyl)benzoate ester prodrugs of

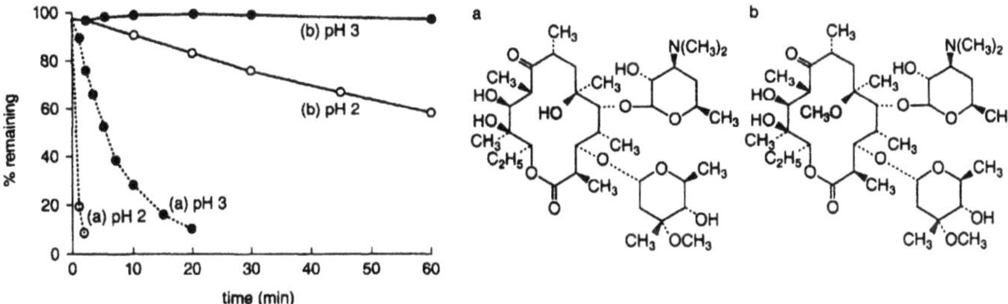

Figure 118. Acid degradation of showing the stabilizing effect of substitution of a methoxy for a hydroxy group. erythromycin (a) and clarithromycin (b) at 37°C. (Reproduced from Ref. 508 with permission.)

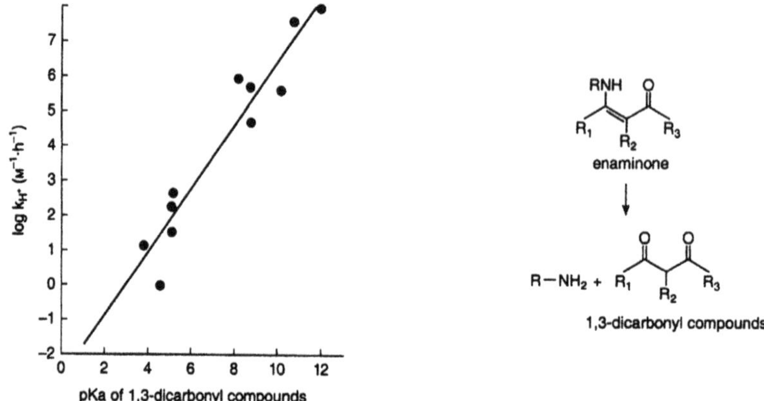

Figure 119. Effect of pK_a of 1,3-dicarbonyl compounds formed by the degradation of various enaminones on the rate of degradation at 25°C. (Reproduced from Ref. 511 with permission.)

acyclovir.[510] As shown in Fig. 119, the acid-catalyzed degradation rate of enaminones, formed between a primary amine and a 1,3-dicarbonyl compound, decreased as the pK_a of the 1,3-dicarbonyl compound decreased, because the rate-determining step in the degradation was proton addition to the vinyl carbon of the enaminone.[511]

Although many studies have been reported on the stabilization of drugs against enzymatic degradation *in vivo*, such as the protection of peptides from metabolism by modifying the N-terminal with an ethyl group,[512] this topic will not be covered here.

2.3.2. Stabilization by Complex Formation

Complex formation between drugs and excipients often leads to stabilization of drugs. The forces involved in complex formation include van der Waals forces, dipole–dipole interactions, hydrogen bonding, Coulomb forces, and hydrophobic interactions.

The effect of complex formation on drug stability can often be described in terms of the processes represented by Scheme 76. If drug D forms a complex with ligand L, the complex (assuming a 1:1 interaction) can be defined by a complexation constant K:

$$K = \frac{[\text{D–L}]}{[\text{D}]_f[\text{L}]_f} \tag{2.117}$$

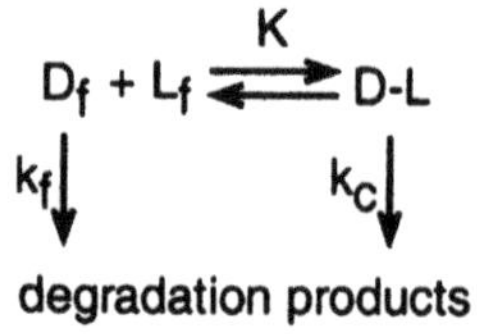

$$\begin{array}{ccc} D_f + L_f & \overset{K}{\rightleftharpoons} & \text{D-L} \\ k_f \downarrow & & \downarrow k_c \\ & \text{degradation products} & \end{array}$$

Scheme 76. Interaction of a drug (D) with a ligand (L). The interaction can lead to either stabilization of the drug ($k_c < k_f$) or catalysis of its breakdown ($k_c > k_f$).

The term [D–L] represents the concentration of the complex, D–L, $[D]_f$ is the concentration of free or uncomplexed drug, and $[L]_f$ is the concentration of free ligand. In Scheme 76, k_f represents the rate constant for the degradation of the drug in the absence of complexation, and k_c is the rate constant for the degradation of the drug in its complexed form. As can be seen, the drug will be stabilized by the presence of L if $k_c < k_f$. The degree of stabilization will also depend on the relative amounts of free and complexed drug, which in turn depends on the concentrations of D and L and the magnitude of K. Conversely, if $k_c > k_f$, complex formation will result in acceleration of the degradation. Differing ligands (L) in a series can affect the degradation rate in two ways: first, by affecting the degree of complexation, as measured by K, and, second, by affecting k_c.

Stabilization of esters such as benzocaine (Fig. 120), procaine, and tetracaine by complex formation with caffeine was reported in the 1950s.[513–515] The stabilization of drugs by caffeine is thought to result from the formation of "stacking" complexes. Thus, attack by water or hydroxide ion on the ester bonds of benzocaine is hindered when the benzocaine molecules are sandwiched between caffeine molecules. Similar stabilization by caffeine has been reported for base-catalyzed degradation of riboflavin.[516]

Ampicillin, cephalexin, and bacampicillin are stabilized by complex formation with aldehydes such as benzaldehyde and furfural,[517–522] although this stabilization involves reversible formation of covalent species. Even though these interactions involve covalent bond formation, they follow Scheme 76, because the covalent association and dissociation (defined by the equilibrium constant K) is fast relative to k_c and k_f. The greater stability of *N*-nitrosoureas in Tris buffers than in carbonate buffers has been ascribed to complex formation with tris(hydroxymethyl)aminomethane.[523]

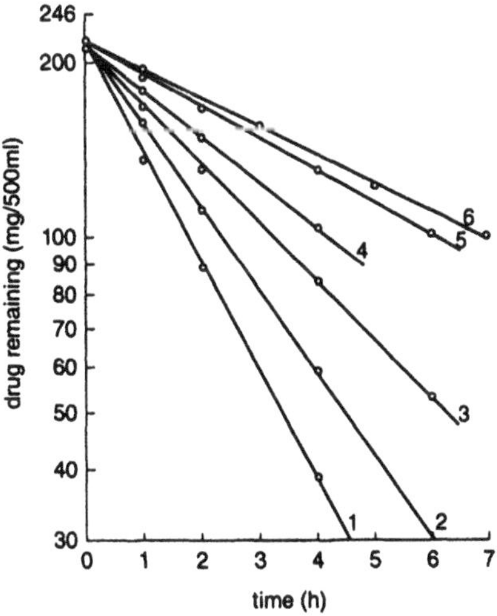

Figure 120. Stabilization of benzocaine by complex formation with caffeine (30°C, 0.04 *N* hydroxide ion). Caffeine concentration: (1) 0.25%; (2) 0.50%; (3) 1.00%; (4) 1.50%; (5) 2.00%; (6) 2.50%. (Reproduced from Ref. 513 with permission of the American Pharmaceutical Association.)

2.3.3. Stabilization by the Formation of Inclusion Complexes with Cyclodextrins

In the caffeine examples given above, the practicality is lost because of the non-inert nature of the ligand, caffeine. Recently, considerable interest has been generated in the use of cyclodextrins (CDs) to improve drug stability. Cyclodextrins are nonreducing cyclic oligosaccharides, consisting of six (α-(CD), seven (β-CD), or eight (γ-CD) dextrose units. Cyclodextrins have a "doughnut" shape, with the interior of the molecule being relatively hydrophobic and the exterior being relatively hydrophilic. Because of their unique chemical structure, cyclodextrins are capable of forming so-called "inclusion" complexes with many drug molecules. Figure 121 best illustrates this concept. If drug D is capable of undergoing chemical degradation in solution, protection of the molecule by inclusion complexation with a cyclodextrin is often possible.

The natural cyclodextrins, α-, β-, and γ-CD, have been chemically modified either to effect stronger complexation or to improve their safety. α-CD and β-CD cannot be used parenterally because of their nephrotoxicity. Recently, a number of modified cyclodextrins, including 2-hydroxypropyl-β-cyclodextrin (HP-β-CD) and sulfobutylether β-cyclodextrins [mainly $(SBE)_{7M}$-β-CD, Captisol^R], have been tested. They show better safety profiles than does β-CD.

The overall loss of drug in the presence of cyclodextrins can be defined by Eq. (2.118). The terms k_c and k_f have the same definitions as in Scheme 76; D–CD and D_f are depicted in Fig. 121. Using Eq. (2.117) and substituting into Eq. (2.118), with the assumption that the cyclodextrin is present in excess of the drug, yields Eq. (2.119).

$$-\frac{d[\mathrm{D}]}{dt} = k_c[\mathrm{D{-}CD}] + k_f[\mathrm{D}]_f \tag{2.118}$$

$$k_{\mathrm{obs}} = \frac{k_f + k_c K[\mathrm{CD}]_f}{1 + K[\mathrm{CD}]_f} = \frac{k_f + k_c K[\mathrm{CD}]_T}{1 + K[\mathrm{CD}]_T} \tag{2.119}$$

The term [CD] is the total concentration of CD present in solution, and k_{obs} is the observed pseudo-first-order rate constant of the degradation of drug. Data such as those shown in Fig. 122 can be fitted to Eq. (2.119) (and variations of this equation) to yield estimates of K and k_c. The value of k_f can also be estimated by curve fitting, but it is best determined by direct measurement of drug degradation in the absence of added cyclodextrin.

Stabilization of prostacyclin and prostaglandins in solution and in the solid state by various CD derivatives is well known. Figure 122 shows the effect of the addition of α-, β-, and γ-CDs on the hydrolysis of prostacyclin in aqueous solution. The methyl ester of

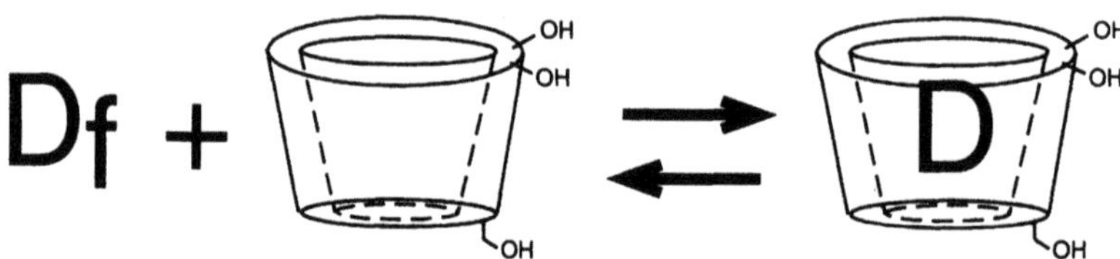

Figure 121. Interaction of free drug molecules (Df) with the cavity of cyclodextrins. The formation of an "inclusion" complex can lead to either stabilization of the drug or catalysis of its breakdown.

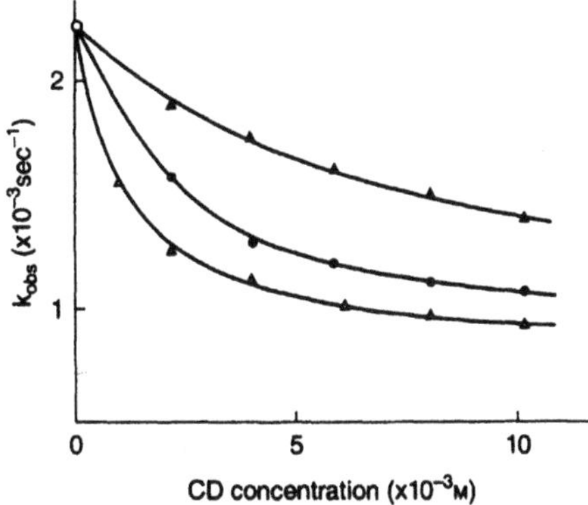

Figure 122. Stabilization of prostacyclin by increasing concentrations of α-CD (●), β-CD (△), and γ-CD (▲) (pH 7.0, 15°C). (Reproduced from Ref. 524 with permission.)

prostacyclin is stabilized to an even larger extent.[524] This differential stabilizing effect has been explained by the fact that ionization of the terminal carboxylic acid of prostacyclin inhibits complex formation. The methylation of the various hydroxyl groups on cyclodextrins has a variable effect on stability. For example, heptakis(2,3-di-*O*-methyl)-β-CD (DM-β-CD) shows a larger stabilizing effect on prostacyclin than does heptakis(2,3,6-tri-*O*-methyl)-β-CD (TM-β-CD) (Fig. 123).[525]

Prostaglandin E_1 in neutral and alkaline solutions is destabilized by β-CD but stabilized by *O*-carboxymethyl-*O*-ethyl-β-CD (CME-β-CD) (Fig. 124).[526] This stabilizing effect has been observed in a fatty alcohol propylene glycol ointment. Dehydration of prostaglandin E_2 and isomerization of prostaglandin A_2 are enhanced by β-CD but inhibited by DM-β-CD and TM-β-CD (Fig. 125).[527] HP-β-CD exhibits an inhibiting effect on degradation only in the acidic pH region.[528]

Stabilization of prostaglandins in the solid state by CDs and their derivatives has also been reported. The stability of 16,16-dimethyl-*trans*-Δ^2-prostaglandin E_1 methyl ester against dehydration and isomerization in the solid state is increased by complex formation with β-CD.[529] As shown in Fig. 126, the stability of prostaglandin E_1 is decreased by the

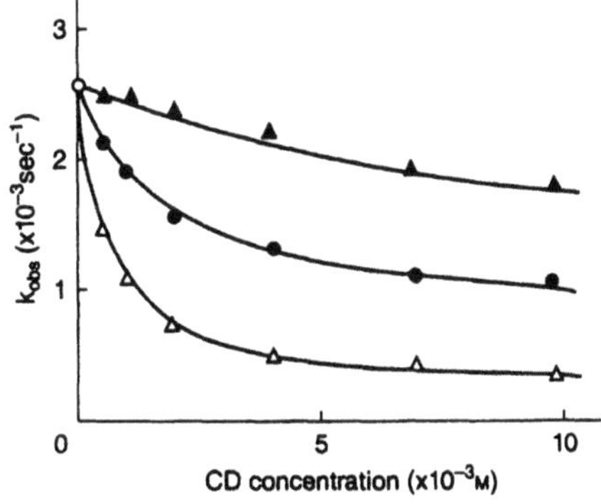

Figure 123. Stabilization of prostacyclin by increasing concentrations of TM-β-CD (▲), β-CD (●), and DM-β-CD (△) (pH 7.0, 15°C). (Reproduced from Ref. 525 with permission.)

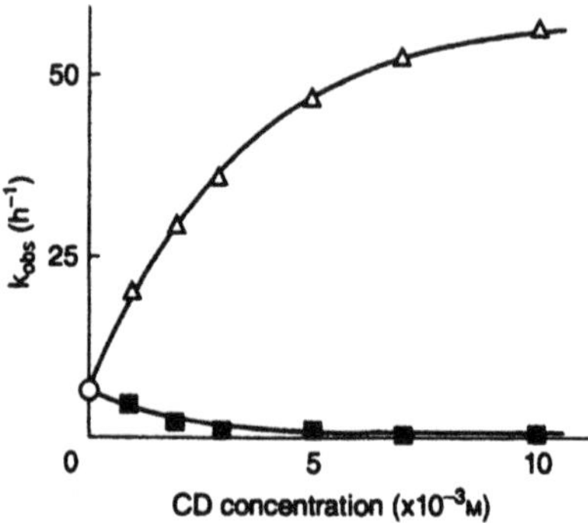

Figure 124. Effect of CD concentration on observed degradation rate constant of prostaglandin E_1 (pH 11.0, 60°C). △, β-CD; ■, CME-β-CD. (Reproduced from Ref. 526 with permission.)

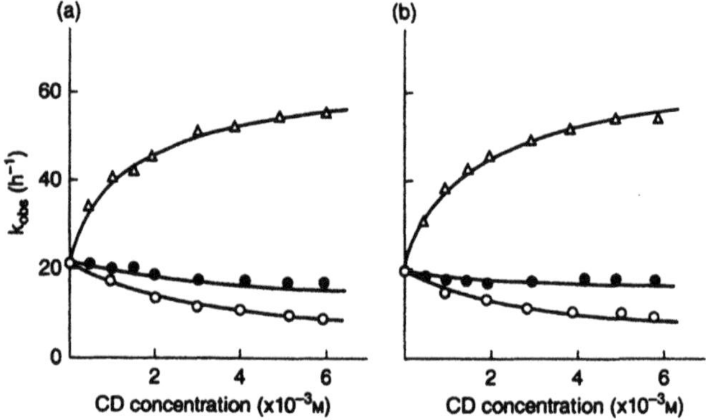

Figure 125. Effect of CD concentration on the rate of dehydration of prostaglandin E_2 (a) and the rate of isomerization of prostaglandin A_2 (b) (pH 11.0, 60°C). △, β-CD; ●, TM-β-CD; ○, DM-β-CD. (Reproduced from Ref. 527 with permission.)

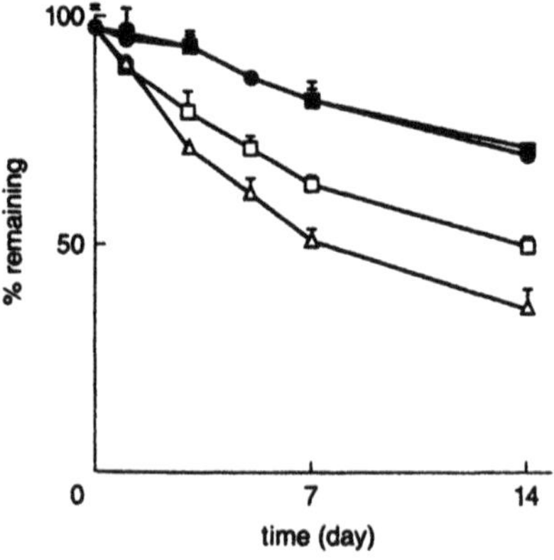

Figure 126. Stabilization of prostaglandin E_1 in the solid state by various CDs versus no CD and a mannitol control at 60°C. [Drug]:[CD]=1:32 (by weight). ■, G_2-β-CD; ●, β-CD; □, no CD; △, mannitol. (Reproduced from Ref. 530 with permission.)

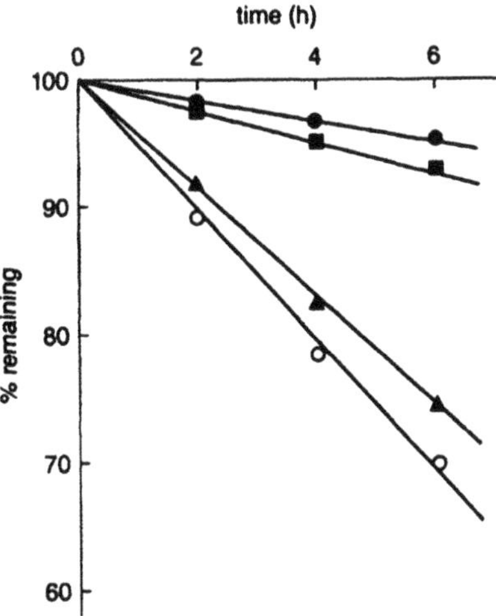

Figure 127. Stabilization of bencyclane fumarate by various CDs (37°C, 0.1*N* HCl). The drug and CD were present in equimolar amounts. ●, β-CD; ■, γ-CD; ▲, α-CD; ○, no CD. (Reproduced from Ref. 531 with permission.)

addition of mannitol upon freezing in the presence of citric acid but increased by the addition of β-CD and 6-*O*-α-D-maltosyl-β-CD (G_2-β-CD).[530] This stabilizing effect has been explained by formation of inclusion complexes, as well as possibly by the fact that moisture, which participates in the degradation, is adsorbed by the CDs.

The stabilizing effect of CDs and their derivatives has been observed with many other drug substances. Hydrolysis of bencyclane fumarate is inhibited by α-, β-, and γ-CD, as shown in Fig. 127.[531] The reduction of the hydrolysis rate of a hydrophobic drug, a dihydropyridine derivative of benzylpenicillin, by HP-β-CD is shown in Fig. 128.[532] Among the numerous other reported examples are the stabilization of doxorubicin[533] and mitomycin C[534] by γ-CD, aspartame by β-CD,[535] daunorubicin in acidic aqueous solution by octakis(2,6-di-*O*-methyl)-γ-CD (DM-γ-CD),[536] estramustine[537] and thymoxamine[538] by DM–

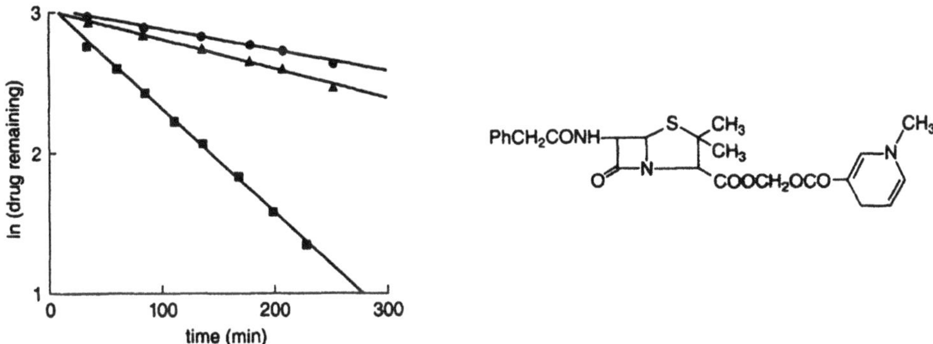

Figure 128. Stabilization of a benzylpenicillin prodrug by HP-β-CD (pH 6.93, 40°C). HP-β-CD concentration: ■, 0; ▲, 0.5%; ●, 1%. (Reproduced from Ref. 532 with permission.)

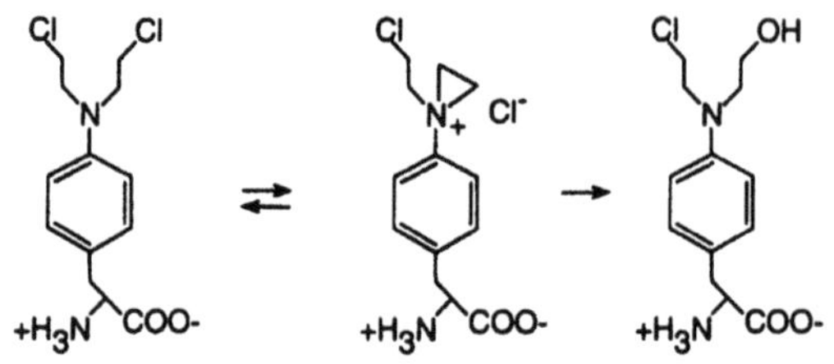

Scheme 77. Reaction pathway leading to melphalan degradation through a polar cyclic ethyleneimmonium intermediate. (Reproduced from Ref. 543 with permission.)

β-CD, thalidomide by HP-β-CD,[539] tauromustine by HP-α-CD,[540] and medroxyprogesterone acetate and megestrol acetate by HP-β-CD and methyl-β-CD.[541] The stabilizing effect of β-CD against the oxidation of 2-[8-methyl-10,11-dihydro-11-oxodibenz[*b,f*]oxepin-2-yl]propionic acid in ointments has also been reported.[542]

Recently, Ma *et al.*[543] studied the stability of the very unstable alkylating agent melphalan in the presence of $(SBE)_{7M}$-β-CD and HP–β-CD. Melphalan undergoes an intramolecular displacement reaction through a very polar cyclic ethyleneimmonium intermediate (Scheme 77). The very polar transition state leading to this intermediate involves considerable charge separation. Both $(SBE)_{7M}$-β-CD and HP-β-CD stabilized melphalan to chemical degradation.[543] What is interesting in this case is that both $(SBE)_{7M}$-β-CD and HP-β-CD have similar binding constants for melphalan and the same K values, as defined in Scheme 76, but k_c was significantly smaller in the case of $(SBE)_{7M}$-β-CD (Table 7). One would intuitively feel that similar binding constants should yield similar positional binding in the cyclodextrin cavity and perhaps similar k_c values. Ma *et al.*[543] were able to demonstrate that the melphalan bound to $(SBE)_{7M}$-β-CD was in a more hydrophobic environment and that there was a small but significant difference in the position of binding of melphalan to $(SBE)_{7M}$-β-CD versus HP-β-CD.

Although the degradation-inhibiting effects of CDs and their derivatives through complex formation can be used as a method for stabilizing pharmaceuticals, CDs and their derivatives may also enhance drug degradation, depending on the reaction mechanisms and the steric arrangement of the drug in the complex. For example, β-CD inhibits the alkaline hydrolysis of benzocaine (ethyl *p*-aminobenzoate)[544] but enhances that of aspirin (Table 8).[545] This has been explained by assuming that complex formation protects the ester bond of ethyl *p*-aminobenzoate from attack by the hydroxide ion nucleophile. In the case of

Table 7. Estimated Values of K, k_f, and k_c for the Degradation of Melphalan at 25°C (pH 7.5) in the Presence of $(SBE)_{7M}$-β-CD and HP-β-CD, in Accordance with Scheme 76[a]

Cyclodextrin	K (M^{-1})	k_f (h^{-1})	k_c (h^{-1})	k_f/k_c
$(SBE)_{7M}$-β-CD	360	0.2	1.8×10^{-3}	114
HP-β-CD	382	0.2	1.1×10^{-2}	20

[a]Reference 543.

Table 8. Effect of β-CD on the Second-Order Rate Constants for the Hydrolysis of Benzocaine and Aspirin[a]

β-CD concentration (%)	Rate constant (M^{-1} h^{-1})	
	Benzocaine 30°C, $[OH^-] = 0.04N$	Aspirin 35°C, pH 10
0.00	0.666	0.258
0.25	0.358	0.530
0.50	0.229	0.807
0.75		1.085
1.00	0.129	

[a]Reference 545.

aspirin, ester bond cleavage is accelerated by the attack of one of the cyclodextrin hydroxyl groups. HP-β-CD, a derivative of β-CD in which some of the hydroxyl groups are masked, stabilizes aspirin (Fig. 129).[546] γ-CD stabilizes anthracyclines such as daunorubicin in the acidic pH region but destabilizes them in the alkaline pH region, as shown in Fig. 130.[547] The destabilizing effect of γ-CD has also been reported for hydrolysis of doxorubicins.[548]

2.3.4. Stabilization by Incorporation into Liposomes, Micelles, or Emulsions

Entrapment of drug substances in liposomes and micelles can lead to changes in their stability. When entrapment reduces the degradation rate, it can be used as a method for stabilizing pharmaceuticals. Aspirin can be partially stabilized by incorporation in L-α-dimyristoylphosphatidylcholine (DMPC)-based liposomes.[489,491] Anesthetics such as procaine are also stabilized by incorporation in liposomes.[490] Physostigmine salicylate in a phospholipid emulsion is stabilized through interaction with phospholipids at the oil–water interface and through incorporation into the internal phase of the emulsion (Fig. 131).[549]

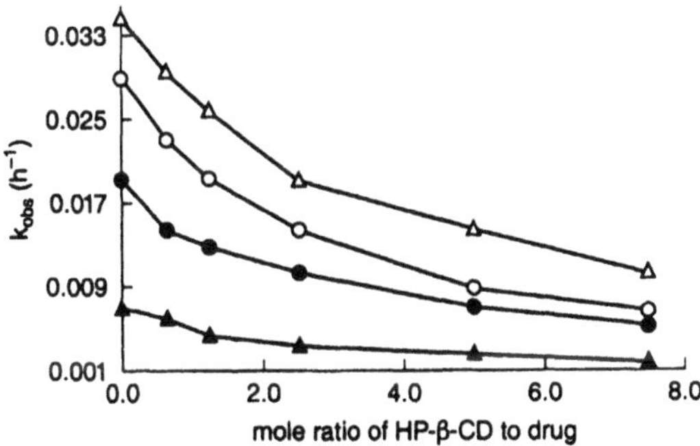

Figure 129. Stabilization of aspirin against degradation with increasing HP-β-CD concentration at different pH values at 40°C. △, pH 6.0; ○, pH 1.3; ●, pH 4.0; ▲, pH 3.0. (Reproduced from Ref. 546 with permission.)

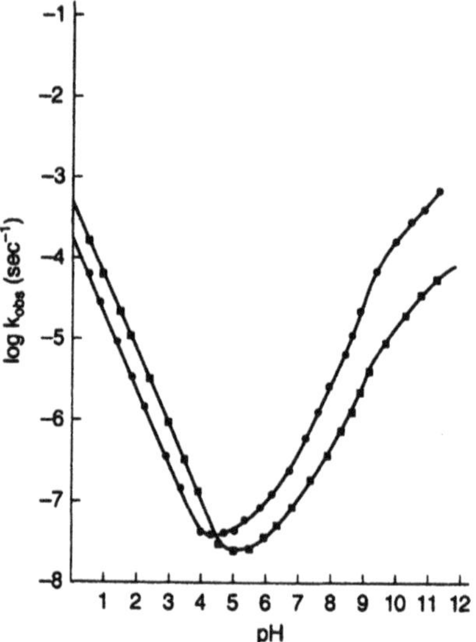

Figure 130. Effect of γ-CD on the degradation of daunorubicin at 50°C, showing stabilization at pH values below 4.5 but accelerated degradation at pH values above 4.5. ●, [γ-CD] = $1.6 \times 10^{-2} M$; ■, no CD. (Reproduced from Ref. 547 with permission.)

Alkaline hydrolysis of forskolin is inhibited in a lipid emulsion.[550] Indomethacin in an oil–water gel is solubilized in the surfactant aggregates and stabilized against acid- and base-catalyzed hydrolysis.[551]

Glycerol, a viscosity-inducing agent, apparently stabilizes drug substances by inhibiting drug diffusion. Photolysis of vitamin B_{12} is inhibited by addition of glycerol.[552]

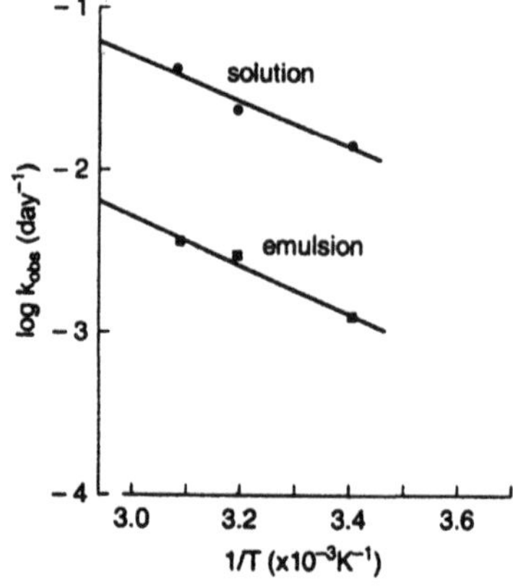

Figure 131. Stabilization of physostigmine salicylate degradation by incorporation into a phospholipid emulsion at pH 5.0. ●, Solution; ■, phospholipid emulsion. (Reproduced from Ref. 549 with permission.)

Table 9. Effect of Antioxidants on Lovastatin Oxidation[a]

Antioxidant[b]	Rate constant ($\times 10^{-3}$ min^{-1})	Inhibition (%)
α-Tocopherol	8.5 ± 0.6	53
BHA[c]	10 ± 1	44
Propyl gallate	12 ± 1	33
None	18 ± 2	—

[a]Reference 553.
[b]Concentration of antioxidant: 0.1% w/w; 40°C.
[c]BHA, Butylated hydroxyanisole.

2.3.5. Addition of Stabilizers Such as Antioxidants and Stabilization through the Use of Packaging

In the previous sections, positive methods for stabilizing pharmaceuticals, involving modification of the molecular structure or the microscopic environment in which the drug is formulated, have been described. On the other hand, the exclusion of some factors adversely affecting drug stability can also lead to the stabilization of pharmaceuticals. Because oxygen, light, and moisture often enhance drug degradation as described earlier, various methods for excluding these factors have been considered as means of stabilizing pharmaceuticals.

Often, the effect of oxygen can be eliminated by the addition of antioxidants. Oxidation of lovastatin in aqueous solution is inhibited by antioxidants such as α-tocopherol and butylated hydroxyanisole (BHA), as shown in Table 9.[553] Similarly, the inhibition of the oxidation of cholecalciferol by α-tocopherol and ascorbic acid[554] and the inhibition of the oxidation of NSC-629243 (an *O*-alkyl-*N*-aryl thiocarbamate; see Scheme 49) by thioglycolic acid[187] have been reported. The ability to inhibit oxidative photodegradation of benzaldehyde has been utilized as a measure of the antioxidant effects of polyhydric phenols.[555] Pharmaceuticals are often stabilized by the utilization of packaging containing an antioxidant.[556] For example, the photooxidation of cianidanol in the solid state was inhibited by lowering the concentration of oxygen with the use of an oxygen absorbent, as illustrated in Figure 132.[402]

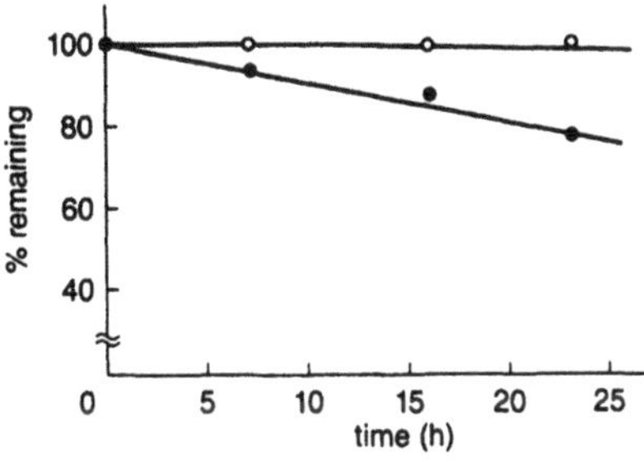

Figure 132. Stabilization of cianidanol against photodegradation (light source: mercury lamp) in the solid state by the use of an oxygen absorbent. ○, With oxygen absorbent; ●, without oxygen absorbent. (Reproduced from Ref. 402 with permission.)

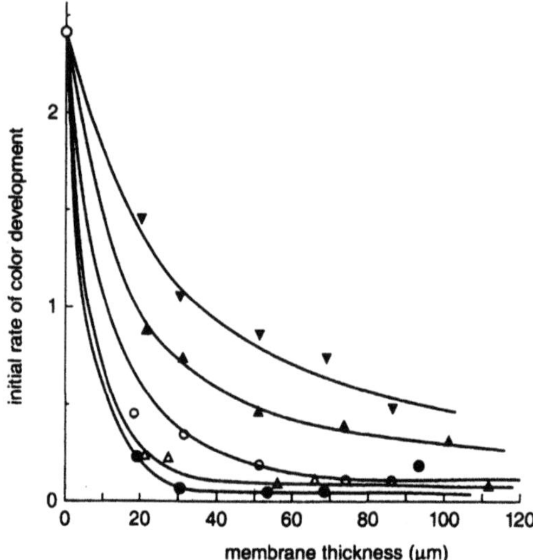

Figure 133. Stabilization of sulfisomidine tablet coloration (light source: mercury lamp) by modifying film coating thickness and oxybenzone content. Samples were irradiated at a temperature below 27°C. Oxybenzone concentration: ▼, 0.5%; ▲, 1%; ○, 2%; △, 5%; ●, 10%. (Reproduced from Ref. 557 with permission.)

The use of photoprotective films generally eliminates the effect of light. As shown in Fig. 133, a film coating containing oxybenzone inhibited coloration and photolytic degradation of sulfisomidine tablets. The extent of inhibition depended on the concentration of the UV absorber and the thickness of the film.[557] Titanium dioxide in a gelatin capsule shell stabilized indomethacin (Fig. 134),[558] and film coatings containing titanium dioxide stabilized nifedipine in a tablet formulation.[559] Incorporation of synthetic iron oxides resulted in the stabilization of uncoated tablets of nifedipine and sorivudine against phododegradation.[560] Stabilization of other pharmaceuticals against photodegradation by the use of

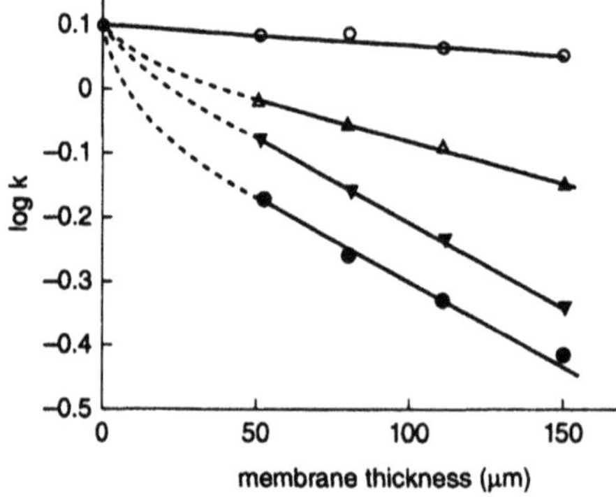

Figure 134. Stabilization of indomethacin coloration (light source: mercury lamp) by the addition of titanium dioxide to gelatin capsule shells. Titanium dioxide concentration: ○, 0; ●, 0.5%; △, 1.0%; ▼, 1.5%. The values of k were calculated from the equation ΔE (color difference) = $k(t^{1/2} - 0.08)$. (Reproduced from Ref. 558 with permission.)

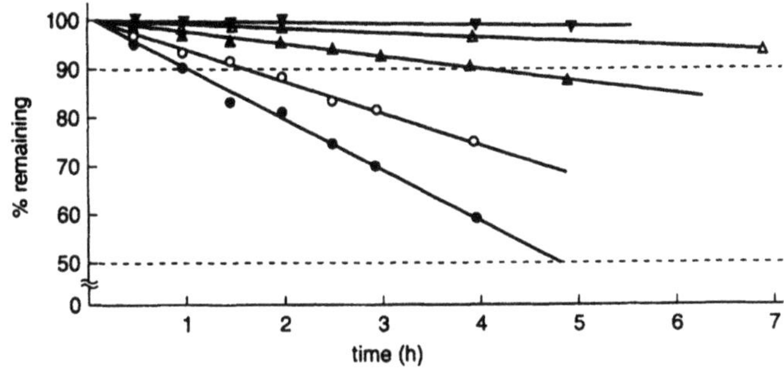

Figure 135. Stabilization of the photochemical degradation (at 290–700 nm) of daunorubicin solutions by the addition of various colorants: ▼, scarlet GN; △, amaranth; ▲, ponceau 6R; ○, tartrazine; ●, no colorant. (Reproduced from Ref. 561 with permission.)

colorants has also been reported; for example, daunorubicin in solution was stabilized by the addition of various colorants (Fig. 135),[561] and ubidecarenone in a dry emulsion was stabilized by the addition of α-(*o*-tolylazo)-β-naphthylamine.[562]

Various methods for protecting pharmaceuticals from the effect of moisture have been considered, with the use of moisture-proof packaging[563,564] and film coatings[565] being the techniques most often utilized. Further discussion of moisture permeability of packaging will be presented in Chapter 4.

Chapter 3

Physical Stability of Drug Substances

Most studies on drug stability have focused on the chemical stability of drug substances, as described in Chapters 1 and 2. However, the physical stability of drugs must also be considered. The physical state of a drug determines its physical properties such as its solubility. Because these properties in turn affect the efficacy and, potentially, the safety of a drug substance, changes in the physical state of a drug substance need to be determined. Traditionally, changes in physical state are assessed by differential scanning calorimetry and X-ray diffraction analysis.

In addition, changes in the physical states of excipients or enabling agents in a dosage form may affect the stability of pharmaceuticals. In this chapter, the physical stability of drug substances and excipients will be briefly described. The physical stability of dosage forms, including changes in dissolution or release rate caused by physical changes, will be described in Chapter 4.

3.1. Physical Degradation

Components of pharmaceuticals (drug substances and excipients) exist in various microscopic physical states with differing degrees of order. Examples are amorphous and various crystalline, hydrated, and solvated states. With time, the drug or the excipient may change from one state, usually unstable or metastable, to a more thermodynamically stable state. The rate of conversion will depend on the chemical potential corresponding to the free-energy difference between the states and the energy barrier (like that for chemical reactions) that must be overcome for the conversion to take place. The following sections address the physical changes that can occur in drug substances and excipients and describe factors affecting these physical changes as well as methods for stabilizing drugs in a fixed defined state.

3.1.1. Crystallization of Amorphous Drugs

Attempts are often made to formulate poorly water-soluble drugs in their amorphous state. This is because the solubility of amorphous materials is generally higher than that of the same substances in their crystalline state. However, because of the lower free energy of the crystalline state, amorphous substances tend to change to their more thermodynamically stable crystalline state with time. Therefore, crystallization of amorphous drug substances

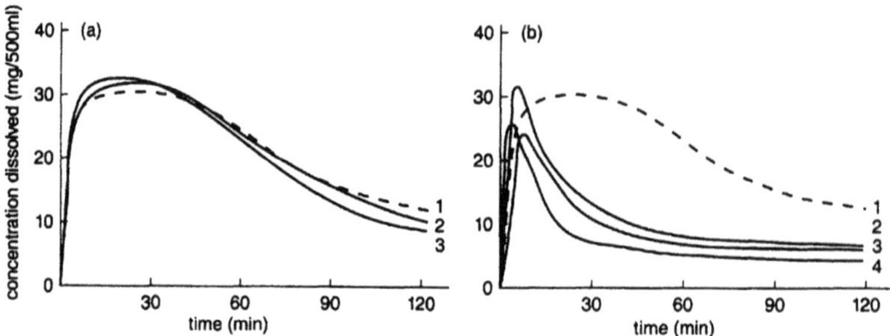

Figure 136. Changes in dissolution behavior of nifedipine from amorphous nifedipine samples exposed to different storage conditions. Storage period at 40°C: (1) 0, (2) 3.5, (3) 6 months; (b) storage period at 21°C and 75% RH: (1) 0, (2) 0.5, (3) 1.5, (4) 4 months. (Reproduced from Ref. 566 with permission.)

may occur during long-term storage and may lead to drastic changes in the release characteristics of the drug and, hence, changes in its clinical and toxicological behavior. Changes in crystal habit during storage have been reported for many drug substances. Some examples are discussed below.

Amorphous nifedipine, coprecipitated with polyvinylpyrrolidone, undergoes partial crystallization during storage under high-humidity conditions. This change from a largely amorphous state to a partially crystalline state resulted in altered dissolution and solubility behavior, as shown in Fig. 136.[566] Amorphous nifedipine prepared by spray drying also exhibited time-dependent crystallization. This crystallization was inhibited by the addition of HP-β-CD.[567] Oxyphenbutazone, which can exist in an amorphous state and three different crystalline states (anhydrous, monohydrate, and hemihydrate), exhibits crystallization and polymorphic transitions during storage depending on humidity, as illustrated in Scheme 78.[568] Amorphous oxyphenbutazone converts to an anhydrous form with lower solubility during storage under conditions of high humidity. A similar crystallization at high humidity was observed with amorphous 6-methylenandrosta-1,4-diene-3,17-dione prepared by grinding with β-CD.[569] Amorphous halopredone acetate prepared by grinding with various excipients, namely, hydroxypropylcellulose (HPC), methyl cellulose (MC), hydroxypropyl methyl cellulose (HPMC), and polyvinylalcohol (PVA), crystallized on subsequent storage at significantly different rates, depending on the excipient polymers, as shown in Fig. 137.[570]

The crystallization rate of amorphous frusemide prepared by spray drying depended on the preparation temperature; higher temperatures apparently provided a more stable amorphous state with a higher glass-transition temperature (T_g).[571] A similar crystallization rate dependency on the spray-drying temperature of macrolide derivatives was seen. Spray

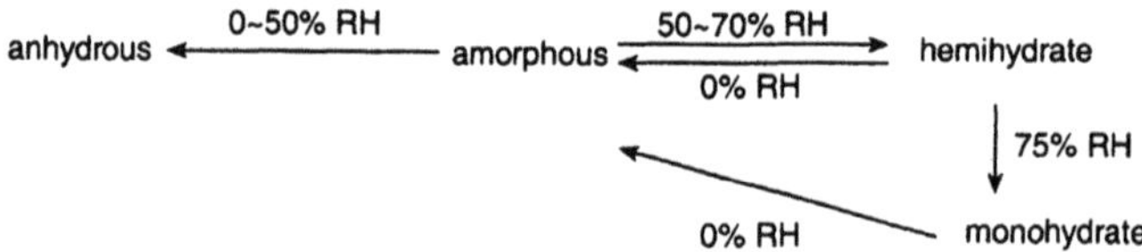

Scheme 78. Schematic representation of the polymorphic transitions of oxyphenbutazone. (Reproduced from Ref. 568 with permission.)

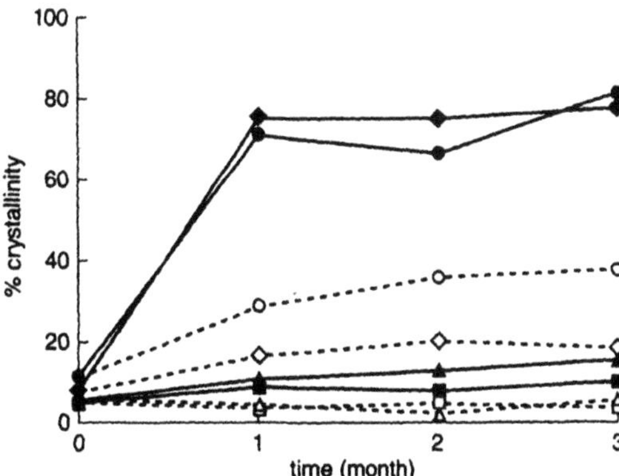

Figure 137. Change in percent crystallinity with time of initially amorphous halopredone acetate prepared by grinding halopredone acetate with various polymeric excipients (40°C, 75% RH). ○, ●, HPC; □, ■, MC; △, ▲, HPMC; ◇, ♦, PVA. ---, aluminum film; ——, cellophane/polyethylene film. (Reproduced from Ref. 570 with permission.)

drying at a temperature between T_g and the temperature at which crystallization started yielded the most stable amorphous materials.[572,573]

Crystallization of amorphous excipients may also occur during the storage of pharmaceuticals. Freeze-dried amorphous sucrose undergoes crystallization at temperatures above its T_g.[574–576] Moisture adsorption upon storage under higher humidity conditions caused crystallization even at temperatures below the T_g owing to the plasticizing effect of the adsorbed moisture. The addition of excipients having a high T_g and low hygroscopicity, such as dextran, raises the T_g and inhibits the crystallization.[577]

3.1.2. Transitions in Crystalline States

Polymorphs are different crystalline forms of the same drug. Because these forms have different free energy or chemical potentials, depending on temperature conditions, transitions between polymorphs occur. Polymorphic transitions during storage may alter critical properties of drugs because the solubility and dissolution rate of drug substances generally

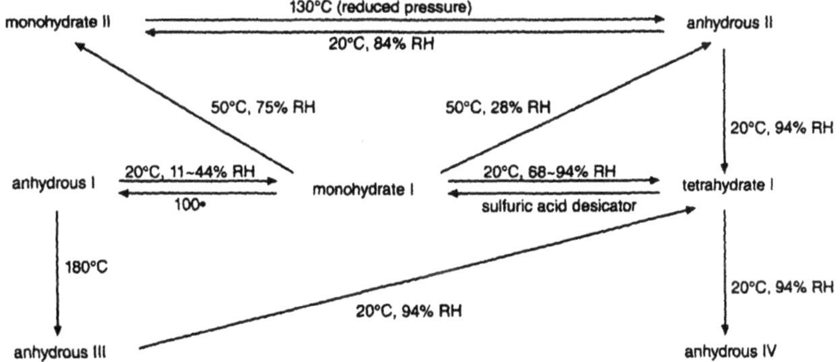

Scheme 79. Schematic representation of the polymorphic transitions of cianidanol. (Reproduced from Ref. 581 with permission.)

vary with changes in their crystalline form. From a storage perspective, temperature and humidity affect polymorphic transitions.

Polymorphic transitions were observed between two crystalline forms of benoxaprofen,[578] three forms of bromovalerylurea,[579] and two forms of pyridoxal hydrochloride.[580] These are just a few examples of the many drug substances exhibiting multiple polymorphic forms. Cianidanol exhibits polymorphic transitions between seven different crystalline forms, depending on temperature and humidity as shown in Scheme 79. However, no difference in dissolution rate was observed among these crystalline forms.[581]

Transitions between anhydrous and hydrated forms have been reported for many drug substances such as raclopride,[582] theophylline,[583,584] nitrofurantoin,[585] sulfaguanidine,[586] and phenobarbital.[587] Again, significant differences in solubility can exist between the anhydrous and hydrated forms of the same drug.

3.1.3. Formation and Growth of Crystals

Molecules in a crystal, and the crystals themselves, should not be considered static. Crystals can grow or decrease in size provided that there is a medium across which the molecules can travel. This could be a liquid phase or a gaseous phase into which the molecules can sublime. For example, drug substances and excipients in solid dosage forms, such as tablets and granules, may recrystallize or sublime onto the surface of the dosage form during storage. So-called "whisker" crystallization was observed in tablets of ethenzamide and caffeine anhydride.[588,589] This crystallization was enhanced in porous tablets and at higher temperatures. Some kinetic studies on the formation and growth of whisker crystals as a function of temperature have been reported. Whisker formation in ethenzamide tablets conformed to apparent zero-order kinetics, and the rate constant followed Arrhenius behavior in the temperature range 20–65°C, as shown in Fig. 138.[588] The effect of humidity on whisker crystallization was complex in that crystallization of ethenzamide and caffeine anhydride tablets was enhanced at higher and lower humidity, respectively, as shown in Fig. 139. Aspirin tablets exhibited whisker crystallization of salicylic acid, a degradation product. This was found to change the tablet strength and to be

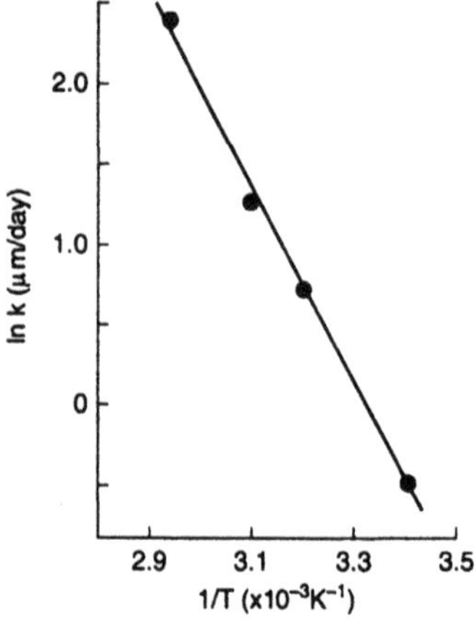

Figure 138. Arrhenius plot of whisker crystal growth kinetics in ethenzamide tablets. (Reproduced from Ref. 588 with permission.)

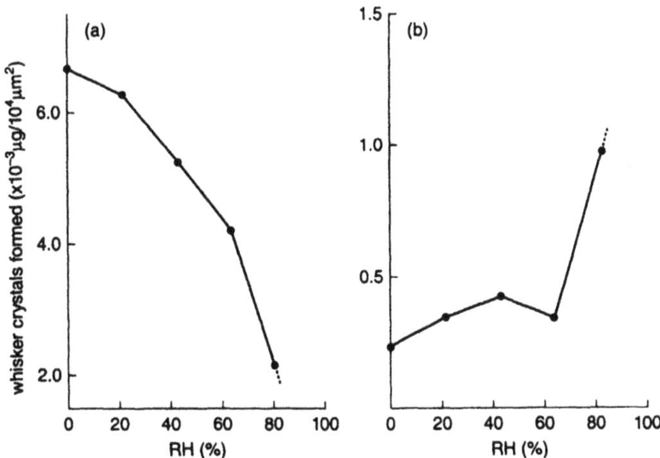

Figure 139. Whisker formation of anhydrous caffeine (a) and ethenzamide (b) in tablet dosage forms at 60°C. (Reproduced from Ref. 589 with permission.)

dependent on tablet pore size.[590,591] Particles of a valproate–synthetic aluminum silicate mixture formed whiskers comprised of valproic acid and sodium valproate (1:1) on their surface.[592] Some excipients may also participate in crystal formation. Tablets containing lactose and mannitol excipients have been shown to form whiskers.[593]

Carbamazepine tablets containing stearic acid formed column-shaped crystals on the tablet surface during storage at high temperature.[594] This crystallization was ascribed to the recrystallization of carbamazepine promoted by the solvent, melted stearic acid.

3.1.4. Vapor-Phase Transfers Including Sublimation

Pharmaceuticals containing components that sublime easily may undergo changes in drug content owing to the sublimation of the drug substances or excipients. In the case of nitroglycerin, which is a liquid with a significant vapor pressure, sublingual tablets exhibited significant variations in drug content during storage owing to intertablet migration through the vapor phase, as shown in Fig. 140.[595,596] This transfer was inhibited by adding water-soluble, nonvolatile fixing agents such as polyethylene glycol.[597]

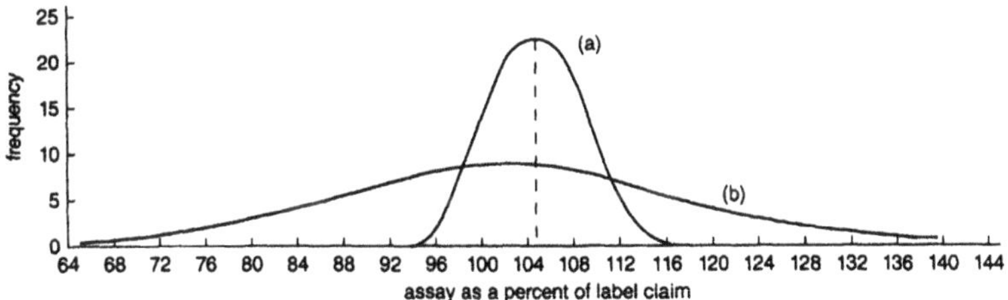

Figure 140. Change in nitroglycerin content uniformity in sublingual tablet due to vapor-phase transfer, before (a) and after storage at 25°C for 5 months (b). (Reproduced from Ref. 596 with permission.)

3.1.5. Moisture Adsorption

Moisture adsorption is generally observed with solid pharmaceuticals. The effect of moisture adsorption on the chemical stability of pharmaceuticals was addressed in Chapter 2. Moisture adsorption during storage can also affect the physical stability of pharmaceuticals, leading to changes in such properties as appearance and dissolution rate. Adsorption of moisture is governed by the physical properties of the drug substance and excipients. For example, the adsorption of moisture by aspirin crystals was enhanced by adding hydrophilic excipients.[598]

Although many books have described the mechanisms of moisture adsorption and adsorption isotherms for drug substances, few reports have dealt with the kinetics of moisture adsorption. Zografi and co-workers reported that the moisture adsorption rate, W', for water-soluble substances can be represented by the following equations, based on a heat-transport control model[599–601]:

$$W' = (C + F)\ln\frac{RH_i}{RH_0} \tag{3.1}$$

RH_i and RH_0 are relative humidity and critical relative humidity, respectively, and C and F are the conductive coefficient and the radiative coefficient, respectively. As shown in Fig. 141,[602] Eq. (3.1) described the adsorption of moisture by a sucrose–potassium bromide mixture.

3.2. Factors Affecting Physical Stability

The physical stability of pharmaceuticals is affected by many of the same variables that affect chemical stability. Specifically, the physical stability of solid pharmaceuticals is affected by the plasticizing effect of water, presumably owing to increases in molecular

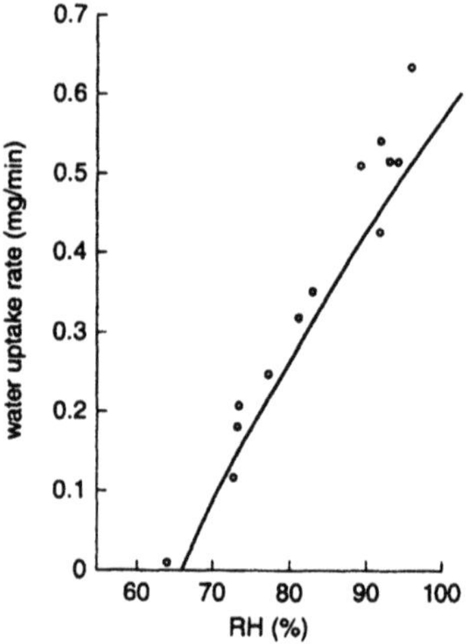

Figure 141. Moisture adsorption of a sucrose–potassium bromide mixture plotted according to Eq. (3.1). (25°C). (Reproduced from Ref. 602 with permission.)

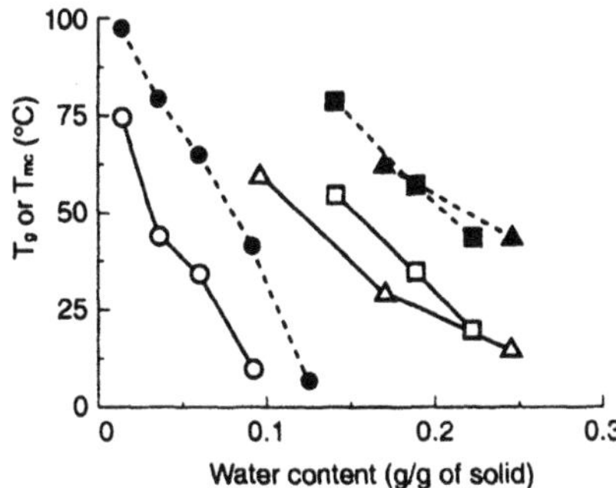

Figure 142. Glass-transition temperature (T_g, ---) and NMR relaxation-based critical mobility temperature (T_{mc}, ——) of lyophilized formulations. ○, ●, α,β-Poly(*N*-hydroxyethyl)-L-aspartamide; △, ▲, polyvinylpyrrolidane; □, ■, dextran. (Reproduced from Ref. 607 with permission.)

mobility. Amorphous indomethacin,[603] nifedipine,[604] and lamotrigine mesylate all show decreased T_g values and increased crystallization in the presence of absorbed moisture.[605]

The decrease in T_g caused by the plasticizing effect of water was explained on the basis of free-volume theory and generally described by Gordon–Taylor/Kelly–Bueche relationships.[606] Although T_g is useful in representing the molecular mobility of amorphous pharmaceuticals, the NMR relaxation-based critical mobility temperature (T_{mc}) is also useful as a measure of molecular mobility. T_{mc} is the critical temperature of appearance of Lorentzian relaxation due to protons in the liquid state in pharmaceutical solids.[607] As shown in Fig. 142, T_{mc} is generally lower than T_g, indicating that glassy pharmaceutical solids exhibit significant molecular mobility even at temperatures below T_g.

Enthalpy relaxation time, determined by differential scanning calorimetry,[608] and mechanical relaxation, determined by dynamic mechanical analysis,[609] can also be used as measures of molecular mobility of amorphous pharmaceutical solids.

3.3. Kinetics of Solid-Phase Transitions

Detailed mechanisms for most physical degradation processes affecting the efficacy and safety of drug products have not been extensively studied because of their complexity. Unlike chemical degradation rates in solution, physical degradation rates usually cannot be predicted on the basis of kinetic parameters estimated from data obtained under accelerated conditions. However, prediction of some physical degradation pathways such as polymorphic changes has been attempted. Several reports dealing with the prediction of polymorphic transitions based on kinetic principles are summarized below.

The Hancock–Sharp equation[610] is often used to describe the kinetics of polymorphic transitions:

$$-\ln\,[\ln(1-\alpha) = \ln B + m \ln t \tag{3.2}$$

where B is a constant. In this equation, α is the fraction of drug in the product state over the fraction in the starting state. By plotting the left-hand side of Eq. (3.2) against the logarithm of time, a linear relationship with a slope of m is obtained. The value of m is then used as an indicator of the transition mechanism. Because each of the mechanisms for polymorphic transitions shown

Table 10. Rate Equations Describing Polymorphic Transitions (Hancock–Sharp Equation)[a]

m value in Eq. (3.2)	Equation	Mechanism
0.62	$\alpha^2 = kt$	$D_1(\alpha)$, one-dimensional diffusion
0.57	$(1-\alpha)\ln(1-\alpha)+\alpha = kt$	$D_2(\alpha)$, two-dimensional diffusion
0.54	$[1-(1-\alpha)^{1/3}]^2 = kt$	$D_3(\alpha)$, three-dimensional diffusion
0.57	$1-2\alpha/3-(1-\alpha)^{2/3} = kt$	$D_4(\alpha)$, three-dimensional diffusion
1.0	$-\ln(1-\alpha) = kt$	$F_1(\alpha)$, first-order kinetics
1.11	$1-(1-\alpha)^{1/2} = kt$	$R_2(\alpha)$, phase boundary (cylindrical)
1.07	$-(1-\alpha)^{1/3} = kt$	$R_3(\alpha)$, phase boundary (spherical)
1.24	$\alpha = kt$	Zero order, zero-order kinetics
2.00	$[-\ln(1-\alpha)]^{1/2} = kt$	$A_2(\alpha)$, two-dimensional growth of nuclei
3.00	$[-\ln(1-\alpha)]^{1/3} = kt$	$A_3(\alpha)$, three-dimensional growth of nuclei

[a]Reference 610.

in Table 10 exhibits a characteristic value of m, determining m according to Eq. (3.2) makes it possible to select a suitable rate equation and to estimate a descriptor rate constant, k.

The polymorphic transition of carbamazepine from form I to form III and that of benoxaprofen from form I to form II exhibited m values of 2.23 and 2.24, respectively, indicating possible mechanisms involving two-dimensional growth of nuclei.[611,578] When m is approximately equal to 2, the reaction conforms to the Avrami–Erofe'ev equation:

$$[-\ln(1-\alpha)]^{1/2} = kt \tag{3.3}$$

The data for carbamazeine and benoxaprofen plotted according to Eq. (3.3) are shown in Figs. 143 and 144, respectively; and the rate constants k obtained from the slopes gave linear Arrhenius plots, indicating that it might be possible to predict the polymorphic transition rates at other temperatures (Fig. 145).

Polymorphic transitions of bromovalerylurea from form I to form II and from form III to form I conformed to mechanisms involving one-dimensional diffusion and two-dimensional nuclei growth processes, respectively. Both transitions also exhibited good Arrhenius behavior in the temperature range studied, as shown in Fig. 146.[579] Transitions of phenyl-

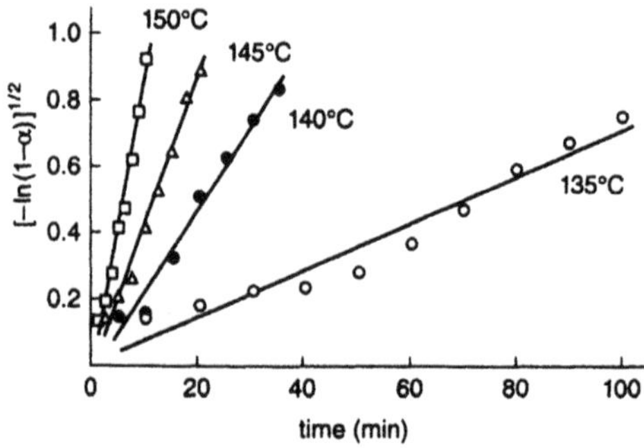

Figure 143. Polymorphic transition kinetics of carbamazepine from form I to form III according to the Avrami–Erofe'ev equation (Eq. 3.3). (Reproduced from Ref. 611 with permission.)

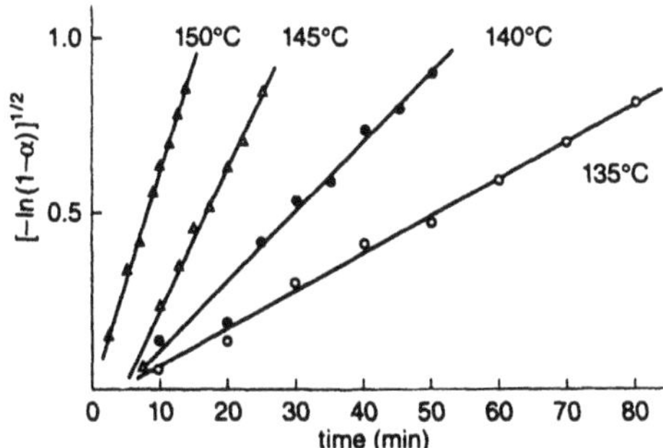

Figure 144. Polymorphic transition kinetics of benoxaprofen from form I to form II according to the Avrami–Erofe'ev equation (Eq. 3.3). (Reproduced from Ref. 578 with permission.)

butazone polymorphs from form α to δ and from form β to δ conformed to two-dimensional patterns and first-order kinetics, respectively (Fig. 147).[612]

The transition of phenobarbital from forms C and E to an anhydrous form conformed to the Jander equation (Eq. 3.4), indicating a three-dimensional diffusion mechanism.[587] Some results are shown in Fig. 148.

$$[1-(1-\alpha)^{1/3}]^2 = kt + C \tag{3.4}$$

The transition of 5-(4-oxo-phenoxy-4*H*-quinolizine-3-carboxamide)-tetrazolate from a tetrahydrate to a monohydrate conformed to zero-order kinetics, and the depend-

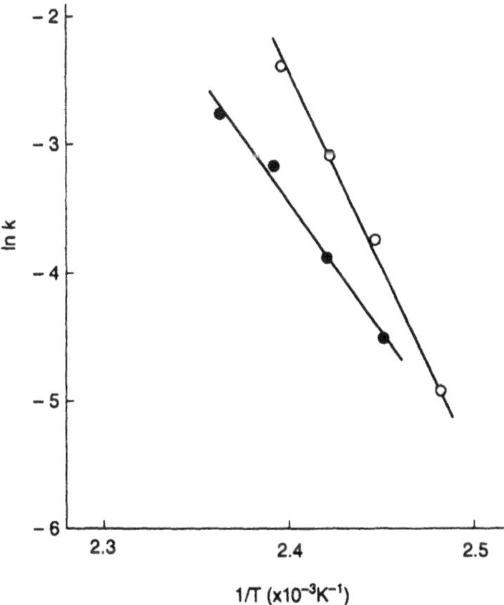

Figure 145. Linear Arrhenius plots for the polymorphic transitions of carbamazepine (○) and benoxaprofen (●). The rate constants k were obtained according to the Avrami–Erofe'ev equation (time unit: min). (Reproduced from Refs. 578 and 611 with permission.)

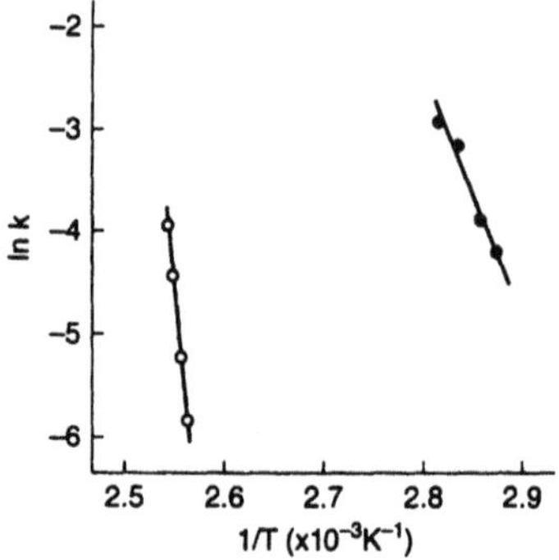

Figure 146. Arrhenius plots of the polymorphic transitions of bromovalerylurea. ○, Transition from form I to form II; ●, transition from form III to form I. The rate constant k is in units of min^{-1}. (Reproduced from Ref. 579 with permission.)

ence of the rate constant on temperature and humidity could be described by Eq. (2.113) (Fig. 149).[613]

The transition of sulfaguanidine from a monohydrate form to an anhydrous form in the presence of differing water vapor pressures[586] conformed to different rate equations among those listed in Table 10. Similarly, the dehydration kinetics of hydrated theophylline[614] of different particle sizes was also described by different rate equations. Polymorphic transitions of anhydrous theophylline in tablets conformed to different rate equations depending on the tablet size and pore size.[583,584]

The rate equations shown in Table 10 have also been used to describe the kinetics of crystallization, that is, the conversion from the amorphous state to a crystalline state. Crystallization of amorphous furosemide dispersed in Eudragit conformed to the rate equation proposed for a three-dimensional diffusion process.[615]

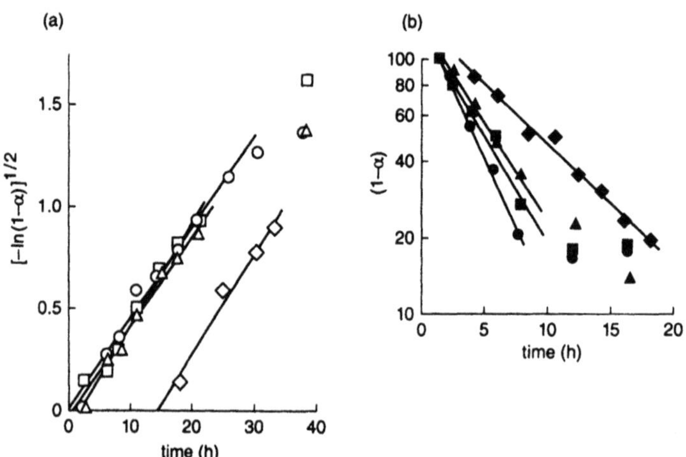

Figure 147. Polymorphic transition kinetics of phenylbutazone according to the Avrami–Erofe'ev equation (a) and by first-order kinetics (b), as a function of exposure to varying relative humidities (60°C). ◇, ♦, 0% RH; △, ▲, 50% RH; □, ■, 70% RH; ○, ●, 80% RH. (Reproduced from Ref. 612 with permission.)

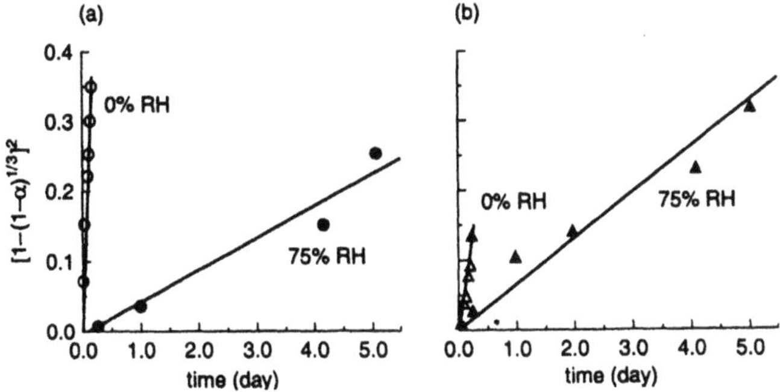

Figure 148. Transition kinetics of two hydrated forms of phenobarbital to an anhydrous species according to the Jander equation (Eq. 3.4). $T = 45°C$. (a) Transition from form C to an anhydrous state; (b) transition from form E to an anhydrous state. (Reproduced from Ref. 587 with permission.)

Some polymorphic transitions can be described by equations other than those listed in Table 10. The transition of nitrofurantoin from its anhydrous form to a monohydrate was described by the following equation[585]:

$$\alpha = \left(1 - \frac{kt + C}{3}\right)^3 \tag{3.5}$$

The Arrhenius equation has been employed as a first approximation in an attempt to define the temperature dependence of physical degradation processes. However, the use of the WLF equation (Eq. 3.6), developed by Williams, Landel, and Ferry to describe the temperature dependence of the relaxation mechanisms of amorphous polymers, appears to have merit for physical degradation processes that are governed by viscosity.

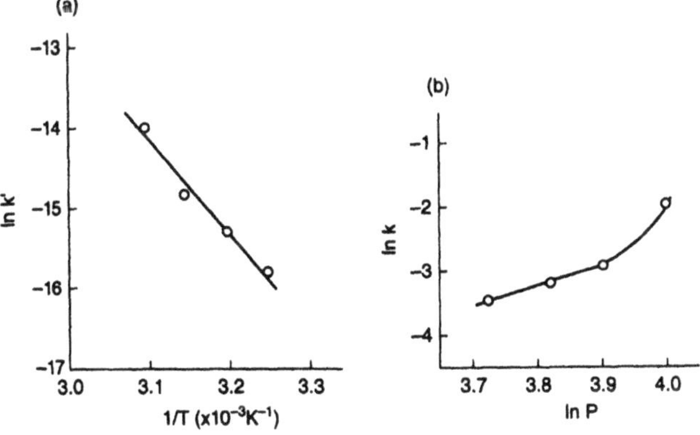

Figure 149. Temperature and humidity dependence of the tetrahydrate-to-monohydrate transition kinetics for 5-(4-oxo-phenoxy-4*H*-quinolizine-3-carboxamide)-tetrazolate. Here k refers to the apparent zero-order rate constant for the process (time unit: h) and $k' = k/P^s$, where P is water vapor pressure. (Reproduced from Ref. 613 with permission.)

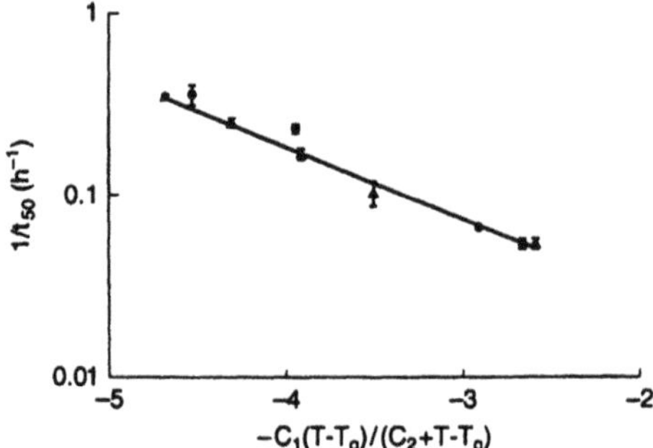

Figure 150. The crystallization rate of nifedipine plotted according to the WLF equation. ▲, Relationship between crystallization rate and temperature; ●, relationship between crystallization rate and T_g. (Reproduced from Ref. 604 with permission.)

$$\ln\left(\frac{k_T}{k_{T_g}}\right) = \frac{-C_1(T - T_g)}{C_2 + (T - T_g)} \tag{3.6}$$

In this equation, k_T and k_{T_g} are the crystallization rates at temperatures T and T_g, respectively, and C_1 and C_2 are constants. The role of viscosity in the crystallization of amorphous sucrose was suggested by the observation that crystallization is enhanced by a lowered T_g resulting from moisture adsorption.[574–577] Also, the crystallization rate of amorphous nifedipine exhibited a temperature dependence best represented by the WLF equation. The increase in the crystallization rate caused by the decrease in T_g under higher humidity conditions was also described by the WLF equation, as shown in Fig. 150.[604]

Chapter 4

Stability of Dosage Forms

The chemical and physical stability of pure drug substances has been described in Chapters 2 and 3, respectively. The stability of pharmaceutical dosage forms is described in this chapter.

Pharmaceutical dosage forms are complex systems composed not only of drug substances but also of various excipients. These excipients, which are non-therapeutic, are intended to contribute desirable, practical properties to the dosage form. Dosage forms may undergo both chemical and physical degradation, as described in previous chapters. This chapter describes preformulation and formulation stability studies, the ways in which stability can affect the functioning of the dosage form, the role of packaging in relation to stability, and the estimation of the shelf life of these complex dosage forms. Requirements for stability testing for New Drug Application submissions to regulatory bodies will be discussed in Chapter 6.

4.1. Preformulation and Formulation Stability Studies

Preformulation studies, such as choice of the crystalline form of the drug and the excipients to be used in the dosage form, are very important for developing stable pharmaceutical products. Preformulation studies provide the initial data that help the formulator decide on possible dosage-form strategies. Because little time is allocated to this stage of drug development, especially in the current environment, it is essential that meaningful results be obtained from simple but rapid screening methods. In this section, the methods for detecting chemical and physical degradation that are generally employed in preformulation studies are described. In addition, a factorial analysis for stability studies is briefly discussed.

4.1.1. Methods for Detecting Chemical and Physical Degradation

Critical for good studies involving the analysis of drugs and their degradants is the establishment and validation of so-called "stability indicating method(s)." Various chromatographic methods are best used to detect chemical changes with time under a variety of stress or nonstress conditions. Validated chromatographic separation techniques such as high-performance liquid chromatography (HPLC) and gas chromatography (GC) coupled to sophisticated detectors provide not only useful quantitative information on drug loss but also insight into the number of degradants formed and their quantitation. Based on the order of elution, insight into the properties of the degradants can also be gleaned. When coupled with detection techniques such as photodiode array UV–visible detection[616] or mass spectrometry, chromatographic methods are invaluable. Some additional techniques and procedures that are used, especially with complex dosage forms, are described in the following sections.

4.1.1.1. *Thermal Analysis*

Differential scanning calorimetry (DSC), differential thermal analysis (DTA), and differential thermogravimetry (DTG) are very useful in formulation screening because calorimetric changes and weight changes caused by chemical and physical degradation of pharmaceuticals can be readily detected. For example, DSC was employed in the preformulation study of a poorly water-soluble drug substance, α-pentyl-3-(2-quinolinyl-methoxy)benzenemethanol (REV5901). As shown in Fig. 151, the free base exhibited an endothermic peak due to melting that was observed at the same position regardless of storage and measurement conditions. On the other hand, the anhydrous and monohydrate hydrochloride salt forms showed different behaviors depending on measurement conditions. The free base was found to be more physically stable than the hydrochloride salt. Based on these results, the free base was chosen for formulation.[617]

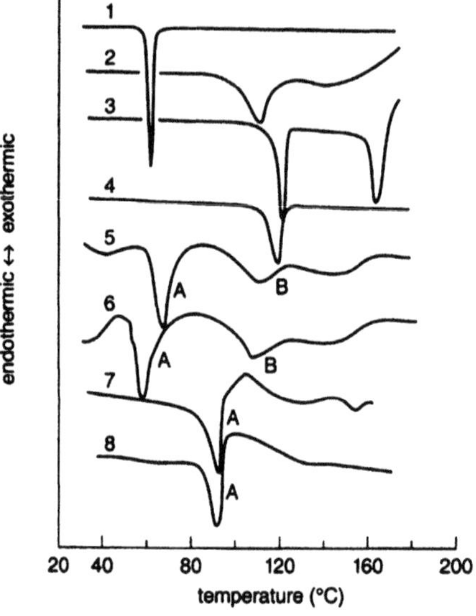

Figure 151. Application of DSC to preformulation studies of REV5901. The DSC thermogram obtained for REV5901 free base (1) showed no significant change with changes in atmospheric conditions. DSC thermograms were recorded for the anhydrous hydrochloride salt on an open pan without purging with N_2 (2), in a pan closed by crimping (3), and in a hermetically sealed pan (4) and for the monohydrate hydrochloride salt on an open pan without purging with N_2 (5), on an open pan with purging with N_2 (6), in a pan closed by crimping (7) and in a hermetically sealed pan (8). (A) Dehydration; (B) melting endotherms. (Reproduced from Ref. 617 with permission.)

Thermal analysis is often capable of easily detecting drug–excipient interactions. For example, accelerated degradation of aspirin caused by physical mixture with silica and aluminum was detected by DSC.[618] Interaction of ibuprofen with magnesium oxide was detected from changes in DSC thermograms (Fig. 152),[619] as was an interaction between enalapril maleate and crystalline cellulose leading to decreased stability.[620] Various other drug–excipient interactions have been detected by this thermal analysis method.[621,622] DSC can also be employed to investigate the stability of finished dosage forms, as was done, for example, with aminophylline suppository formulation.[623]

DSC, DTA, and DTG are useful for detecting physical changes in addition to chemical degradation. Crystallization of amorphous drugs and polymorphic transitions (see Chapter 3) have been extensively studied using these methods.[568,580,581]

The kinetics of degradation can be studied using isothermal calorimetry, that is, calorimetry performed at constant temperature. Recently, sensitive thermal conductivity microcalorimeters useful for detecting even small amounts of degradation at room temperature have become available. For example, the slow solid-state degradation of cephalosporins at a rate of approximately 1% per year was successfully measured by microcalorimetry.[624]

Microcalorimetry has been employed in studying the kinetics of chemical degradation of various drug substances. Heat flow produced from the hydrolysis of aspirin in acidic solution decreased according to first-order kinetics as shown in Fig. 153, indicating that degradation can be measured by microcalorimetry.[625,626] Apparent first-order rate constants for ampicillin degradation in aqueous solution measured by microcalorimetry exhibited a pH–rate profile similar to that obtained from iodometric titrations (Fig. 154).[627] The total heat flow produced from oxidation of ascorbic acid in aqueous solution measured in various vessels was proportional to the amount of degraded ascorbic acid, indicating that the degradation can be easily followed (Fig. 155).[628] The apparent enthalpy change of this oxidation was calculated to be 224 kJ/mol. Initial heat flow measurements utilizing microcalorimetry at several elevated temperatures have been used to calculate the energy of activation for the degradation of drug substances such as tetracycline and phenytoin. The degradation rate at 25°C was predicted from the rate constant obtained by HPLC and the activation energy obtained by microcalorimetry.[629] Microcalorimetry has also been used for determining degradation order and mechanism.[630,631]

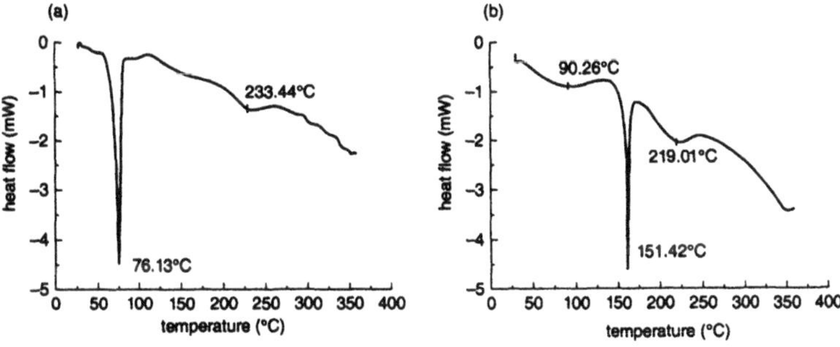

Figure 152. DSC thermograms showing the interaction between ibuprofen and magnesium oxide. (1:1 mixture). (a) Before storage; (b) after 1-day storage at 55°C. (Reproduced from Ref. 619 with permission.)

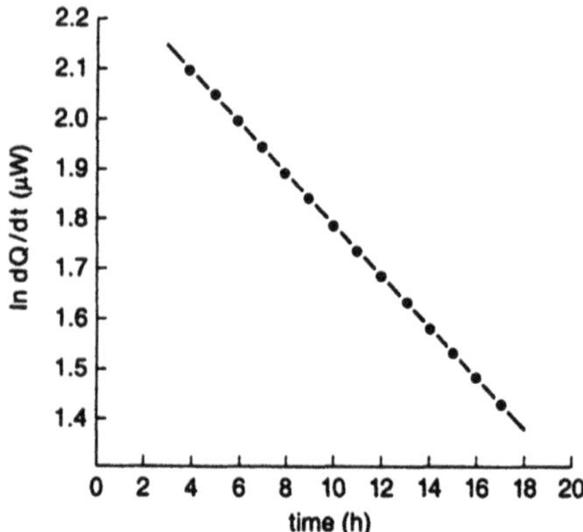

Figure 153. A natural logarithm plot of heat changes with time produced during the hydrolysis of aspirin at pH 1.1 and 45°C. (Reproduced from Ref. 625 with permission.)

Microcalorimeters are capable of measuring very small amounts of heat flow. This advantage of microcalorimetry was demonstrated in the measurement of the oxidation rate of α-tocopherol saturated with oxygen gas.[632] As shown in Fig. 156, the rate constants for the temperature range 23–40°C determined by microcalorimetry were consistent with those extrapolated from the rate constants determined by HPLC at temperatures above 50°C. The rate constant at room temperature was determined rapidly by microcalorimetry, whereas its determination by HPLC would require long-term stability testing over several months. In this case, no change in activation energy was observed in the temperature range studied, indicating that the degradation rate at room temperature could be estimated by extrapolating accelerated data. Microcalorimetry is most effective, however, for degradation exhibiting nonlinear Arrhenius plots. For example, it can be used to determine degradation rates at room temperature when the degradation mechanism and apparent activation energy for degradation vary with temperature. In such cases, extrapolating stability data obtained at elevated temperatures can lead to overestimation or underestimation of the stability at room temperature.

Whereas microcalorimetry is most suitable for the study of degradations that result in relatively large enthalpy changes, such as those seen in the examples of oxidation and

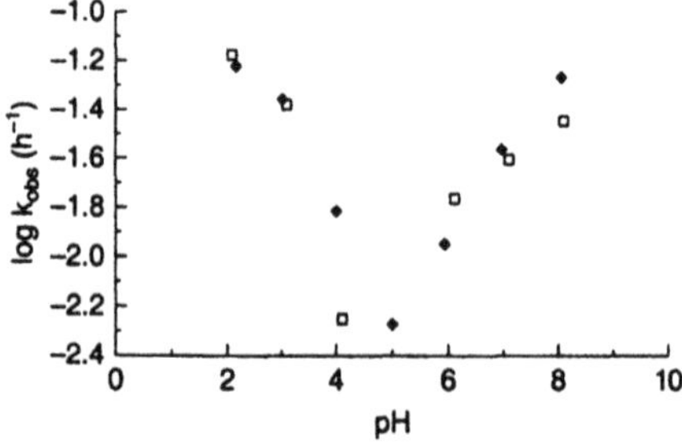

Figure 154. pH–rate profiles for the degradation of ampicillin measured by microcalorimetry (□) and iodometric titration (♦) at 37°C. (Reproduced from Ref. 627 with permission.)

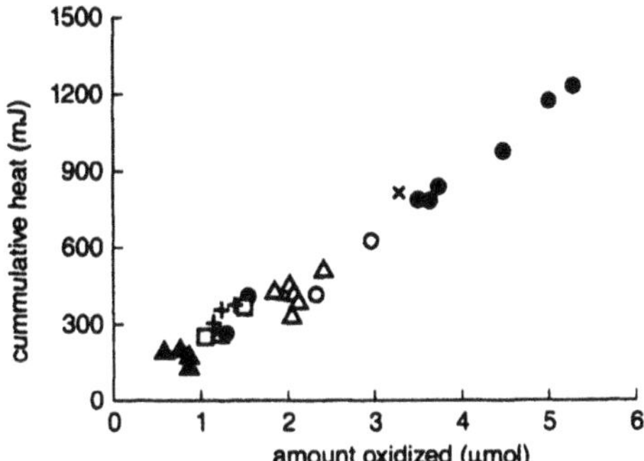

Figure 155. Linear relationship between heat flow and degradation of ascorbic acid in aqueous solution (pH 3.0–6.5, 25°C). △, ●, Glass vials, ascorbic acid concentration: 0.10% w/v; ▲, glass vials, N_2-purged; □, glass vials, EDTA added; ○, stainless steel vessels; ×, stainless steel vessels, EDTA added; +, glass vials, ascorbic acid concentration: 0.01% w/v. (Reproduced from Ref. 628 with permission.)

solid-state degradation given above, the technique may provide erroneous results for degradations accompanied by little heat flow. An additional limitation is that the technique is nonspecific and provides little information on the molecular mechanisms of degradation.

In addition to chemical degradation, microcalorimetry has been applied to the detection of physical changes in drug substances and excipients. An example is the change in the hydration of lactose.[633]

4.1.1.2. Diffuse Reflectance Spectroscopy

Diffuse reflectance spectroscopy (DRS), established by Kortum and co-workers in the 1950s,[634,635] was employed by Lach and co-workers to detect the solid-state interactions between various drug substances such as oxytetracycline and various excipients such as

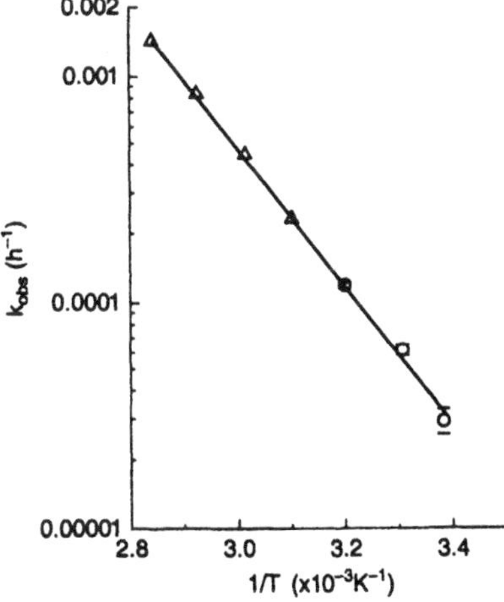

Figure 156. Arrhenius plots of the oxidation rate constant of α-tocopherol measured by microcalorimetry (○) and HPLC (△). (Reproduced from Ref. 632 with permission.)

magnesium trisilicate.[636–645] The DRS spectrum of an isoniazid–magnesium oxide mixture exhibited a decrease in reflectance r_∞ with increasing isoniazid content, as shown in Fig. 157. The remission function, calculated by the Kubelka–Munk equation (Eq. 4.1), was proportional to isoniazid content (Fig. 158).[640]

$$f(r_\infty) = \frac{(1 - r_\infty)^2}{2r_\infty} \tag{4.1}$$

Thus, the solid-state degradation could be followed quantitatively by DRS. A difficulty with the technique, especially when performed at short wavelengths, is spectral interference from the degradation products. On the other hand, visible color development of solid dosage forms alters spectra at relatively long wavelengths such that quantitative analysis by DRS is possible.[640] Color development in an ascorbic acid–lactose mixture exhibited a remission function proportional to the ratio of colored to intact sample for both powders and tablets (Fig. 159), indicating that quantitative analyses are possible.[646] The slope of the linear relationship depended on sample density, suggesting that measurement under constant conditions is necessary. DRS is especially useful for detecting small changes occurring locally on solid surfaces.

4.1.1.3. *Miscellaneous Methods*

NMR and infrared (IR) spectroscopy are also used to investigate the chemical stability of drug substances. Determination of the hydrolysis rate of esters such as atropine by NMR,[647] a nondestructive near-IR analysis of aspirin tablets,[648] and determination of the hydrolysis rate of diltiazem by polarimetry[649] have been reported. Unusual methods, such as measurement of the dielectric properties of dosage forms like gelatin and methylcellulose microcapsules (Fig. 160), have been used to detect physical changes.[650,651] These changes

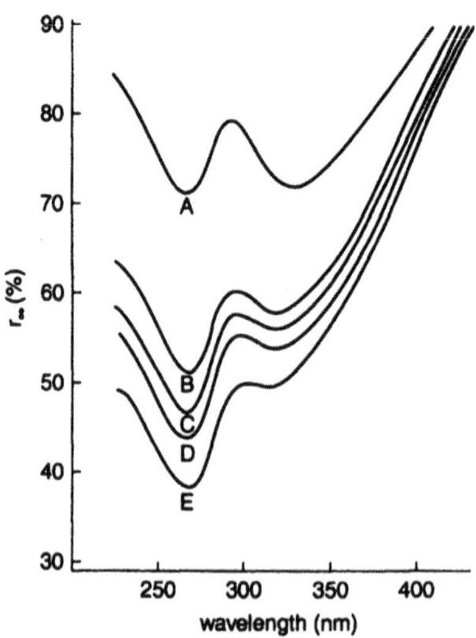

Figure 157. Diffuse reflectance spectroscopy of isoniazid–magnesium oxide mixtures. Isoniazid concentration (mg/g of MgO): (A) 3, (B) 7, (C) 10, (D) 13, (E) 16. (Reproduced from Ref. 640 with permission.)

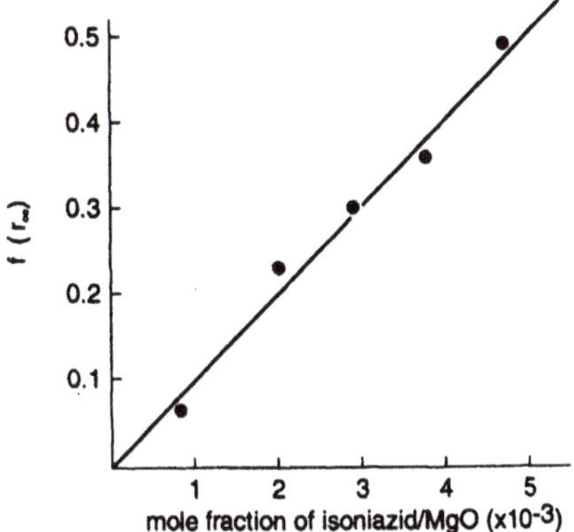

Figure 158. Remission function of isoniazid–magnesium oxide mixture at 268 nm as a function of isoniazid content. (Reproduced from Ref. 640 with permission.)

were then related to changes in drug release rates from these dosage forms. Measurement of weak chemiluminescence has also been applied to a stability study.[652]

In the formulation screening of solid dosage forms, chemical compatibility is sometimes evaluated using suspensions or slurries.[653–656] Although this information may be difficult to relate to the stability of the dosage forms, it may provide some preliminary information on the stability of formulation components.

4.1.2. Factorial Analysis

In formulation studies, all the factors affecting the stability of the pharmaceutical product have to be considered. Because the stability of pharmaceuticals is generally affected by numerous and complex factors, quantitative analysis of the role of each would involve a very large and complex series of experiments. The effect of each individual factor would have to be tested under conditions in which all other factors are maintained constant. Factorial analysis attempts to minimize the number of experiments needed to get meaningful results and, therefore, to save time and labor.

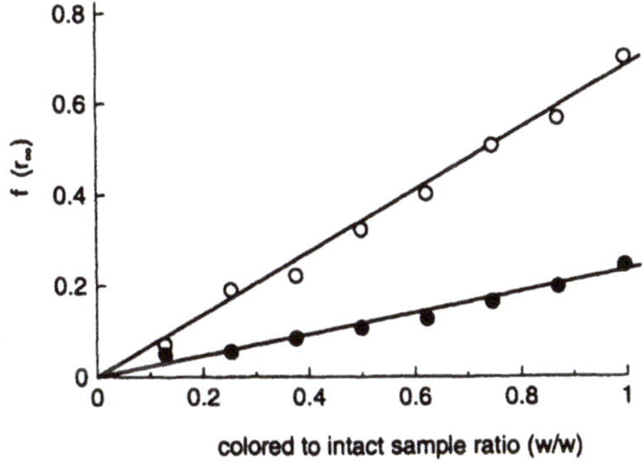

Figure 159. Remission function of ascorbic acid–lactose mixture as a function of colored sample content. ●, Powder; ○, tablet. (Reproduced from Ref. 646 with permission.)

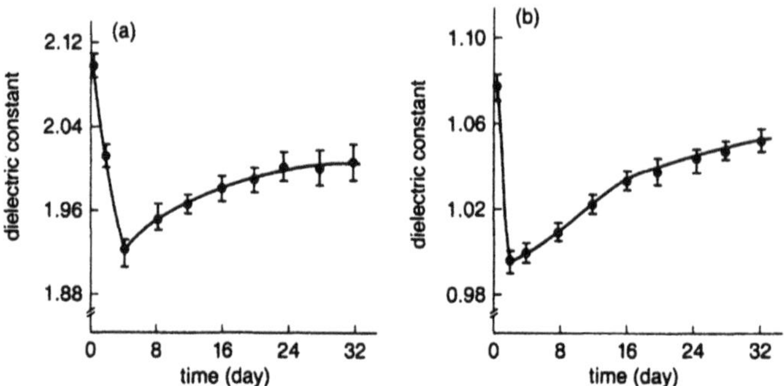

Figure 160. Changes in dielectric constant of gelatin (a) and methylcellulose microcapsules (b) during aging at 45°C. (Reproduced from Ref. 650 with permission.)

Consider the following example. The dependence of degradation rate on reactant concentrations ([A], [B], [C], . . .) is described by additive terms for each degradation pathway, as represented by the rate equation (Eq. 2.4).[657] Therefore, k_{obs} [the terms inside brackets in Eq. 2.4] is described by Eq. (4.2) when, for example, hydroxide ion (B) and phosphate ion (C) catalyze the degradation independently.

$$k_{\text{obs}} = k_b[\text{B}] + k_c[\text{C}] + \cdots \tag{4.2}$$

On the other hand, a factor such as ionic strength that directly affects the rate constant k [Eqs. (2.4), (2.6), and (2.7)] enters into the rate equation as a product term:

$$k_{\text{obs}} = f(\text{A}) \{k_b[\text{B}] + k_c[\text{C}] + \cdots\} \tag{4.3}$$

To obtain k_b, k_c, . . ., and $f(\text{A})$, experiments are performed at two levels for each factor, such that 2^n experiments are required when the number of factors is n. For example, 2^3 experiments are performed at low and high levels of each factor, as shown in Fig. 161, when three factors

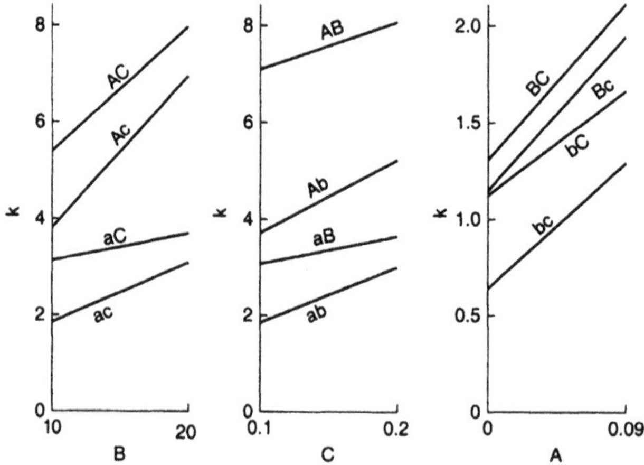

Figure 161. 2^3-factorial analysis. Upper-case letters (A, B, C) represent the high level of each factor, and lower-case letters (a, b, c) represent the low level. (Reproduced from Ref. 657 with permission.)

(A, B, and C) affect the degradation rate. Unless factor A interacts with the additive factors (B and C), k_b, k_c and f(A) can be obtained from the experimental results.[657] The Plackett–Burman method (n factors are analyzed by performing $n + 1$ experiments)[658] and the randomized block method are employed to perform factorial analysis experiments less than 2^n.

The application of factorial analysis in stability studies of various pharmaceuticals has been reported. For example, the effects of temperature, ionic strength, buffer concentrations, and other factors on complex formation between polyvinylpyrrolidone and drug substances such as salicylic acid were analyzed.[659] Factorial analysis was performed to elucidate the effects of excipients, light, humidity, temperature, and other factors on the stability of chlorpromazine hydrochloride,[660] aspirin,[661,662] and amphotericin B.[663] Factorial analysis has also been performed in investigations of the physical stability of dosage forms. The effects of humidity and packaging on the dissolution rate of controlled-release theophylline tablets were analyzed using such a design.[664]

4.2. Functional Changes in Dosage Forms with Time

Dosage forms are designed to perform certain functions. For example, a specific dosage form such as a traditional tablet might be designed for rapid release of the active drug upon exposure to an aqueous medium. This is usually assessed *in vitro* through measurement of the aqueous dissolution rate of the drug from the dosage form. If large changes in the dissolution characteristics of the drug on long-term storage of the dosage form are observed, this would indicate that changes are occurring in the dosage form that may compromise the performance of the dosage form in patients. Such functional stability studies are as important as the studies involving the chemical and physical stability of the drug substance. Altered functioning of the dosage form with time may be related to changes in the chemical or physical properties of the drug, excipients, coating materials etc., or it may be related to complex interactions between various components of the dosage form.

The chemical stability of components of dosage forms can be assessed in ways similar to those described in Chapter 2. For the assessment of physical changes to dosage forms, changes specific to each dosage form should be evaluated, in addition to the assessment of the physical degradation of drug substances and excipients described in Chapter 3. This section gives examples of physical degradation of dosage forms that affects the functioning of the dosage form and considers the functional stability of the dosage forms, as well as the ability to predict changes in functionality.

4.2.1. Changes in Mechanical Strength

Storage of solid dosage forms under humid conditions may bring about adsorption of moisture and lead to changes in the mechanical strength of tablets.[665–669] Adsorption of moisture by tablets in blister packages increased with increasing humidity, and resulted in decreased mechanical strength (Fig. 162).[668] The change in mechanical strength was described as a function of the moisture sensitivity of the tablet, the moisture permeability of the package, and the humidity conditions. Knowledge of the dependency on each of these factors allowed prediction of the long-term storage properties of the product. A close correlation was seen between calculated and observed properties.[668]

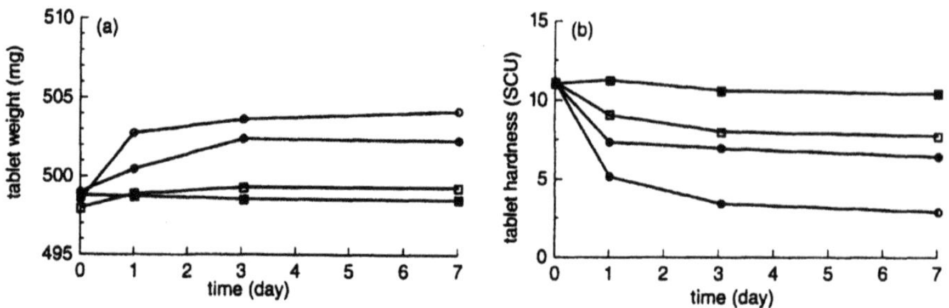

Figure 162. Moisture adsorption (a) and strength change (b) of model tablets stored in blister packages maintained at 21–22°C and varying relative humidities. ■, 25% RH; □, 60% RH; ●, 70% RH; ○, 95% RH. (Reproduced from Ref. 668 with permission.)

4.2.2. Changes in Drug Dissolution from Tablets and Capsules

Dissolution (or release) of a drug substance from a dosage form, such as a tablet or a capsule, is a very important characteristic. Dissolution characteristics have been known to change upon storage. For example, the dissolution rate of carbamazepine tablets decreased markedly when they were stored at room temperature and 100% RH for a period as short as 6 days (Fig. 163).[670] Such dramatic changes in dissolution rate may alter the bioavailability of the drug. However, changes in *in vitro* dissolution behavior do not necessarily mean that changes in bioavailability will occur. Only in those cases in which dissolution limits bioavailability or in which *in vitro* dissolution reasonably mimics what happens *in vivo* is there likely to be a correlation. Although the *in vitro* dissolution rate of soft gelatin capsules containing digoxin and polyethylene glycol 400 decreased during storage, no significant changes in bioavailability were observed.[671] Similarly, a change in the *in vitro* dissolution rate of etodolac capsules during storage was not reflected in a change in bioavailability.[672] However, a nitrofurantoin capsule formulation exhibited a decrease in *in vitro* dissolution rate and in *in vivo* absorption rate upon storage.[673]

4.2.2.1 *Effect of Formulation on Changes in Dissolution*

The stability of the dissolution characteristics of dosage forms during storage can be affected by formulation components and processing. A phenobarbital tablet containing

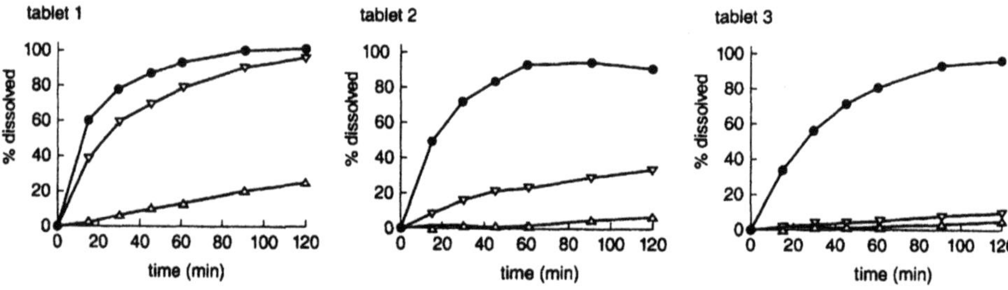

Figure 163. Changes in dissolution rate of carbamazepine tablets during storage. ●, Before storage; △, after 6-day storage at 100% RH; ▽, dried at 85°C after 6-day storage at 100% RH. (Reproduced from Ref. 670 with permission.)

gelatin as a binder exhibited a marked decrease in dissolution rate during storage at 98% RH.[674] The dissolution rate of hydrochlorothiazide tablets containing acacia increased during storage at high temperatures.[675] A decrease in dissolution rate during storage was also observed with a hydrochlorothiazide bead formulation containing sodium starch glycolate,[676] whereas the dissolution rate of nitrofurantoin capsules decreased during storage at high humidity as the content of carbomer increased, leading to a decrease in bioavailability as shown in Fig. 164.[677] This was consistent with gelation of the carbomer during storage.

Alginic acid, a tablet disintegrant, exhibited a decrease in swelling force generation after storage under a variety of conditions, indicating the possibility of decreased tablet dissolution (Fig. 165).[678] Changes in the rate of dissolution of phenytoin sodium capsules after storage depended on the excipients used; for example, dissolution rate increased with time when calcium sulfate was added but decreased with time when lactose was added.[679] Decreased dissolution rate during storage at high humidity has been reported for theophylline pellets and prednisone tablets containing microcrystalline cellulose.[680,681]

As excipients affect the dissolution characteristics of dosage forms over time, the water content of dosage forms can also affect stability. Higher water content generally causes a larger change in dissolution behavior.[666,680] Storage of calcium 4-aminosalicylate tablets at high humidity prolonged the disintegration time and decreased the dissolution rate as shown in Fig. 166.[682] Even though the decrease in dissolution was correlated with an increase in water content, the mechanism for this change was not clear. A tablet containing polyvinylpyrrolidone as a disintegrant exhibited a larger change in drug release when stored at 23°C compared to 65°C, suggesting that water evaporation at higher temperatures leads to stabilization of drug release characteristics.[683] It has been reported for both prednisone and erythromycin tablets that moisture-permeable packaging is likely to change the drug release from the tablets.[684,685]

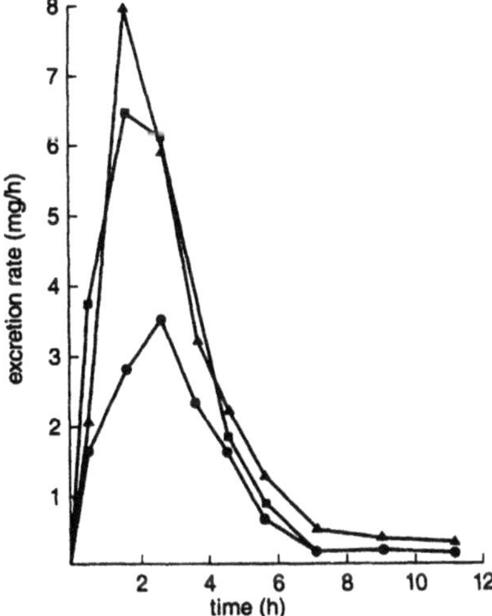

Figure 164. Changes in nitrofurantoin excretion rate from nitrofurantoin capsules containing different amounts of carbomer and stored under different conditions. ▲, Before storage; ■, formulation containing less carbomer after 1-year storage at 40°C and 30% RH; ●, formulation containing more carbomer after 1-year storage at 40°C and 60% RH. (Reproduced from Ref. 677 with permission.)

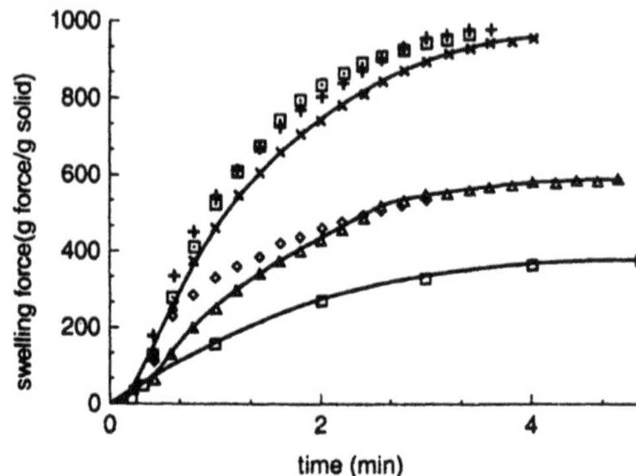

Figure 165. Changes in the swelling force of alginic acid after one-year storage under a variety of conditions. ▣, Before storage. Storage conditions: +, 25°C; ×, 30°C; ◇, 30°C and 75% RH; △, 40°C; □, 50°C. (Reproduced from Ref. 678 with permission.)

Sustained-release, wax-based nifedipine tablets exhibited a change in drug release after storage. Formation of nifedipine microcrystals and structural changes in the wax vehicle were given as explanations for the observed change.[686] Solid dispersions of griseofulvin prepared with polyethylene glycol 3000 (PEG 3000) showed a decreased dissolution rate of griseofulvin after storage, which was ascribed to the crystallization of PEG 3000.[687] This change in drug release was inhibited by addition of the surfactant sodium dodecyl sulfate to the formulation.

4.2.2.2. *Changes in Drug Release from Coated Dosage Forms*

The stability of the drug release characteristics of film-coated tablets and pellets is affected by the stability of the films. This should be the case especially when the film contributes significantly to the rate-limiting step in the release. The stability of a film coat

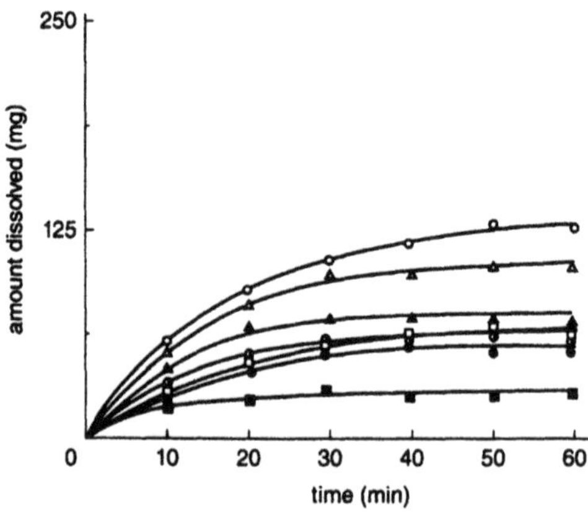

Figure 166. Changes in the dissolution rate of calcium 4-aminosalicylic acid after storage. The curves represent the dissolution rates before storage (○), after storage for 8 (△), 16 (◑), and 24 days (□) at 23°C and 32.9% RH, and after storage for 8 (▲), 16 (●), and 24 days (■) at 23°C and 92.9% RH. (Reproduced from Ref. 682 with permission.)

prepared from an aqueous polymeric dispersion was affected by the curing process.[688] Drug release from enteric-coated and sugar-coated tablets was more susceptible to the effect of humidity than that from film-coated tablets. For example, storage of sugar-coated tablets changed the disintegration time, leading to increased[689] or decreased dissolution rate.[690,691] Enteric-coated aspirin tablets exhibited a decrease in dissolution rate during storage at 33°C and 60% RH, as shown in Fig. 167.[692] A similar decrease in dissolution rate is reported for sugar-coated chlorpromazine tablets.[693]

Although drug release from film-coated tablets is generally more stable than that from enteric-coated and sugar-coated tablets, the release rates may change depending on storage conditions. For example, film-coated chlorpromazine tablets exhibited a change in drug release at temperature conditions cycling between 30°C and room temperature.[693] Storage of tableted microencapsulated aspirin granules prepared with polyacrylate–polymethacrylate-based polymers resulted in decreased drug release with storage time, an effect ascribed to cross-linking of the polymer leading to prolonged disintegration time.[694]

4.2.2.3. Changes in Capsule Shells with Time and Storage Conditions

Capsules prepared from gelatin are physically unstable at water contents outside the range of 12–18%. Storage of two chloramphenicol capsules at high humidity prolonged the disintegration time and decreased the drug release rate, as shown in Fig. 168.[695] Decrease in the drug release from ampicillin capsules during storage at high humidity was suggested to be due to the agglomeration of drug particles caused by moisture.[696]

Drug release from capsules may change owing to the reaction of the capsule shells with the contents. A decrease in the drug release rate from capsules containing polysorbate 80 was explained by assuming that cross-linking of gelatin was promoted by formaldehyde formed from the oxidation of polysorbate 80.[697] Interaction of dyes and gelatin in capsule shells, especially under light, could change the rate of drug release from capsules.[698,699] These

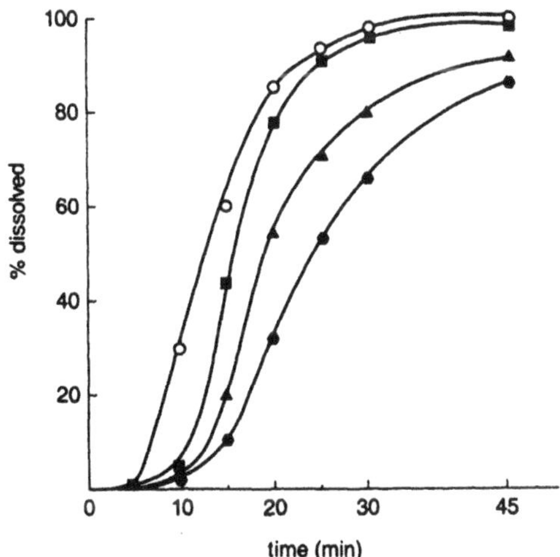

Figure 167. Changes in the dissolution rate of enteric-coated aspirin tablets after storage. The curves represent the dissolution rates before storage (○) and after storage at 33°C and 60% RH for 10 (■), 20 (▲), and 42 days (●). (Reproduced from Ref. 692 with permission.)

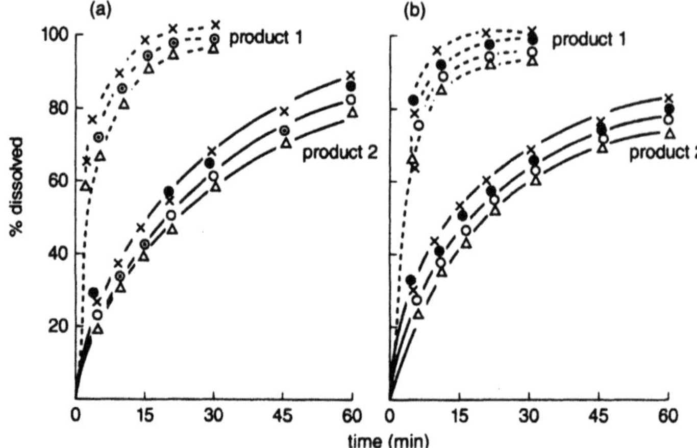

Figure 168. Changes in the dissolution rate of two different chloramphenicol capsules at 49% RH (a) and 66% RH (b). The curves represent the dissolution rates before storage (×) and after storage for 2 (●), 8 (○), and 16 weeks (△). (Reproduced from Ref. 695 with permission.)

changes were enhanced by combinations of the effects of high humidity and light, suggesting a contribution of photoreaction(s) to the changes seen.[698] These detrimental effects were eliminated when dissolution was tested in simulated gastric and intestinal fluids with pepsin and pancreatin, respectively (Fig. 169).[700] Gelatin-coated acetaminophen tablets exhibited a marked decrease in dissolution rate during storage at high humidity (30°C, 80% RH), which was moderated by the addition of pancreatin to the dissolution medium (Fig. 170).[701] Presumably, the addition of pancreatin helps to cleave the cross-linked gelatin. Decreases in disintegration and release rate were also reported for ketoprofen rectal capsules upon storage.[702]

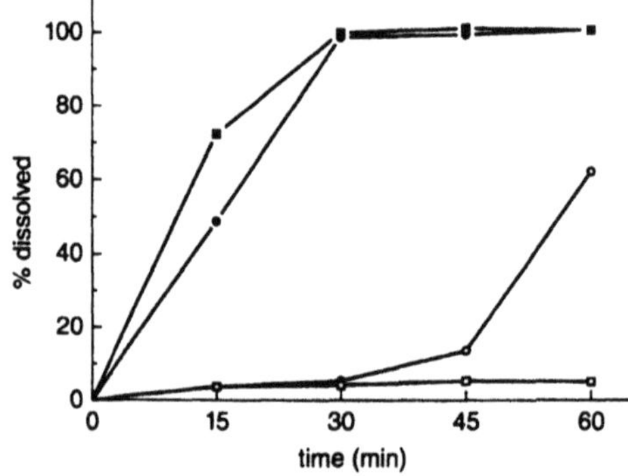

Figure 169. Effect of added enzymes on the decrease in the dissolution rate of gelatin capsules containing dyes after storage under fluorescent light for 2 weeks at 80% RH. Open symbols represent capsules without enzyme; filled symbols represent capsules with added pancreatin (0.9%). □, ■, Chocolate brown opaque capsules; ○, ●, bright blue opaque capsules. (Reproduced from Ref. 700 with permission.)

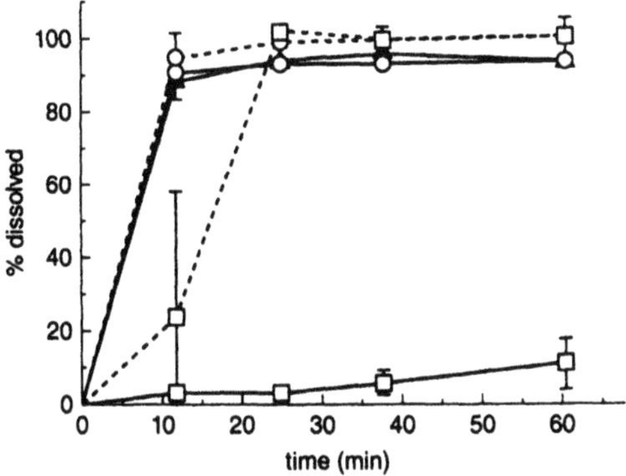

Figure 170. Effect of added enzymes on the decrease in the dissolution rate of gelatin-coated acetaminophen tablets during storage. Solid curves represent tablets without enzyme; dashed curves represent tables with added pancreatin 1%). ▲, Before storage; ○, after 7-month storage at room temperature; □, after 7-month storage at 37°C and 80% RH. (Reproduced from Ref. 701 with permission.)

4.2.2.4. Prediction of Changes in Dissolution

It is difficult to describe changes in dissolution or drug release rates during storage by kinetic equations because of the complicated and varied mechanisms involved. However, some attempts have been made, and various empirical relationships noted. The disintegration time of tablets containing gelatin as a binder followed an unusual relationship with time of storage, as shown in Fig. 171.[703] Plotting the logarithm of a dissolution rate constant for prednisolone tablets versus storage time yielded a linear relationship (Fig. 172), thus allowing one to predict the dissolution rate for this formulation after any given storage time. Similarly, a relationship was seen between dissolution rate and water content such that

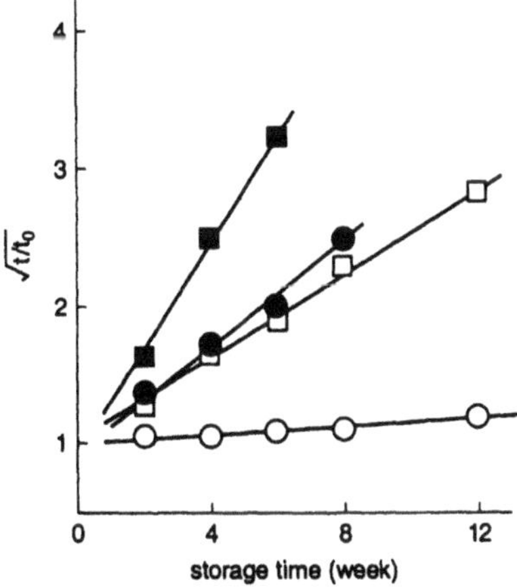

Figure 171. Changes in the disintegration time of difenamizole tablets containing gelatin as a binder during storage at 40°C. Water content: ○, 5.02%; □, 5.70; ●, 6.01%; ■, 6.48%. The ratio of the disintegration time after storage (t) to that before storage (t_0) is plotted versus storage time. (Reproduced from Ref. 703 with permission.)

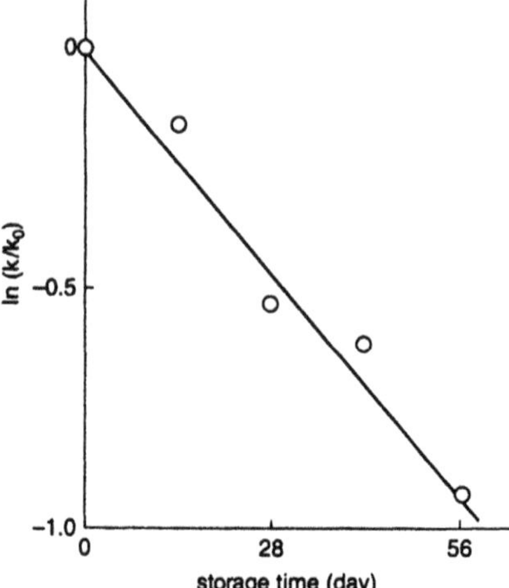

Figure 172. An apparent log-linear relationship between the change in the dissolution rate of prednisolone tablets and storage time at 4°C. Water content: 5.41%. The ratio of the dissolution rate after storage (k) to that before storage (k_0) is plotted versus storage time. (Reproduced from Ref. 704 with permission.)

changes in water content with time in various packagings permitted prediction of dissolution changes upon storage.[704]

Changes during storage in the moisture permeability of cellulose acetate films used to effect controlled release (Fig. 173) were predicted from changes in mechanical properties of the films.[705] The mechanical properties measured were relaxation time versus mechanical stress.

For model tablets coated with polymer films composed of ethyl cellulose and hydroxypropyl methyl cellulose phthalate, plotting the logarithm of moisture permeability and dissolution rate versus the logarithm of physical aging time yielded a linear relationship (Fig. 174). This suggested that long-term stability of moisture permeability and dissolution rate can be estimated from this empirical algorithm.[706]

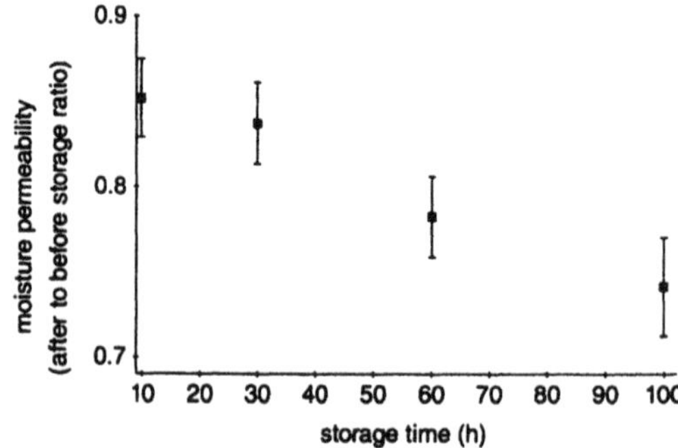

Figure 173. Changes in the moisture permeability of cellulose acetate film during storage at 100°C. (Reproduced from Ref. 705 with permission.)

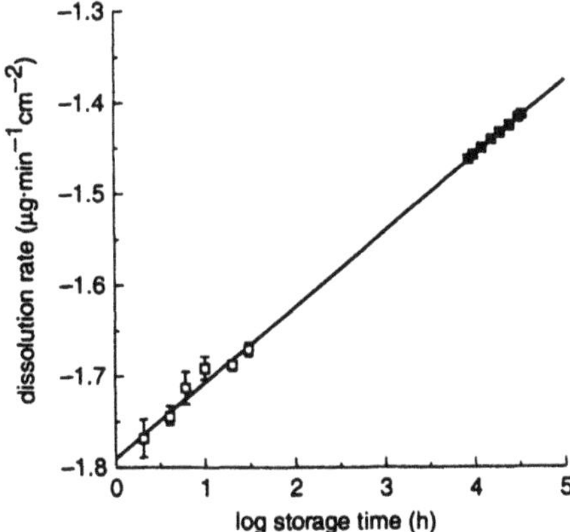

Figure 174. A log–log relationship between the change in dissolution rate of hydroxypropyl methyl cellulose phthalate-coated tablets and storage time at high temperature (80°C). □, Observed; ■, predicted. (Reproduced from Ref. 706 with permission.)

It is generally accepted that the stability of dissolution rate during room-temperature storage cannot be predicted from shorter-term storage under accelerated conditions of high temperature and humidity. This was confirmed by the observation that tablets containing polyvinylpyrrolidone exhibited a marked change in dissolution rate during storage at 23°C, whereas no change was observed at 65°C.[683] On the other hand, short-term accelerated testing is considered to be somewhat useful. Changes in the dissolution rate of hydrochlorothiazide tablets at room temperature, for example, were correlated to changes observed at 37, 50, and 80°C, suggesting that stability evaluation by accelerated testing may be possible in some cases.[675] That no change was seen in dissolution rate during short-term storage of film-coated, enteric-coated, and sugar-coated tablets under accelerated conditions provided some confidence that dissolution at room temperature should not change significantly with time.[693]

4.2.3. Changes in Melting Time of Suppositories

Suppositories are designed to melt after rectal administration, and this process is crucial to the release of active ingredients. If storage results in hardening of the suppositories such that the time required for them to melt is prolonged, this could result in an inferior product. As shown in Fig. 175, long-term storage of some products, even at 20°C, resulted in a prolongation of melting times.[707,708] The hardening effect increased with increased storage temperature up to 25°C but decreased at higher temperatures owing to partial melting of the suppository base (Fig. 176). Thus, in this case, accelerated testing at a higher temperature would not have been useful.

Many suppository bases are made up of various acylglycerols. Hardening of suppositories is considered to result from various phase transitions, crystallization, and transesterification reactions in these lipids. The DSC thermograms of a semisynthetic, hard base triglyceride shown in Fig. 177 indicate that a polymorphic phase transition occurred during storage.[709]

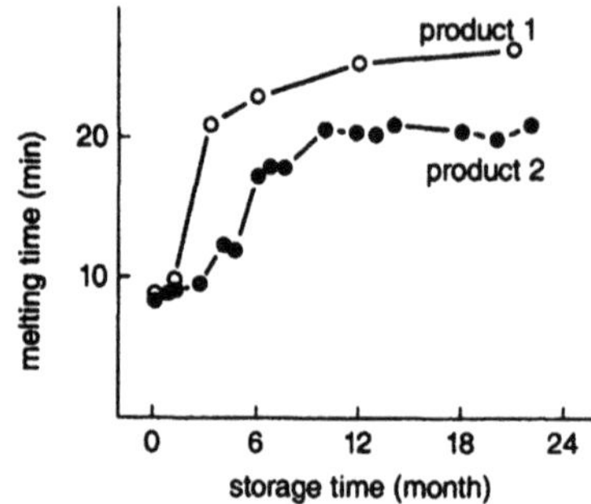

Figure 175. Changes in the melting time of suppositories following storage at 20°C. (Reproduced from Ref. 707 with permission.)

Changes in melting properties during storage were also reported for effervescent suppositories even when stored at 25°C; this indicated that these products should be stored below 15°C (Fig. 178).[710]

4.2.4. Changes in Drug Release Rate from Polymeric Matrix Dosage Forms, Including Microspheres

Polymeric matrix dosage forms intended for controlled release may undergo changes in drug release rate upon storage. For example, poly(*d,l*-lactic acid) microspheres exhibited shrinkage and decreased release rates of phenobarbitone after 6-month storage at 37°C.[711] Various physical properties of the matrix such as the glass-transition temperature (T_g) and the crystalline states of polymers affect drug release from polymeric matrix dosage forms. Changes in these properties during storage can lead to changes in the drug release rate.

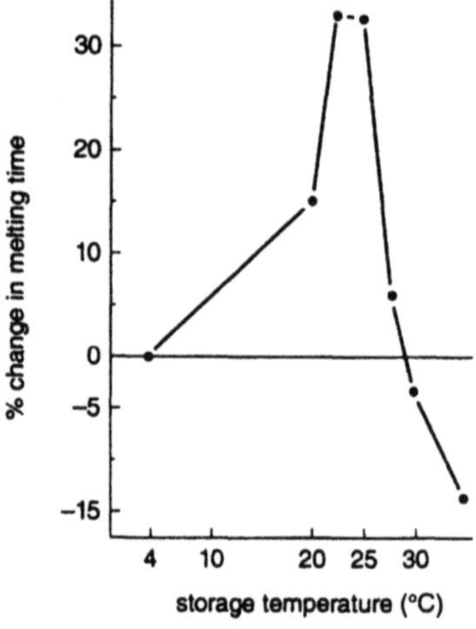

Figure 176. Effect of storage temperature on the charge in melting time of suppositories following 6-month storage. (Reproduced from Ref. 707 with permission.)

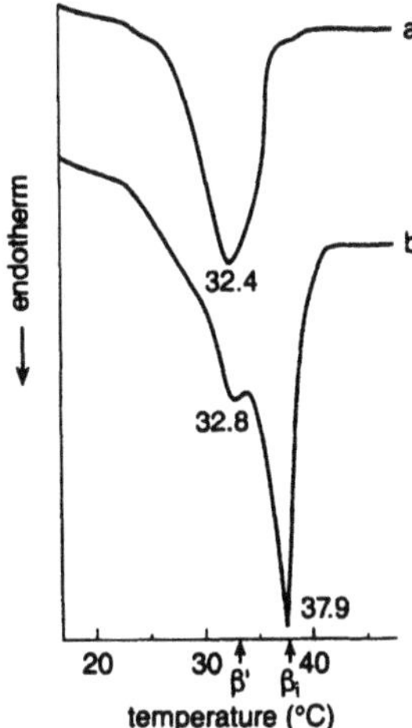

Figure 177. DSC curves showing polymorphic phase transitions of a triglyceride on storage. (a) Before storage; (b) after 6-month storage at 25°C. (Reproduced from Ref. 709 with permission.)

An increase in the T_g of poly(*d,l*-lactide-*co*-glycolide) microspheres was observed during storage at 40°C.[712] The T_g of biodegradable microspheres may also decrease as a result of polymer decomposition during storage. The lowered T_g of poly(*l*-lactide) microspheres due to the lowered molecular weight of the polymer resulted in an increase in drug release rate from the microspheres.[713] Particularly important may be the acceleration of hydrolysis in matrices containing basic drug substances.[714]

Changes in the crystalline state of polymer matrices during storage may result in changes in drug release from microspheres. Amorphous poly(*l*-lactide) microspheres containing progesterone exhibited an increased release rate following storage due to polymer crystallization (Fig. 179).[713,715] Storage at temperatures above the T_g increased crystallization, as shown by a diminished exothermic peak in the DSC thermogram recorded following storage, compared to that of a fresh sample (Fig. 180). It was suggested that drug

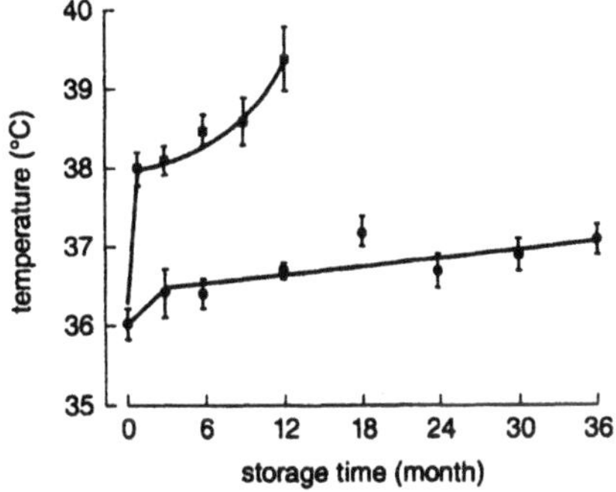

Figure 178. Changes in the dropping points (the temperature required for suppositories to melt and fall from a grease cup orifice) of suppositories following storage at 15°C (●) or 25°C (■) and 75% RH. (Reproduced from Ref. 710 with permission.)

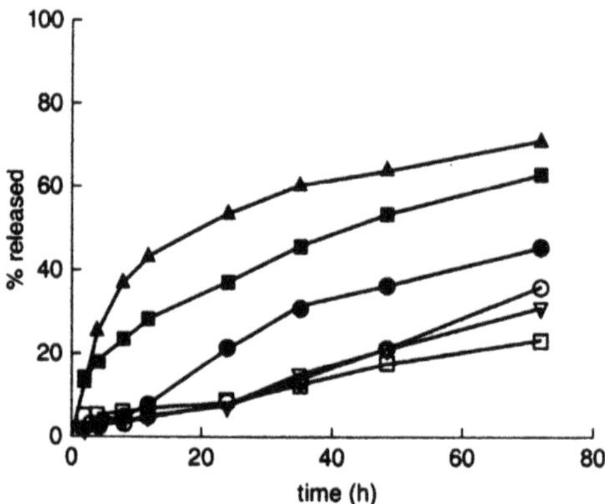

Figure 179. Changes in drug release rate due to crystallization upon storage. Drug release rates were measured before storage (○), after 7-month storage at 30°C and 0% RH (▽), at 30°C and 50% RH (●), and at 30°C and 75% RH (▲), and after 6-month storage at 50°C and 0% RH (□) and at 50°C and 11% RH (■). (Reproduced from Ref. 713 with permission.)

distribution within the microspheres changed with crystallization, resulting in the increased release rate.[713,715] It was proposed that increased crystallization might have occurred even at temperatures below the T_g in the presence of increased humidity, thereby lowering the T_g through polymer hydrolysis.

4.2.5. Drug Leakage from Liposomes

During storage, liposomes may exhibit physical instability, leading to leakage of intraliposomal entrapped drugs. In addition, chemical degradation of lipid membrane components resulting from oxidation and hydrolysis also changes drug release rates from liposomes. For example, phospholipid hydrolysis increased the permeability of a liposome membrane, resulting in increased leakage.[716]

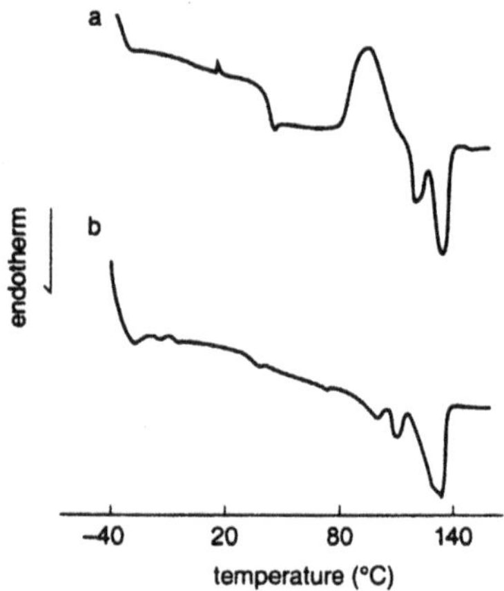

Figure 180. DSC curves demonstrating crystallization of poly(*l*-lactide) in microspheres following storage. (a) Before storage; (b) after 6-month storage at 50°C and 11% RH. (Reproduced from Ref. 713 with permission.)

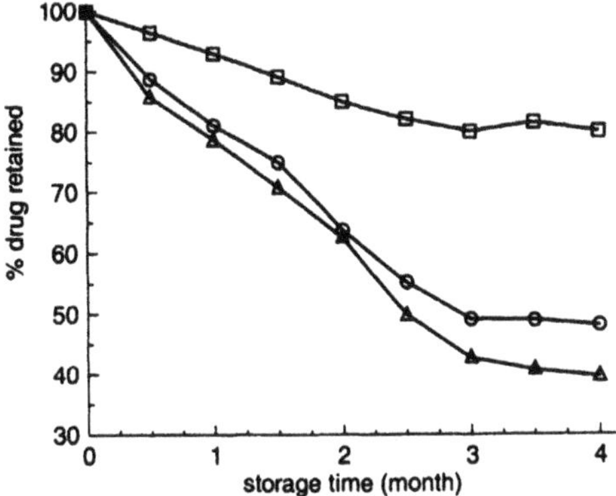

Figure 181. Effect of membrane components on the leakage of 5-fluorouracil from liposomes during storage at 4°C. ○, LUV (PC/PS/CH 7:4:5); △, LUV (MC/PS/CH 7:4:5); □, MLV (PC/PS/CH 7:4:5). LUV, Large unilamellar vesicle, PC, 1,2-dipalmitoyl-sn-glycero-3-phosphocholine monohydrate, PS, dipalmitoyl-DL-α-phosphatidyl-L-serine, CH, cholesterol, MC, 1,2-dimyristoyl-sn-glycero-3-phosphocholine monohydrate, MLV, multilamellar vesicle. (Reproduced from Ref. 717 with permission.)

Drug leakage from liposomes following storage depends on liposomal structure and membrane components, as shown in Fig. 181.[717,718] Optimization of membrane components and excipients to reduce drug leakage during storage has been attempted. Liposomes made from egg yolk lecithin exhibited drug leakage following storage; however, this effect was reduced by storage at low temperature in an oxygen-free atmosphere or by including antioxidants such as α-tocopherol in the formulation (Table 11).[719] Drug leakage was diminished in collagen-containing solutions, suggesting that collagen produced a decrease in liposome permeability through an antioxidant effect (Fig. 182).[720]

Aggregation of liposomes upon storage also depends on membrane components. Liposomes that included taurine as an isotonic solute were most stable at an optimal content of benzalkonium chloride.[721] Repulsive energy forces between particles described by the Deryaguin–Verwey–Overbeek theory appeared to account for the stabilization.

Table 11. Release of Carboxyfluorescein from Liposomes Following Storage[a]

	Time for 50% release (days)		
Liposome composition[b]	Room temperature	Room temperature, oxygen-free	4°C
EYL	13	30	100
EYL:Chol (5:1 molar ratio)	50	—	180
EYL: α-T (20:1)	100	140	>400
EYL: α-T (5:1)	120	140	>400
EYL: Chol: α-T (5:1:0.3)	190	—	>400

[a]Reference 719.

[b]Abbreviations: EYL, egg yolk lecithin; Chol, cholesterol; α-T, α-tocopherol.

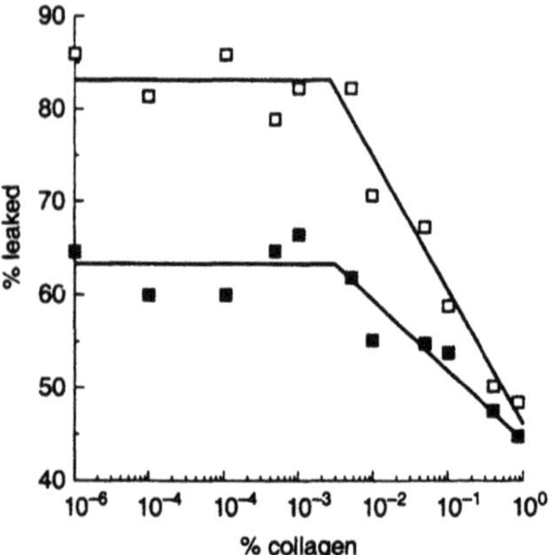

Figure 182. 5(6)-Carboxyfluorescein leakage from liposomes during incubation at 20°C for 70 h in collagen-containing solutions as a function of collagen concentration. Lipid concentration: □, 0.04%; ■, 0.4%. (Reproduced from Ref. 720 with permission.)

4.2.6. Aggregation in Emulsions

Aggregation is a normal physical phenomenon in emulsion formulations. Oxygen-transporting emulsions of perfluorodecalin needed stabilizing additives to prevent aggregation.[722] A series of total parenteral nutrition admixtures exhibited changes in the droplet size during storage, which was detected by Coulter counter and laser diffractometry measurements (Fig. 183).[723] The emulsion stability was dependent on the emulsion zeta potential and was predicted by the Deryaguin–Landau–Verwey–Overbeek theory.[724]

Increasing the storage temperature from 25 to 40°C markedly reduced the stability of a clofibride emulsion for oral administration, whereas storage at 4°C caused rapid phase separation owing to decreased solubility.[725] In this study, the physical stability of emulsions under stressed conditions was evaluated by subjecting them to ultracentrifugation (25,500 × *g* over 1 h), repeated freeze–thaw cycles (16 h of freezing at −18°C and 8 h of thawing at 25°C), and excessive shaking (150 strokes/min at 25°C over 48 h).

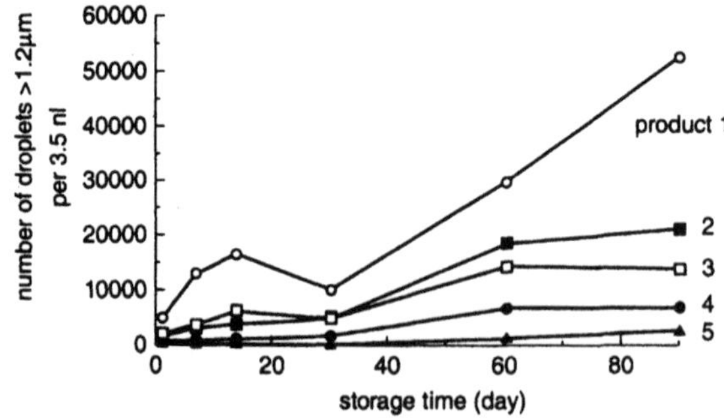

Figure 183. Changes in the droplet size of emulsions with time at 4°C. (Reproduced from Ref. 723 with permission.)

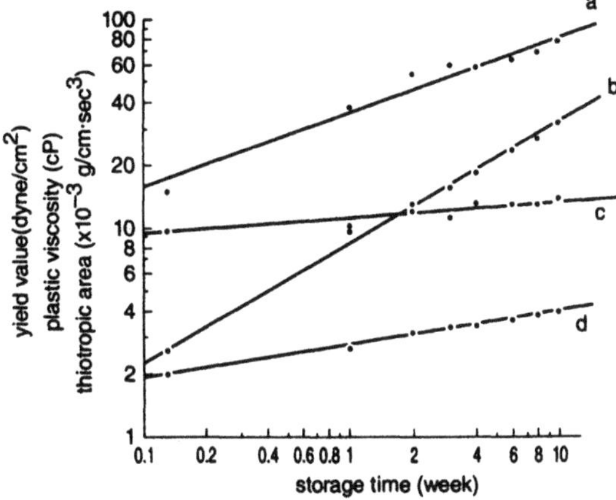

Figure 184. Rheologic aging of a lotion formulation with time. (a) yield value; (b) dynamic yield value; (c) plastic viscosity; (d) thixotropic area. (Reproduced from Ref. 727 with permission.)

Photomicrographs were taken with a specialized low-temperature scanning electron microscope[726] to assess the phase inversion of a cream formulation during long-term storage. Rheologic aging of lotion formulations was represented as a function of time using the following equation[184]:

$$P = at^b \tag{4.4}$$

In this equation, P represents a rheologic parameter, and a and b are constants. Figure 184 shows the plots obtained for four rheologic parameters.

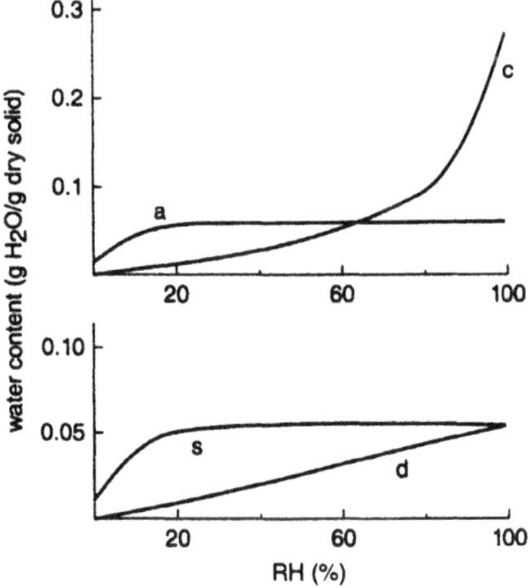

Figure 185. Moisture adsorption by gelatin capsules. a, Monomolecular layer of water molecules bound to surface; c, multimolecular layers of water; d,adsorption from dried state; s, desorption from saturation. (Reproduced from Ref. 728 with permission.)

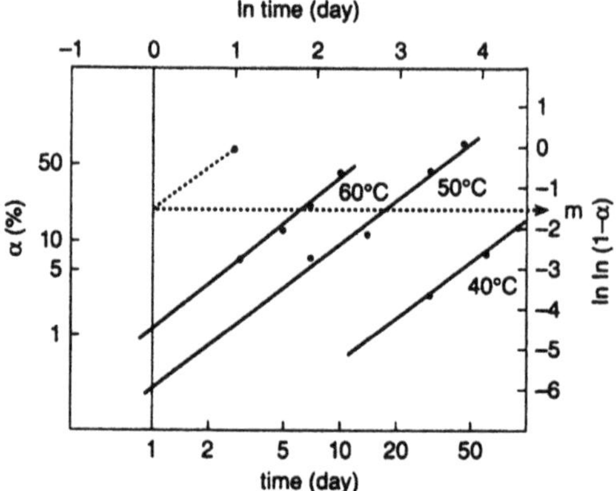

Figure 186. Discoloration of parenteral ascorbic acid formulation plotted according to the Weibull equation. α: Decrease in percent transmittance. (Reproduced from Ref. 730 with permission.)

4.2.7. Moisture Adsorption

Moisture adsorption by solid dosage forms can result not only in increased chemical drug degradation but also in changes in the functional stability of dosage forms.

A hard gelatin capsule exhibited moisture sorption depending on humidity, the hysteresis of which was analyzed by the Young–Nelson hypothesis. This allowed the formation of a monolayer of adsorbed moisture to be distinguished from normal condensation of moisture and from absorption of moisture (Fig. 185).[728]

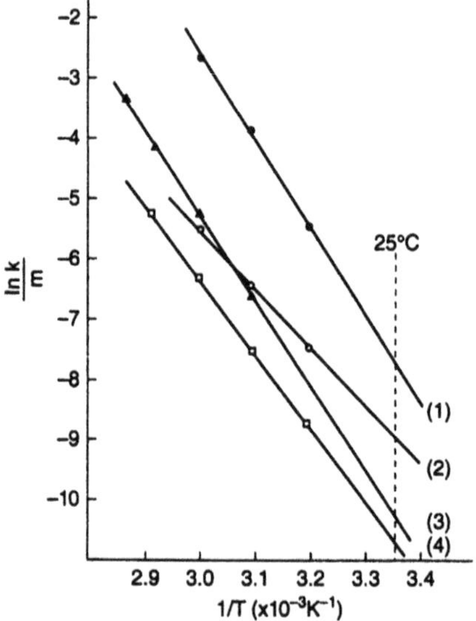

Figure 187. Arrhenius plots for discoloration of parenteral formulations of ascorbic acid (1), reserpine (2), thiamine hydrochloride (3), and ATP (4). The parameters k and m represent the constants obtained using the Weibull equation (time unit: day). (Reproduced from Ref. 730 with permission.)

The moisture permeation rate of a sugar coating composed of sucrose, talc, and other minor components was reported to conform to Fick's equation.[729] The permeation rate appeared to be rate-controlling for moisture adsorption by sugar-coated tablets.

Moisture adsorption of dosage forms in relation to the moisture permeation of packaging will be described in Section 4.3.1.

4.2.8. Discoloration

Although the discoloration of dosage forms may result from chemical degradation, the mechanisms are usually unclear. Thus, discoloration is generally considered to be a physical degradation (degradation of "appearance"). Empirical equations such as the Weibull equation have been used to predict discoloration of some dosage forms. Discoloration of a parenteral formulation of ascorbic acid was described by the Weibull equation (Eq. 2.69), and the constants representing discoloration rate were obtained from the slopes (Fig. 186).[730] The fact that the kinetics also conformed to Arrhenius behavior (Fig. 187) suggested that it would be possible to predict discoloration rates under a variety of conditions.

Changes in crystallinity during storage of solid-state emulsions from which oil-in-water emulsions are prepared was reported.[731]

4.3. Effect of Packaging on Stability of Drug Products

The role that packaging plays in the overall perceived and actual stability of the dosage form is well established. Packaging plays an important role in quality maintenance, and the resistance of packaging materials to moisture and light can significantly affect the stability of drugs and their dosage forms. It is crucial that stability testing of dosage forms in their final packaging be performed.[732]

The primary role of packaging, other than its esthetic one, is to protect the dosage forms from moisture and oxygen present in the atmosphere, light, and other types of exposure, especially if these factors affect the overall quality of the product on long-term storage. Protection from light can be achieved using primary packaging (packaging that is in direct contact with the dosage forms) and secondary packaging made of light-resistant materials. Incorporating oxygen adsorbents such as iron powder in packaging units can reduce the effect of oxygen. Details on the contributions of packaging to the stability of dosage forms have been extensively presented elsewhere. This section will emphasize the effect of packaging on moisture adsorption as it affects the stability of dosage forms and consider the interaction between dosage forms and packaging.

4.3.1. Moisture Penetration

Many studies have been conducted on predicting the role of packaging in moisture adsorption by dosage forms. Adsorption of moisture by tablets contained in polypropylene films was successfully modeled from storage temperature and the difference in water vapor pressure between the inside and outside of the packaging, as shown in Fig. 188.[733] Similarly, the moisture adsorption of moisture by cloxazolam tablets through press through packaging (PTP) composed of polyvinyl chloride and on aluminum film under nonisothermal conditions was predicted from the moisture permeability coefficient of the packaging as well as from temperature and humidity conditions inside and outside the packaging.[734]

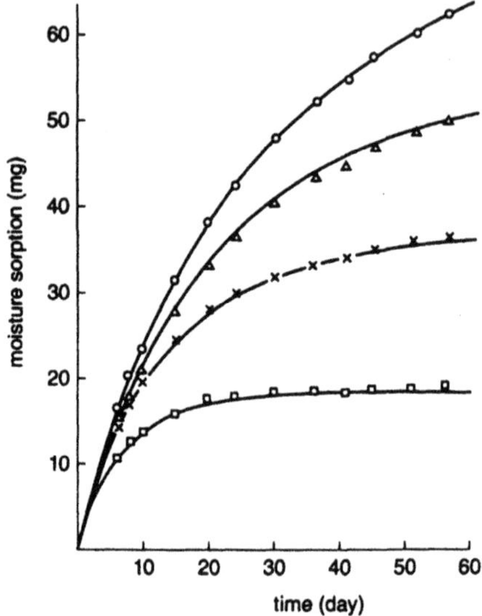

Figure 188. Prediction of moisture adsorption by tablets in polypropylene films (25°C and 75% RH). Number of tablets placed in a permeability cup: □, 3; ×, 6; △, 9; ○, 12. Lines represent predicted sorption curves. (Reproduced from Ref. 733 with permission.)

Chemical and physical degradation of packaged dosage forms caused by moisture adsorption has been predicted from the moisture permeability of the packaging. For example, strength changes of lactose–corn starch tablets in strip packaging (SP) and PTP,[735] discoloration of sugar-coated tablets of ascorbic acid,[736,737] and hydrolysis of aspirin aluminum tablets in PTP and glass bottles[738] were predicted using the moisture permeability coefficient of the packaging.

Desiccants are often used to eliminate moisture in packaging when the moisture resistance of the packaging itself is not sufficient to prevent exposure. The utility of desiccants has been assessed based on a sorption–desorption moisture transfer model.[739]

4.3.2. Adsorption onto and Absorption into Containers and Transfer of Container Components into Pharmaceuticals

Pharmaceuticals may interact with packaging and containers, resulting in the loss of drug substances by adsorption onto and absorption into container components and the incorporation of container components into pharmaceuticals. Diazepam in intravenous fluid containers and administration sets exhibited a loss during storage due to adsorption onto glass and adsorption onto and absorption into plastics.[740,741] Nitroglycerin, a liquid with a significant vapor pressure, is also significantly adsorbed onto and absorbed into containers. Decreases in the drug content of nitroglycerin tablets in SP[742] and nitroglycerin solutions in glass and plastic containers[743,744] due to adsorption/absorption were analyzed by a diffusion model and a model consisting of adsorption onto the surface followed by partitioning into the plastic.

Polyvinyl chloride (PVC), a polymer often used for pharmaceutical containers, is known to interact with various drug substances. Adsorption and absorption of nitroglycerin onto

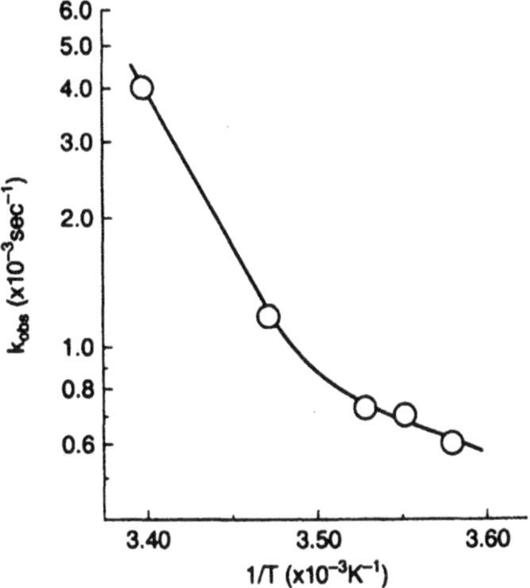

Figure 189. An attempted Arrhenius plot for the rate of absorption of nitroglycerin into a PVC container (pH 5.6). (Reproduced from Ref. 745 with permission.)

and into PVC conformed to apparent first-order kinetics, and the apparent rate constant describing uptake showed nonlinear Arrhenius behavior (Fig. 189).[745] Absorption of clomethiazole edisylate and thiopental sodium into PVC infusion bags was observed.[746] The pH dependence of adsorption/absorption of acidic drug substances such as warfarin and thiopental and basic drug substances such as chlorpromazine and diltiazem indicates that only the un-ionized form of the drug substance is adsorbed onto or absorbed into PVC infusion bags.[747] The absorption was correlated to the octanol–water partition coefficients of the drugs, suggesting that prediction of absorption from partition data is possible.[748,749] A parameter referred to as sorption number (S_n) was used to predict drug loss to PVC bags.[750,751] S_n is defined by the plastic–infusion solution partition coefficient of the drug, the diffusion coefficient of the drug in the plastic, the fraction of the drug un-ionized in solution, the volume of the infusion solution, and the surface area of the plastic. As shown in Fig. 190,

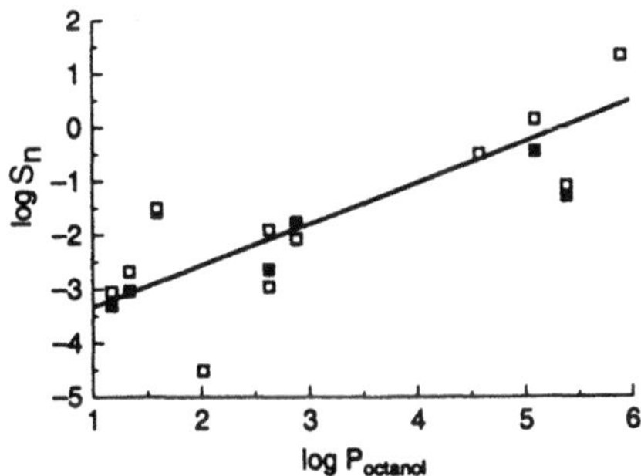

Figure 190. A log–log plot of sorption number (S_n) into a PVC container versus octanol–water partition coefficient ($P_{octanol}$) for a number of drugs. Drugs were stored at 15–20°C for 1, (□) and 24 h (■). (Reproduced from Ref. 750 with permission.)

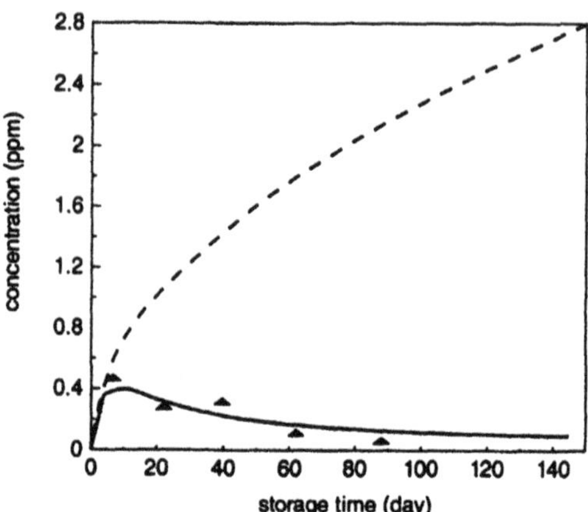

Figure 191. Accumulation of carboxylic acid alkyl ester from polypropylene container at 25°C and pH 2. ——, Predicted from diffusion and degradation rate equation; ---, diffusion-controlled accumulation; ▲, observed. (Reproduced from Ref. 748 with permission.)

the logarithm of this parameter was correlated with the logarithm of the octanol–water partition coefficients of various solutes.

Polymers such as nylon 6 (polycaprolactam) are known to adsorb drug substances such as benzocaine.[752] Glass surfaces are also known to adsorb drug substances. Chloroquine solutions in glass containers decreased in concentration owing to adsorption of the drug onto the glass.[753]

Rubber closures are also known to absorb materials, including drugs. Absorption of preservatives such as chlorocresol into the rubber closures of injectable formulations has been studied extensively.[754–756] Water permeability of rubber closures used in injection vials is considered an important parameter in assessing the closures, but quantitative prediction of water permeability through rubber closures is difficult because the diffusion coefficient of water is dependent on relative humidity.[757]

In addition to absorption onto and absorption into containers, transfer of container components into pharmaceuticals may affect the perceived stability/quality of drug dosage forms. Adsorption of volatile components from rubber closures onto freeze-dried parenterals during both dosage form processing and storage brought about haze formation upon reconstitution.[758–760] Leaching of dioctyl phthalate, a plasticizer used especially in PVC plastics, into intravenous solutions containing surfactants was observed.[761,762] The time course of dissolution of an alkyl ester of a carboxylic acid, which originated in a polymer composite packaging material, from polypropylene containers was predicted by a diffusion rate and degradation rate equation, as shown in Fig. 191.[763]

4.4. Estimation of the Shelf Life (Expiration Period) of Drug Products

Shelf life is best defined as the time span over which the quality of a product remains within specifications. That is, it is the time period over which the efficacy, safety, and esthetics of the product can be assured. When the degradation of the essential components cannot be adequately described by a rate expression, shelf life cannot be easily estimated or projected. When a quality-indicating parameter changes with time via complex kinetics that cannot be adequately explained or predicted, one must determine stability solely from experimental observations. Many physical degradation processes exhibit this kind of behavior. In contrast, estimation of shelf life is often possible when the shelf life is governed by a degradation process that can be adequately described by a rate expression (like many chemical degradation processes).

Estimation of product shelf life is done by two methods—estimation from data obtained under the same conditions as those that the final product is expected to withstand and estimation from tests conducted under accelerated conditions. This section describes these two methods for estimating the shelf life of pharmaceuticals when chemical degradation is the major contributor to the degradation process and the degradation can be adequately described by a rate expression.

4.4.1. Extrapolation from Real-Time Data

The Woolfe equation has been used to estimate the shelf life of a product from data obtained at the same temperature/conditions as those expected for the final product.[764] The time at which drug content diverges from its specifications is estimated by extrapolating the time course of degradation at a specific temperature/condition. When the time course of drug content (C) is represented by

$$t = \bar{t} = \frac{C - \bar{C}}{b} \tag{4.5}$$

where $\bar{t}$ is the average of t, $\bar{C}$ is the average of C, and b is constant, the confidence interval of the time at which drug content diverges from specifications, determined by regression analysis of n sets of time-content data, is described by

$$t_L = t + \left\{ (t - \bar{t})g \pm \frac{t'\sqrt{V_e}}{b} \sqrt{\frac{(t-\bar{t})^2}{S_{tt}} + \left(1 + \frac{1}{n}\right)(1 - g)} \right\} \frac{1}{1-g} \tag{4.6}$$

where t' is the one-sided Student's t value with degrees of freedom and $n - 2$.

$$g = \frac{t'^2 V_e}{b^2 S_{tt}} \qquad V_e = \frac{S_{cc} - S_{ct}^2/S_{tt}}{n-2}$$

The shelf life (the lower confidence limit of the time at which the drug content diverges from the specification range) can thus be estimated. When the lower limit of content specification is 90%, the shelf life corresponds to t_L at C of 90%. If V_e is large or b is small, shelf life cannot be estimated because the value of g must be not less than 1. Because the confidence interval becomes narrowest at $\bar{t}$, as shown in Fig. 192, a more precise estimate of shelf life can be obtained by extrapolating the regression curve determined from a larger number of time–content data points at larger t.

The Carstensen equation is used to calculate the confidence limit of C at a specific t[765]:

$$C_L = \bar{C} + b(t - \bar{t}) \pm t' \sqrt{\left[1 + \frac{1}{n} + \frac{(t-\bar{t})^2}{S_{tt}}\right] V_e} \tag{4.7}$$

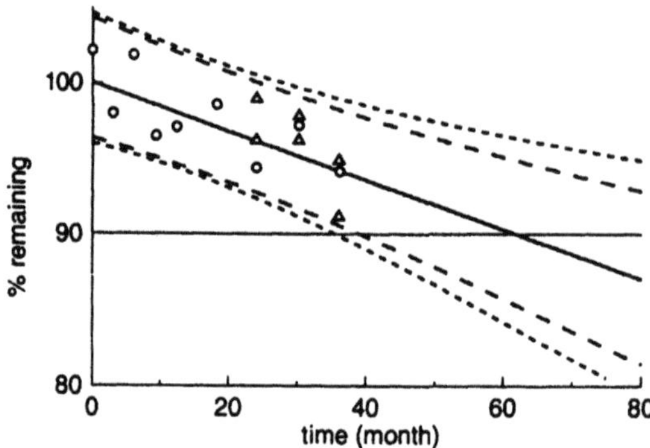

Figure 192. Time–drug content curve with 95% confidence intervals. A zero-order degradation of 2%/year was assumed. Assay error was assumed to be 2% standard deviation. -----, 95% significant limit calculated from data represented by open circles; – – –, 95% significant limit calculated from all the data including data represented by open triangles.

where k is the average annual rate constant. Thus, whereas the Woolfe equation allows one to estimate the confidence limit of t as a function of C, the Carstensen equation permits estimation of the confidence limit of C as a function of t.

4.4.2. Shelf-Life Estimation from Temperature-Accelerated Studies

In temperature-accelerated studies, shelf life at a storage temperature T_1 is estimated from the shelf life at an elevated temperature T_2, according to

$$\ln \frac{t_{90(T1)}}{t_{90(T2)}} = \frac{E_a}{R}\left(\frac{1}{T_1} - \frac{1}{T_2}\right) \tag{4.8}$$

Shelf life is referred to as $t_{90(T_1)}$ when the lower specification limit of content is 90%. Shelf life exhibits a log-linear relationship versus $1/T$ in a given temperature range when the activation energy is constant (Fig. 193). The latter condition usually is only met when the degradation mechanism is the same across the temperature range of exposure. For example, a shelf life of 6 months at 40°C corresponds to a shelf life of 3 years at 25°C when an activation energy of 22.1 kcal/mol is assumed.

4.4.2.1. Experimental Design of Accelerated Testing

Experimental designs for accelerated temperature testing were proposed in the 1960s by Tootill,[766] Kennon,[767] and Lordi and Scott,[768] as well as many others. A practical chart for estimating shelf life from the data from accelerated testing at 41.5 and 60°C, which was proposed by Lordi and Scott (Fig. 194), is introduced here because of its historical value in drug stability studies, although it is no longer useful because the advent of computers has made more complex numerical calculations trivial. Shelf lives at 25, 41.5, and 60°C, $t_{90(25)}$, $t_{90(41.5)}$, and $t_{90(60)}$, respectively, are represented by

$$\ln t_{90(60)} = 2 \ln t_{90(41.5)} - \ln t_{90(25)} \tag{4.9}$$

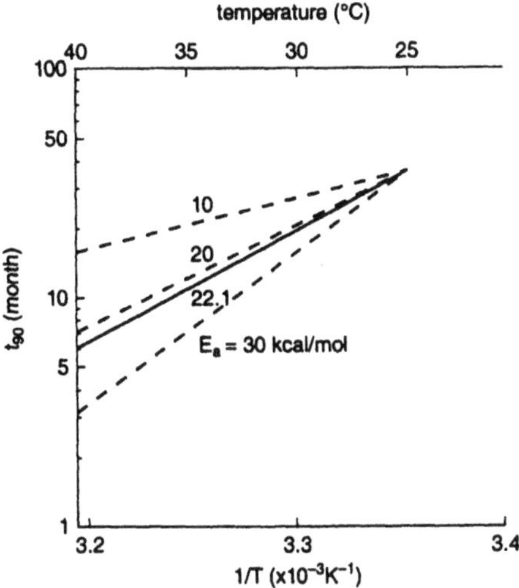

Figure 193. Temperature dependence of shelf life.

The bold, solid descending lines in Fig. 194 represent $t_{90(60)}$ as a function of $t_{90(25)}$ at a specific $t_{90(41.5)}$ represented on the right-hand y-axis. The intersection of a horizontal line representing $t_{90(41.5)}$ and a bold, solid descending line representing $t_{90(60)}$ corresponds to $t_{90(25)}$. Thus, $t_{90(25)}$ can be estimated from $t_{90(41.5)}$ and $t_{90(60)}$. Activation energy can be determined from the

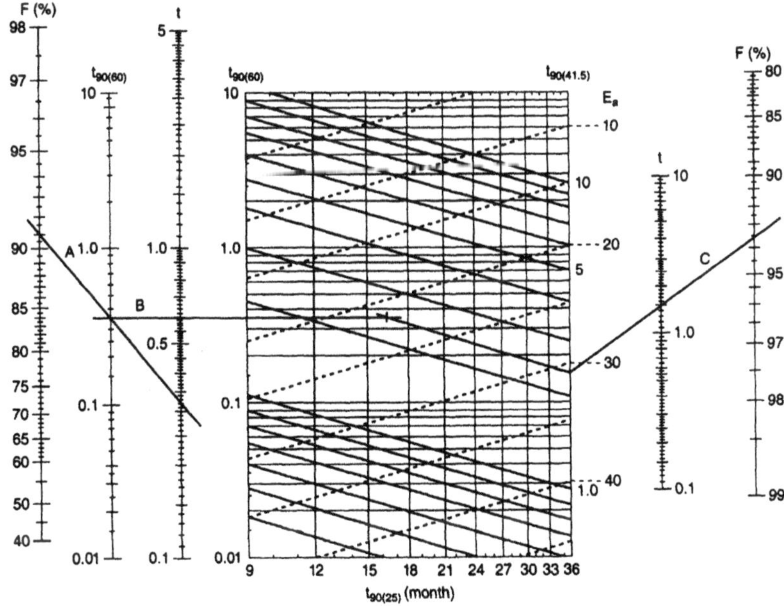

Figure 194. A Lordi chart for estimating shelf life from accelerated data. (Reproduced from Ref. 768 with permission.)

dashed lines. The measures shown on both sides of the figure are used to calculate $t_{90(60)}$ and $t_{90(41.5)}$ from the percent of drug remaining, F, at time t for a first-order reaction.

Presently, shelf life is usually estimated by regression analysis according to Eq. (4.8) and its variants using computers. A simplified method for shelf-life estimation, regardless of reaction order, has been proposed,[769–772] and various computer programs have been developed. However, it should be noted that the application of Eq. (4.8) is limited to a temperature range in which E_a can be regarded as constant (as discussed in Chapter 2).

Using data for the rate of production of degradants in addition to data for drug loss significantly enhances the precise estimation of shelf life.[773,774] This is especially the case when the formation of degradation products can be measured with higher precision than the drug loss, and only degradation data at the initial stage are used for the estimation.

4.4.2.2. Estimation of Shelf Life Using Accelerated-Test Data at a Single Level of Temperature

It is theoretically possible to estimate the shelf life of a product from a single measurement of drug degradation at a single time point and temperature if the activation energy for the degradation is known. Of course, the quality of the estimate is strongly affected by assay error. For example, when a pharmaceutical having a shelf life of $t_{90(25)}$ is stored at a temperature T for a time span t, the probability that degradation percentage is determined to be x using an assay method having a standard deviation of σ is represented by Eq. (4.10) (for zero-order degradation kinetics).[775] Thus, when degradation percentage is determined to be x, the true value of shelf life is $t_{90(25)}$ at a probability represented by Eq. (4.11).

$$P(x) = \frac{1}{\sqrt{2\pi}\sigma} \exp\left[-\left(100 - \frac{C}{t_{90(25)}} - x\right)^2 \frac{1}{2\sigma^2}\right] \tag{4.10}$$

where

$$C = 10t \exp\left[\frac{-E_a}{R}\left(\frac{1}{T} - \frac{1}{298}\right)\right]$$

$$P(t_{90(25)}) = \frac{1}{\sqrt{2\pi}\sigma} \cdot \exp\left[-\left(100 - \frac{C}{t_{90(25)}} - x\right)^2 \frac{1}{2\sigma^2}\right] \cdot \frac{C}{t_{90(25)}{}^2} \tag{4.11}$$

The distribution of the true value of the shelf life is shown as a function of the observed value for the degradation percentage in Fig. 195. As the observed value of degradation percentage decreases, the mean and range of the shelf-life estimate increase. Figure 196 shows the probability that the shelf life is longer than 2 years as a function of degradation percentage after a storage at 40°C for 6 months. The probability depends on activation energy and assay error.

Figure 197 shows the relationship between the observed degradation percentage and estimated shelf life at a probability of 95%. If the activation energy is known, the shelf life can be estimated from only the value of the degradation percentage observed after storage at 40°C for 6 months with the use of this figure. When the value of the activation energy is unknown, the shelf life should be estimated assuming a smaller value of E_a than expected in

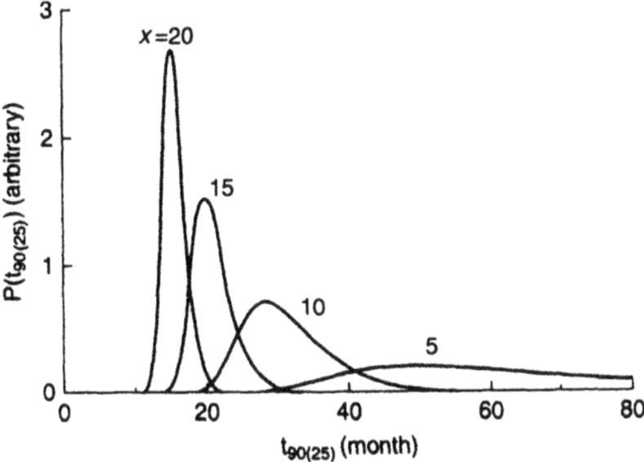

Figure 195. Distribution of true shelf life predicted from degradation percentage observed at 40°C. *x*: Percent degraded during 6-month storage at 40°C. Assay error (standard deviation): 2%; activation energy: 20 kcal/mol. (Reproduced from Ref. 775 with permission.)

order to obtain a conservative value for the shelf life. The actual shelf life may, in fact, be longer.

The shelf life estimated from more than three values of degradation percentage observed at an elevated temperature exhibits a smaller distribution range than that estimated from a single value shown in Fig. 195.[775] This is because the use of more than three observed data values provides information on the data variation and enables a Monte Carlo simulation of degradation data.[776] This simulation method yields a longer shelf-life estimate because of the smaller distribution range.

As described in Chapter 6, stability testing guidelines recommend 40°C as the temperature for accelerated testing. Figure 198 shows the probability that all the observed values of degradation percentage after storage at 40°C for 2, 4, and 6 months are less than 10% (assuming a 90% specification) as a function of the true shelf life. This probability depends largely on activation energy and assay error. For example, for an energy of activation of 10

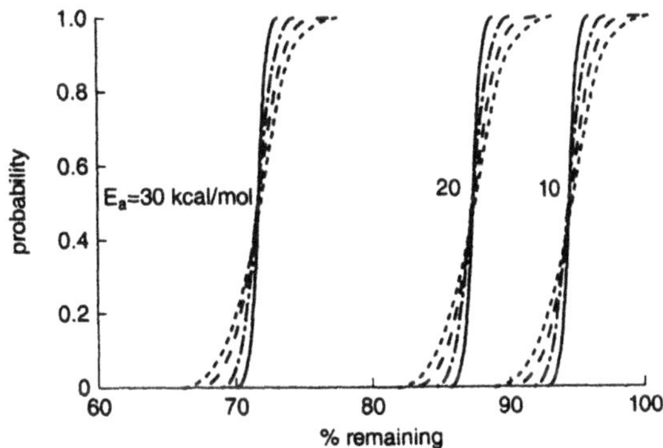

Figure 196. Relationship between degradation percentage observed at 40°C and probability of shelf life longer than 2 years. Assay error (standard deviation): ——, 0.5; — · —, 1; –––, 1.5; ---, 2%. (Reproduced from Ref. 775 with permission.)

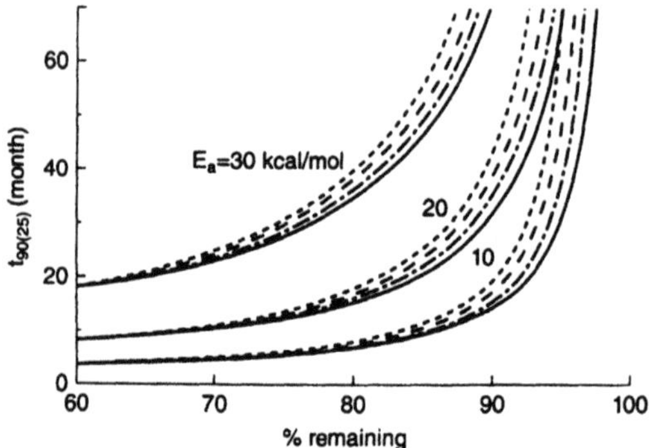

Figure 197. Relationship between degradation percentage observed at 40°C and shelf life estimated at 95% probability. Assay error (standard deviation): ——, 2; — · —, 1.5; ---, 1; ---, 0.5%. (Reproduced from Ref. 775 with permission.)

kcal/mol, even a product with a shelf life of less than 1 year exhibits degradation percentage data less than 10% at an increasing probability as assay error increases.[777]

4.4.3. Estimation of Shelf Life under Temperature-Fluctuating Conditions

Actual storage temperature of pharmaceuticals fluctuates with time. The degradation rate at fluctuating temperatures is higher than that at the average temperature. The difference is determined by the fluctuation pattern and activation energy for the degradation reaction.[778–780] For example, the degradation rate of a reaction under a temperature cycle represented by a sine curve with an average of 20°C and a range of 10°C (Fig. 199) is 1.08 times larger than that at a constant temperature of 20°C (for a reaction with an activation energy of 20 kcal/mol).

The concept of average kinetic temperature (T_k) was introduced by Haynes[781] to predict the stability of pharmaceuticals stored under fluctuating temperature conditions over the course of a year. T_k is represented by Eq. (4.12) using month-average temperatures from

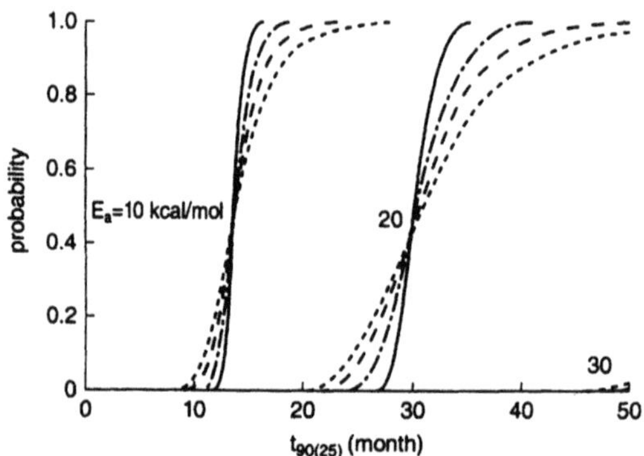

Figure 198. Effect of activation energy and assay error on the probability that all the values of degradation percentage observed after a storage at 40°C for 2, 4, and 6 months are less than 10%. Assay error (standard deviation): ——, 0.5; — · —, 1; ---, 1.5; ---, 2%. (Reproduced from Ref. 777 with permission.)

Figure 199. A model of fluctuating temperature.

January to December (T_1 to T_{12}) and corresponds to a "virtual" temperature at which a product would undergo degradation at the same rate as a product exposed to the fluctuating temperature pattern. Table 2 shows the values of T_k for various cities as a function of E_a values. Note that these critical temperature values vary most significantly from the average temperature values for those cities with the biggest temperature fluctuations and for reactions with higher E_a values.

$$T_k = \frac{-E_a/R}{\ln k - \ln A}$$

$$= \frac{-E_a/R}{\ln\left\{\dfrac{\exp(-E_a/RT_1) + \exp(-E_a/RT_2) + \cdots + \exp(-E_a/RT_{12})}{12}\right\}} \quad (4.12)$$

Table 12. Average Kinetic Temperatures of Various Cities for Reactions with Different Activation Energies

		Average kinetic temperature for a reaction with an activation energy (kcal/mol) of:		
City	Average temperature[a]	10	20	30
Sapporo	8.2	10.4	12.5	14.0
Sendai	11.9	13.6	15.2	16.5
Tokyo	15.6	17.0	18.5	19.7
Osaka	16.3	17.9	19.5	20.8
Kagoshima	17.6	18.9	20.2	21.3
Naha	22.4	22.9	23.4	23.9
Oslo	3.8	5.6	7.4	8.7
Berlin	9.4	10.7	11.9	13.0
London	9.7	10.3	10.9	11.5
Beijing	11.8	14.7	17.2	18.8
Washington, D.C.	14.3	16.2	18.1	19.4
Rome	15.5	16.3	17.2	18.0
Sydney	17.9	18.2	18.6	19.0
Rio de Janeiro	23.8	23.9	24.0	24.1
Djakarta	27.2	27.2	27.2	27.2
Manila	27.4	27.4	27.4	27.5
Bombay	27.5	27.5	27.6	27.7
Bankok	28.4	28.5	28.5	28.6
Djibouti	29.5	29.8	30.2	30.6

[a]Average over the period 1961–1990.

An alternative to the concept of average kinetic temperature is the kinetic ratio (α), a ratio of a rate constant at a fluctuating temperature to that at a standard temperature, T_{ref}:

$$\alpha = \frac{\int_0^{\tau} \exp\left(\frac{-E_a}{RT}\right) d\theta}{\tau \cdot \exp\left(\frac{-E_a}{RT_{ref}}\right)} \tag{4.13}$$

The kinetic ratio can be used to represent the effect of temperature fluctuations.[782]

In addition, the stability prediction under fluctuating conditions, many papers have dealt with the use of artificial climate laboratories[783] and methods for shelf-life calculation.[778,779,784,785]

Chapter 5

Stability of Peptide and Protein Pharmaceuticals

Chapters 2, 3, and 4 concerned the stability of pharmaceuticals containing pharmacologically active ingredients of relatively low molecular weight. This chapter addresses the stability of peptide and protein drugs. Peptides and proteins can undergo some of the same degradation processes seen in small molecules. However, the stability of protein and peptide pharmaceuticals can be affected by additional reactions that alter their tertiary or higher structures.

Like drugs of low molecular weight, peptides and proteins undergo chemical degradation pathways such as hydrolysis and racemization. Depending on their molecular weight, they are also susceptible to physical degradation by denaturation, aggregation, and precipitation. Because of the complicated degradation mechanisms, it is generally more difficult to predict the stability of peptide and protein pharmaceuticals.

Chemical and physical properties of peptides and proteins have been studied extensively. The thermodynamics of protein structure have also been studied in detail and reported in many excellent reviews and books.[786–788] The present chapter focuses on the stability of peptide and protein pharmaceuticals upon storage.

5.1. Degradation of Peptide and Protein Pharmaceuticals

Degradation observed with peptide/protein pharmaceuticals is classified into chemical and physical mechanisms. The former involve changes in covalent bonds, and the latter involve changes in noncovalent interactions such as hydrophobic bonding/associations. For a specific peptide/protein, degradation usually includes both chemical and physical pathways as well as interactive pathways that might result when a molecule undergoes intermolecular disulfide exchange accompanied by precipitation. This section outlines each of the major chemical and physical degradation pathways.

5.1.1. Chemical Degradation

The major known chemical degradation pathways for peptides and proteins are deamidation, racemization, isomerization, hydrolysis, disulfide formation/exchange, β-elimination, and oxidation.

5.1.1.1. Deamidation

Asparagine residues in peptides and proteins undergo deamidation via cyclic imide formation followed by subsequent hydrolysis to form the corresponding aspartic and *iso*-aspartic acid peptides. This mechanism occurs primarily under neutral-to-basic pH conditions. Deamidation of an asparagine residue to the corresponding aspartic acid residue may also occur via a mechanism that does not involve cyclic imide formation, as shown in Scheme 80. Glutamine residues also undergo deamidation, but at much slower rates.

Adrenocorticotropic hormone (ACTH), which has 38 amino acid residues, exhibited pseudo-first-order deamidation in the neutral-to-alkaline pH region. The deamidation rate increased with increasing pH and buffer concentration. Deamidation via the cyclic imide of an asparagine residue was suggested since both the aspartic acid and *iso*-aspartic acid peptides were detected as deamidation products. As shown in Table 13, the rate of disappearance of ACTH showed good mass balance with the rates of appearance of deamidated ACTH and ammonia, indicating that the rate-determining step for the deamidation is not degradation of the cyclic imide but its formation.[789] A model hexapeptide with an asparagine residue (Asn-hexapeptide) exhibited a similar deamidation reaction. The deamidation rate was higher for asparagine residues having a smaller amino acid at the C-terminal side of the residue, as shown in Table 14, indicating that steric factors may influence cyclic imide formation.[790,791]

Deamidation of ACTH under acidic pH conditions is considered to be direct deamidation to the aspartic acid peptide since the *iso*-aspartic acid peptide was not observed as a

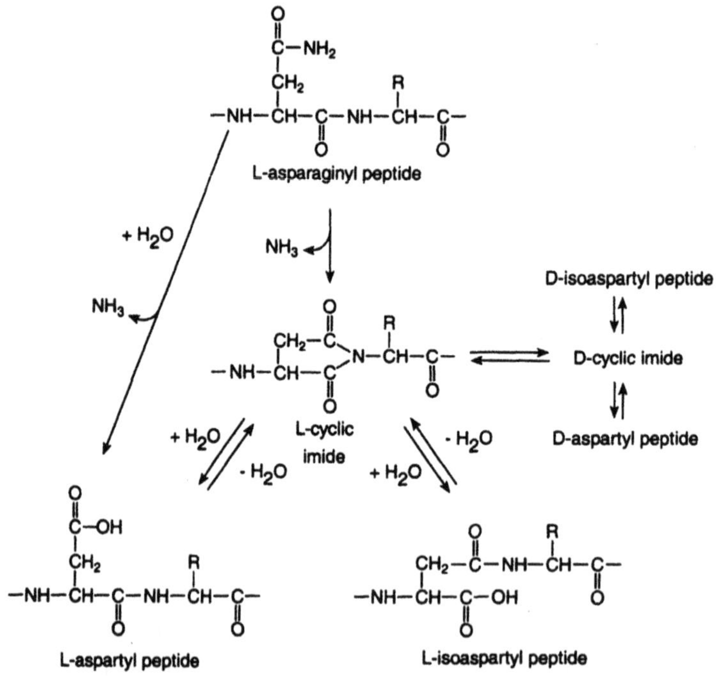

Scheme 80. Scheme showing the deamidation, isomerization, and racemization of peptides having asparagine or aspartic acid residues.

Table 13. The Effect of Glycine Buffer Concentration on the Deamidation of ACTH at pH 9.6 and 37°C[a]

	Apparent rate constant (h^{-1}) at pH 9.6, 37°C and glycine buffer concentration of:		
	10m*M*	50m*M*	100m*M*
Disappearance of ACTH	6.6×10^{-2}	1.4×10^{-1}	2.3×10^{-1}
Appearance of deamidated ACTH	4.7×10^{-2}	1.4×10^{-1}	2.6×10^{-1}
Appearance of ammonia	6.5×10^{-2}	1.6×10^{-1}	2.9×10^{-1}

[a]Reference 789.

degradation product.[789] Similar direct deamidation of the model Asn-hexapeptide resulted in 100% formation of the aspartic acid peptide.[790,791]

Insulin has two asparagine residues that undergo deamidation. At acidic pH values, Asn A-21 undergoes deamidation, whereas at neutral pH and in suspensions, deamidation at residue Asn B-3 predominates.[792] Deamidation of insulin at pH 2 and 3 was also enhanced by self-association.[793]

5.1.1.2. Isomerization and Racemization

Peptides and proteins having an aspartic acid residue undergo hydrolysis, isomerization, and racemization via cyclic imide formation. As shown in Scheme 80, L-aspartic acid peptide can isomerize to L-*iso*-aspartic acid peptide via its L-cyclic imide. The L-cyclic imide intermediate is capable of undergoing racemization to the D-cyclic imide and thus forms the D-aspartic acid peptide and the D-*iso*-aspartic acid peptide on hydrolysis.

Following storage of a secretin solution, aspartoyl[3] secretin (cyclic imide) and β-aspartyl[3] secretin (isomer) were detected in the solution by reversed-phase HPLC, indicating that isomerization occurred via the cyclic imide.[794,795] An Asp-hexapeptide also exhibited isomerization via cyclic imide formation at pH values above 4.[796] The rate of formation of the cyclic imide was affected by the size of the amino acid on the C-terminal side of the aspartic acid residue.[797] A cyclic imide was also detected as a major degradation product of basic fibroblast growth factor at pH 5.[798]

Table 14. The Effect of the Amino Acid Residue on the C-terminal Side of Asn on the Deamidation of Asn-Hexapeptides[a]

Asn-hexapeptide	t_{50} (days)
Val-Tyr-Pro-Asn-Gly-Ala	1.89 (pH 7.5)
Val-Tyr-Pro-Asn-Ser-Ala	5.55 (pH 7.5)
Val-Tyr-Pro-Asn-Ala-Ala	20.2 (pH 7.4)
Val-Tyr-Pro-Asn-Val-Ala	106 (pH 7.5)
Val-Tyr-Pro-Asn-Pro-Ala	70 (pH 7.4)
Val-Tyr-Pro-Asn-Pro-Ala	106 (pH 7.4)

[a]Reference 791.

Racemization has also been observed with many peptides and proteins. Casein exhibits racemization at aspartic acid, phenylalanine, glutamic acid, and alanine residues.[799] Racemization of serine and histidine residues has been reported for histrelin (a nonapeptide)[800] and a decapeptide,[801] agonists of luteinizing hormone-releasing hormone (LH-RH). As shown in Fig. 200, the main degradation pathway of decapeptide (an antagonist of LH-RH) above pH 7 was epimerization.[802]

5.1.1.3. Hydrolysis

Hydrolysis is a pathway often observed during peptide and protein degradation. As shown in Scheme 81, aspartic acid residues in particular are susceptible to hydrolysis in the acidic pH range. Secretin, apart from undergoing isomerization, also undergoes degradation by hydrolysis of its aspartic acid residues at position-3 and position-15.[794,795] Hydrolysis of aspartic acid residues under acidic conditions has also been observed with recombinant human macrophage colony-stimulating factor,[803] recombinant human interleukin-11,[804] and a hexapeptide.[796] Hydrolysis may also occur at serine and histidine residues.[799–802]

5.1.1.4. Cross-Linking through Disulfide Bond Formation and Other Covalent Interactions

Oxidation of cysteine residues of peptide and protein molecules yields intra- and intermolecular disulfide bonds (Scheme 82), leading to changes in tertiary structure. Also, normal disulfide bonds in peptide and protein molecules can undergo thiol-catalyzed intra- and intermolecular exchange reactions, leading to changes in secondary and tertiary structures (Scheme 83). Furthermore, the disulfide bond itself is susceptible to cleavage via β-elimination and forms dehydroalanine residues and persulfides (Scheme 84). These

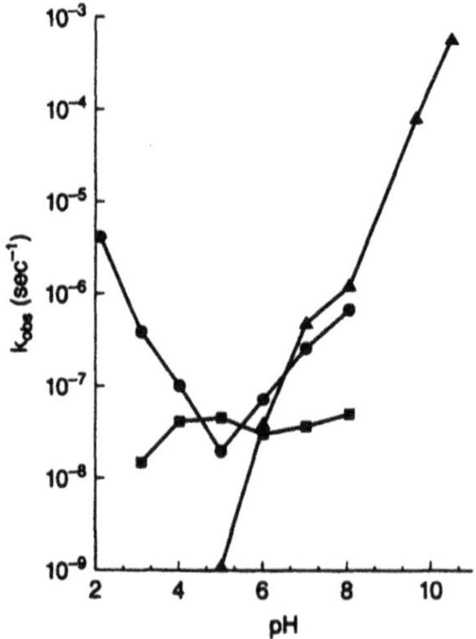

Figure 200. pH–rate profiles for deamidation (●), epimerization (▲), and hydrolysis (■) of a decapeptide LH-RH antagonist at 80°C. (Reproduced from Ref. 802 with permission.)

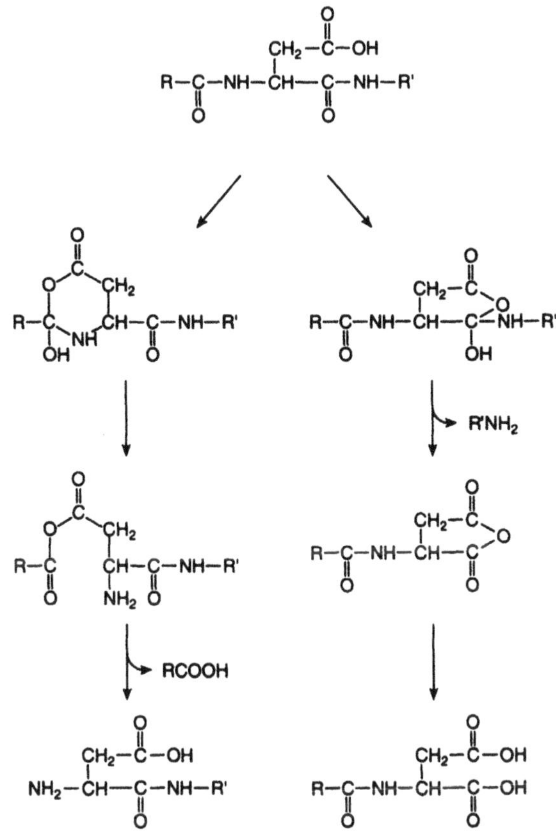

Scheme 81. Pathways proposed for the hydrolysis of peptides at aspartic acid residues.

products may also participate further in disulfide exchange reactions, resulting in formation of new cross-linkages.

Lysozyme exhibits deamidation at pH 6 and deamidation and hydrolysis at pH 4, whereas cleavage of disulfide residues and formation of new disulfide bonds were observed at pH 8.[805] The half-lives for reaction at the disulfide residues due to β-elimination were similar for 14 proteins, including insulin, indicating that cleavage of disulfide bonds is relatively independent of both primary and higher structures.[806] Intermolecular formation of new disulfide bonds leads to aggregation of peptides and proteins. For example, lyophilized bovine serum albumin and insulin undergo aggregation via intermolecular disulfide bond formation at a rate dependent on the water content of the lyophile.[807–810] Lyophilized

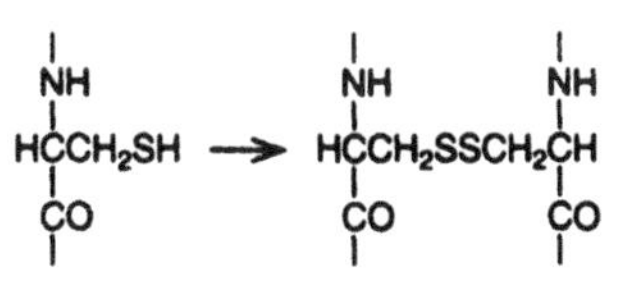

Scheme 82. Formation of a disulfide bond through oxidation of cysteine residues.

$$RS^- + R'S\text{-}SR' \rightleftharpoons RS\text{-}SR' + R'S^-$$

$$R'S^- + RS\text{-}SR \rightleftharpoons R'S\text{-}SR + RS^-$$

Scheme 83. Disulfide exchange reactions.

β-galactosidase also exhibited aggregation via disulfide bond formation at relatively low water content. The participation of disulfide bond formation in this case was confirmed by size-exclusion chromatography. The formed aggregate was not dissociated by guanidine hydrochloride but was dissociated by dithiothreitol, a disulfide bond reductant.[811]

Covalent bond formation, other than disulfide bond formation, is also involved in other intermolecular cross-linkages. The covalent linkages in the aggregates of freeze-dried ribonuclease A appeared to result from the participation of lysine, asparagine, and glutamine residues as suggested by amino acid analysis of the aggregates.[812,813] Lyophylized formulations of recombinant tumor necrosis factor-α formed dimers and oligomers that were nonreducible.[814] Upon storage, insulin formulations yielded covalent dimers; the extent of dimerization was highly dependent on the formulation.[815] The deamidated A-21 asparagine of one insulin molecule and the B-1 phenylalanine residue of another were found to be involved in the formation of the dimeric species.[816] The aggregates of basic fibroblast growth factor have been characterized by a systematic approach using UV spectroscopy, size-exclusion HPLC, and reversed-phase chromatography.[817]

5.1.1.5. Oxidation

Formation of disulfide bonds from cysteine residues is an oxidation reaction. A cysteine residue in α-amylase is oxidized at pH 8.0.[818] Methionine and histidine residues are also susceptible to oxidation. Oxidation of methionine residues has been observed during storage of parathyroid hormone[819] and relaxin.[820] Degradation of freeze-dried ribonuclease A was ascribed to oxidation because molecular oxygen was involved in the degradation process.[821]

Oxidation of methionine to methionine sulfoxide in small peptides was catalyzed by Fe^{3+} and promoted by ascorbic acid. Intramolecular catalysis by a histidine residue was involved in this oxidation, and its effect was maximal when the histidine and methionine residues were separated by one residue.[822]

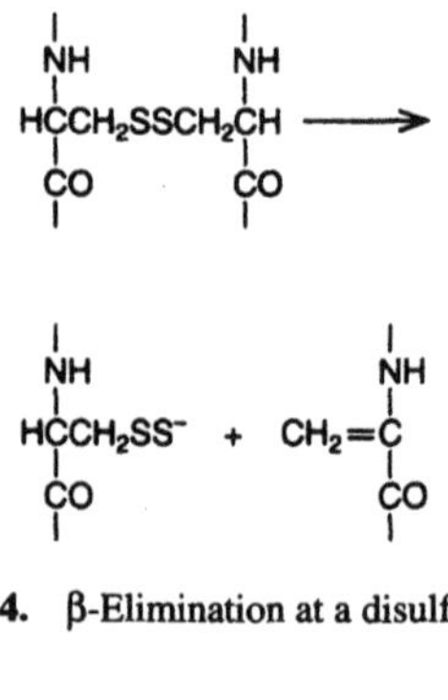

```
  |           |
  NH          NH
  |           |
 HCCH2SSCH2CH   ——>
  |           |
  CO          CO
  |           |

  |                  |
  NH                 NH
  |                  |
 HCCH2SS-  +  CH2=C
  |                  |
  CO                 CO
  |                  |
```

Scheme 84. β-Elimination at a disulfide bond.

Lyophilized porcine pancreatic elastase exhibited denaturation during storage at 40°C and 75% relative humidity in which oxidation of tryptophyl groups was probably involved.[823] Reducing sugar impurities in mannitol used as an excipient[824] induced solid-state oxidative degradation of a cyclic heptapeptide.

5.1.2. Physical Degradation

Larger peptides and proteins are susceptible to noncovalent or physical changes (so-called 'physical degradation') in addition to chemical degradation. Physical degradation includes denaturation, aggregation, adsorption, and precipitation. Denaturation, that is, an alteration of tertiary (and/or quaternary) structure, generally results in loss of bioactivity. Furthermore, exposure of hydrophobic groups upon denaturation often leads to adsorption onto surfaces, aggregation, and precipitation. Denaturation may also prompt chemical degradation pathways often not seen with the native or natural tertiary (and/or quaternary) structure. Therefore, there is considerable interest in preventing denaturation while formulating protein drugs.

Cross-linkages via disulfide bond formation cause aggregation of peptides and proteins, as described in Section 5.1.1.4. Hydrophobic bond formation, on the other hand, causes aggregation without covalent changes. Freeze-dried human growth hormone exhibited noncovalent aggregation in addition to chemical degradation via methionine oxidation and deamidation of asparagine residues.[825] Noncovalent aggregation was also observed with β-galactosidase in aqueous solution, although freeze-dried β-galactosidase, having limited moisture, exhibited aggregation via disulfide bond formation. Size-exclusion chromatograms indicated that the aggregates formed in solution were dissociated by guanidine hydrochloride, a hydrophobic-bond-breaking reagent (Fig. 201).[811]

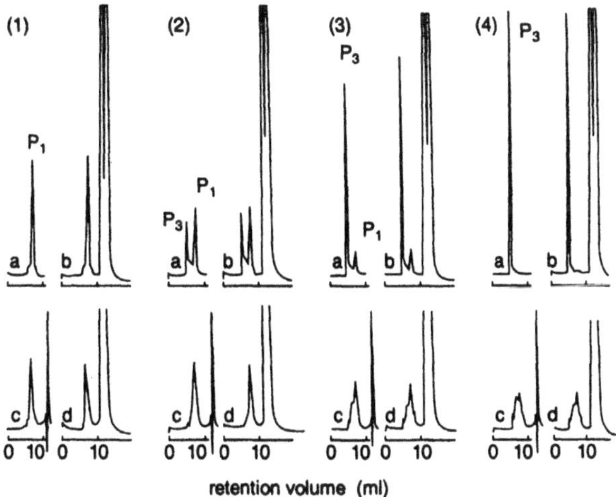

Figure 201. Size-exclusion chromatograms of aggregates formed during storage of β-galactosidase solution at 50°C. Duration of storage: (1) 0, (2) 30, (3) 240, (4) 360 min. (a) No additive; (b) in the presence of dithiothreitol; (c) in the presence of guanidine hydrochloride; (d) in the presence of dithiothreitol and guanidine hydrochloride. (Reproduced from Ref. 811 with permission.)

Noncovalent aggregation has been suggested for many other proteins, but not always confirmed. A conjugate formed between a vinca alkaloid and a monoclonal antibody exhibited aggregation in solution, the mechanism of which (covalent or noncovalent) was not clear.[826] Aggregates formed upon agitation of insulin solutions in the presence of hydrophobic surfaces (Teflon) were dissociated with urea, suggesting noncovalent aggregation.[827]

5.1.3. Degradation in Peptide and Protein Formulations

Degradation of peptides and proteins in formulations is complex because various factors may be involved in the degradation. Therefore, it is usually not easy to elucidate degradation mechanisms. Analytical methods such as electrophoresis and gel permeation chromatography are useful in assessing the stability of peptide and protein formulations and in studying degradation mechanisms.[828,829] Peptide mapping procedures were used to elucidate the mechanism of degradation of monoclonal antibody formulations containing polysorbate 80; they indicated that the major routes of degradation were deamidation, oxidation, and formation of cross-linkages.[830] Temperature-gradient gel electrophoresis is useful for investigating whether protein denaturation is reversible or irreversible.[831] Quasi-elastic light scattering is useful for determining changes in size distribution upon peptide or protein aggregation.[832]

Adsorption of peptides and proteins onto the walls of containers has been reported for various formulations.[833–835] General solutions to this problem remain to be found, although surfactants appear to be effective in reducing drug binding to surfaces.[836,837] Freeze-dried formulations of peptides and proteins may degrade via moisture adsorption from the headspace of the containers and rubber stoppers.[838]

5.2. Factors Affecting the Degradation of Peptide and Protein Drugs

Degradation of peptide and protein drugs is affected by various factors. Chemical degradation of peptide and protein drugs is often dependent on pH, buffer components and concentration, and the presence of excipients.[839–845] As with small-molecule drugs, stabilization by incorporation into microspheres has been observed with peptide and protein drugs.[846] This section describes mainly the factors affecting physical degradation such as denaturation and aggregation, which are specific to peptide and protein drugs. Although physical degradation may occur during formulation processing, including steps such as stirring, filtration, dilution, pressure loading, freezing, and drying, this section focuses on degradation during storage of peptide and protein pharmaceuticals.

5.2.1. Moisture Content and Molecular Mobility

The chemical and physical stability of peptides and proteins in the solid state is largely affected by moisture content. Freeze-dried ribonuclease A with a higher water content exhibited a greater extent of aggregation, as shown in Fig. 202.[847] The effect of water content on protein stability has also been reported for lyophilized human growth hormone[825,848] and a lyophilized monoclonal antibody–vinca conjugate[849] among others.

Moisture absorption often decreases storage stability of lyophilized proteins; however, extremely low moisture content can also decrease storage stability. The optimum residual

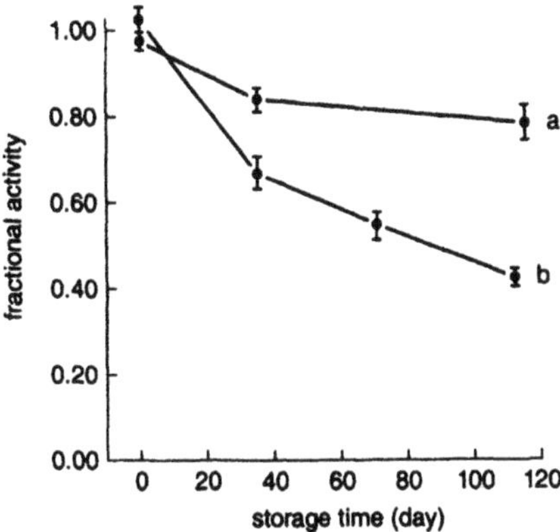

Figure 202. Effect of water content on the aggregation of freeze-dried ribonuclease A at 45°C. Moisture content: approximately 125 mol (a) and 700 mol (b) of H_2O per mole of enzyme. (Reproduced from Ref. 847 with permission.)

moisture in lyophilized tissue-type plasminogen activator was consistent with the minimum amount of moisture necessary to shield strongly polar groups in protein molecules.[850] A similar effect of water content on protein stability was reported for lyophilized billirubin oxidase.[851]

Destabilization caused by excess moisture absorption has been demonstrated with various lyophilized proteins. Storage of freeze-dried insulin at high relative humidity caused, as expected, a higher moisture absorption, which then resulted in increased aggregation, as shown in Fig. 203.[810] Aggregation of freeze-dried bovine serum albumin via disulfide bond formation increased with increasing water content (Fig. 204).[807,852] Aggregation of lyophilized β-galactosidase and bovine serum albumin in solution also increased with increasing water content.[809,853] Destabilization caused by moisture absorption can be ascribed to the plasticizing effect of water, which increases the molecular mobility of lyophiles. Plasticization of lyophilized β-galactosidase and bovine serum albumin due to moisture absorption was accompanied by an increase in the mobility of water molecules as measured by the spin–lattice relaxation time of $H_2{}^{17}O$.[809,853]

Plasticization due to moisture absorption also increases the mobility of protein molecules themselves. The molecular mobility of freeze-dried bovine serum γ-globulin, as

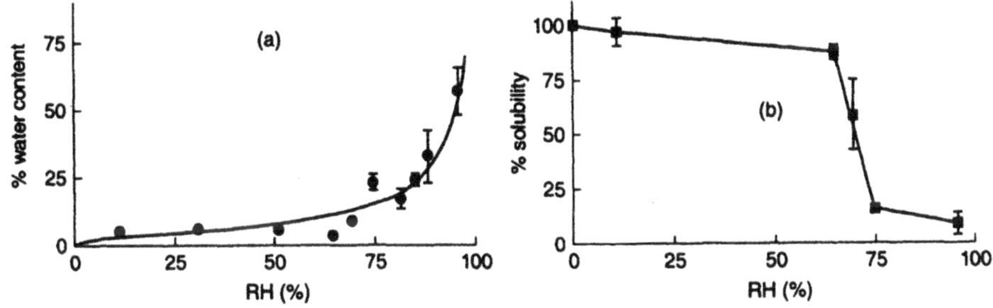

Figure 203. Relationship between moisture adsorption (a) and aggregation (b) for freeze-dried insulin stored at 50°C for 24 h. (Reproduced from Ref. 810 with permission.)

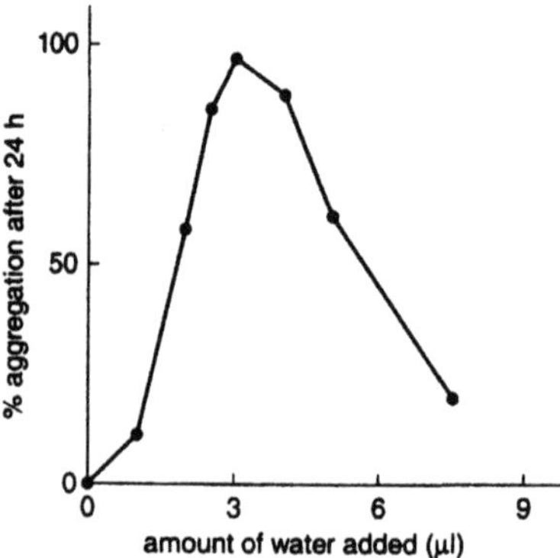

Figure 204. Effect of added water content on the aggregation of freeze-dried bovine serum albumin stored at 37°C for 24 h. [From W. R. Liu, R. Langer, and A. M. Klibanov, Moisture-induced aggregation of lyophilized proteins in the solid state, *Biotechnol. Bioeng.* **37**, 177–184 (1991). Reproduced by permission of Wiley-Liss, Inc., a subsidiary of John Wiley & Sons, Inc.]

measured by the spin–spin relaxation time of protein protons, increased with increasing water content, resulting in enhanced aggregation.[854] The critical temperature of appearance of liquidlike protons with higher mobility in lyophilized proteins (T_{mc}), detected by NMR relaxation measurements, was closely related to protein stability, such that protein aggregation became measurable on the experimental time scale at temperatures above the critical temperature.[855] A decrease in T_{mc} upon moisture absorption was also observed with lyophilized protein formulations containing polymer excipients such as dextran and polyvinyl alcohol.[856,857]

5.2.2. The Role of Excipients

Excipients used in peptide and protein formulations have major effects on denaturation and aggregation during storage. Freeze-dried proteins are most stable in viscous glassy states,[858] and their stability is increased by interactions such as hydrogen bonding between proteins and surrounding molecules.[859] Therefore, excipients that contribute to the maintenance of a glassy state or interact with the proteins can stabilize the proteins. Storage stability of alkaline phosphatase,[860] recombinant tumor necrosis factor-α,[814] a monoclonal antibody,[861] and a recombinant human interleukin-1 receptor antagoist[862] was increased by excipients such as sugars and HP-β-CD. Freeze-dried β-galactosidase was stabilized against inactivation during storage by adding excipients, which remained in an amorphous state.[863,864]

The glass transition temperature (T_g) is considered to be a critical variable in estimating the molecular mobility of amorphous materials, and the addition of excipients with a high T_g is believed to increase the storage stability of lyophilized formulations. Stabilization by excipients that have high T_g values has been demonstrated with lyophilized bovine somatotropin and lysozyme[865] and with lyophilized interleukin-2.[866] The NMR relaxation-based critical mobility temperature (T_{mc}), described above, can also be used as a measure of storage stability of lyophilized protein formulations. The T_{mc} of a lyophilized bovine serum γ-globulin formulation containing dextran increased with increasing molecular weight of the dextran, resulting in increased storage stability.[856] In contrast, addition of polyvinyl alcohol with higher mobility lowered the T_{mc} of the formulation, resulting in decreased storage stability.[857]

The stability of proteins in aqueous solutions is improved by excipients exhibiting preferential exclusion, such as sugars.[867,868] Porcine growth hormone was stabilized by HP-β-CD.[869] Denaturation during storage of urease and interleukin-2 was inhibited by nonionic surfactants such as poloxamers.[870] Inhibition of protein aggregation in solutions by various sugars and surfactants has also been reported for acidic fibroblast growth factor.[871] Stabilizing effects of polymeric additives have been reported for low-molecular-weight urokinase,[872] human IgM monoclonal antibody,[873] and human keratinocyte growth factor.[874,875]

5.3. Degradation Kinetics of Peptide and Protein Pharmaceuticals

5.3.1. Quantitative Description of Peptide and Protein Degradation

Chemical degradation of peptide and protein pharmaceuticals can also be analyzed kinetically in the same manner as for small-molecule drugs. Specifically, chemical degradation of small peptides in aqueous solutions generally conforms to simple first-order kinetics. For example, first-order kinetics have been reported for the hydrolysis in aqueous solution of secretin, which has 27 amino acid residues (Fig. 205).[795] Deamidation, hydrolysis, and epimerization of an LH-RH antagonist having 10 amino acid residues (Fig. 206),[802] deamidation of calcitonin, having 32 amino acid residues (Fig. 207),[876] and degradation of gonadorelin[877] and growth hormone-releasing hexapeptide[878] also follow first-order kinetics.

Kinetic analysis has even been attempted for the degradation of peptide and protein pharmaceuticals for which the mechanism and pathways are unknown. Apparent inactivation of α-chymotrypsin and bromelain in aqueous solutions was described by monoexponential

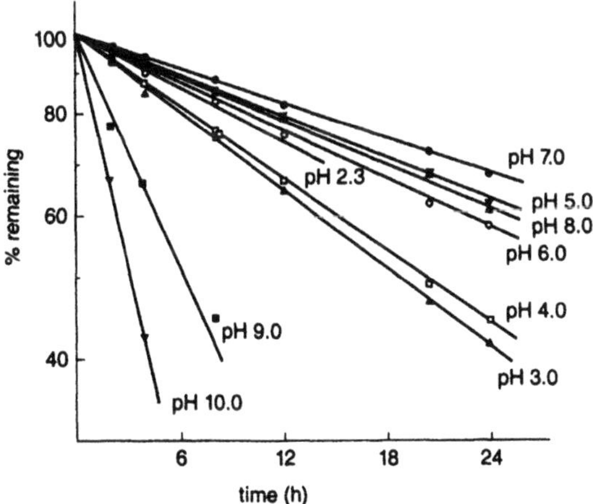

Figure 205. First-order plots of the hydrolysis of secretin in aqueous solution at varying pH values (60°C). (Reproduced from Ref. 795 with permission.)

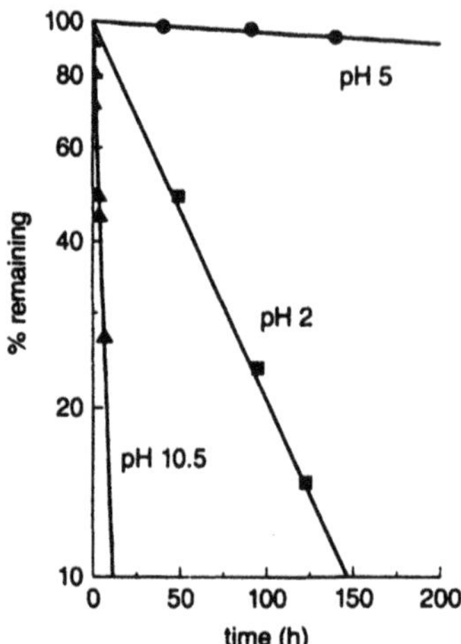

Figure 206. First-order plots of the degradation of an LH-RH antagonist in aqueous solution at varying pH values (80°C). (Reproduced from Ref. 802 with permission.)

decay, apparent first-order kinetics (Fig. 208).[879] Apparent inactivation of kallikrein was described according to a kinetic model representing a reversible and an irreversible reaction (Eq. 5.1) or an alternative model representing independent irreversible inactivation of two isoenzymes (Eq. 5.2).[879]

$$N \underset{k_2}{\overset{k_1}{\rightleftarrows}} D \xrightarrow{K_3} \tag{5.1}$$

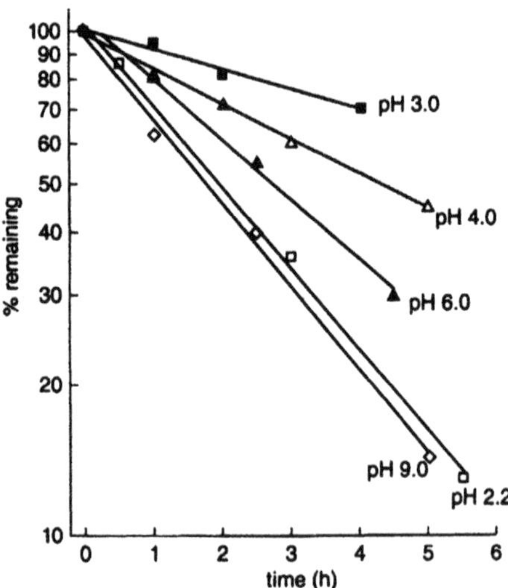

Figure 207. First-order plots of the deamidation of calcitonin in aqueous solution at varying pH values (70°C). (Reproduced from Ref. 876 with permission.)

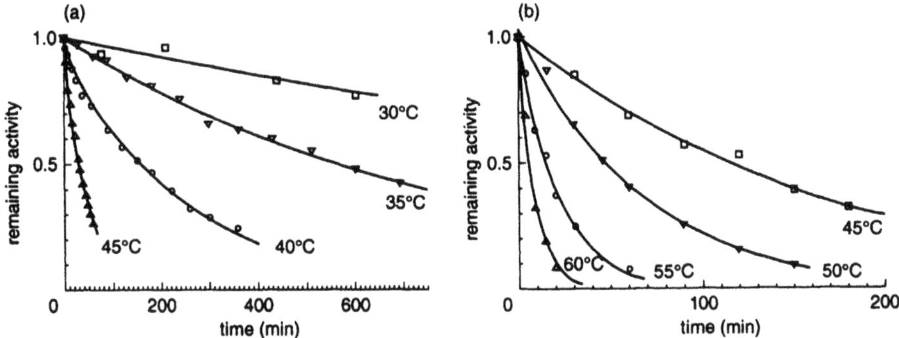

Figure 208. First-order plots of inactivation of α-chymotrypsin (a) and bromelain (b) in aqueous solution at various temperatures (pH 7.4). (Reproduced from Ref. 879 with permission.)

$$N_1 \xrightarrow{k_1} D \qquad N_2 \xrightarrow{k_2} D \tag{5.2}$$

The kinetics of solid-state degradation of peptide and protein pharmaceuticals is difficult to describe for many of the same reasons that it is difficult to describe solid-state inactivation of small molecules. Apparent inactivation of digestive enzymes such as lipase was analyzed empirically using the Weibull equation (Eq. 2.69), as shown in Fig. 209,[880,881] whereas apparent inactivation of dry horse serum cholinesterase was adequately described by a first-order equation even for inactivation in the solid state.[882]

5.3.2. Temperature Dependence of the Degradation Rate of Peptide and Protein Drugs

Stability prediction for peptide and protein drugs under accelerated testing conditions is possible if the temperature dependence of the degradation rate is determined and found to be well behaved. The temperature dependence can often be represented by the Arrhenius equation, as was seen with small-molecule drugs. Linear Arrhenius plots and the values of apparent activation energy calculated from the slopes have been reported for chemical degradation of various peptides in aqueous solutions. Values of approximately 20 kcal/mol

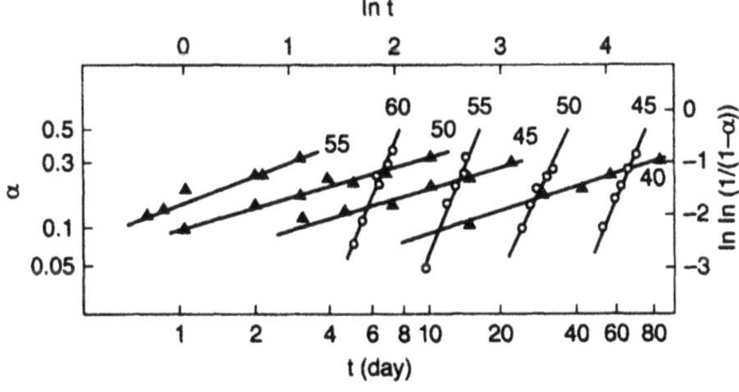

Figure 209. Weibull plots for the inactivation of two different lipases in the solid state. α: Fraction inactivated. ▲, Pancreatic lipase; ○, Rhizopus lipase. (Reproduced from Ref. 881 with permission.)

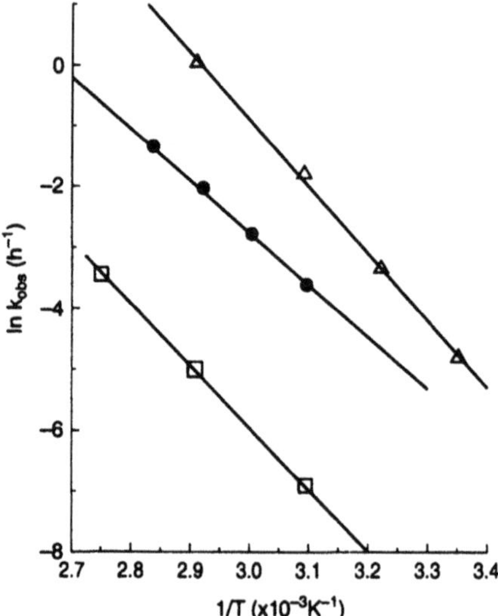

Figure 210. Arrhenius plots for chemical degradation of two Asn-hexapeptides and salmon calcitonin in aqueous solution. △, Val-Tyr-Pro-Asn-Gly-Ala, pH 7.5; □, Val-Tyr-Pro-Asn-Val-Ala, pH 7.5; ●, salmon calcitonin, pH 6.0. (Reproduced from Refs. 790, 791, and 876 with permission.)

for deamidation of a hexapeptide having an asparagine residue in a neutral region[790,791] and 96 kJ/mol (22.9 kcal/mol) for hydrolysis of gonadorelin and triptorelin,[883] each having 10 amino acid residues, were reported. Other values include 24 kcal/mol for the hydrolysis of LH-RH[801]; 20–33 kcal/mol for alkaline racemization of casein[799]; 15 kcal/mol for deamidation of calcitonin at pH 6[876]; and 21 kcal/mol for hydrolysis of cholecystokinin-B receptor antagonist in the acidic pH region.[884] Figure 210 shows some typical Arrhenius plots for

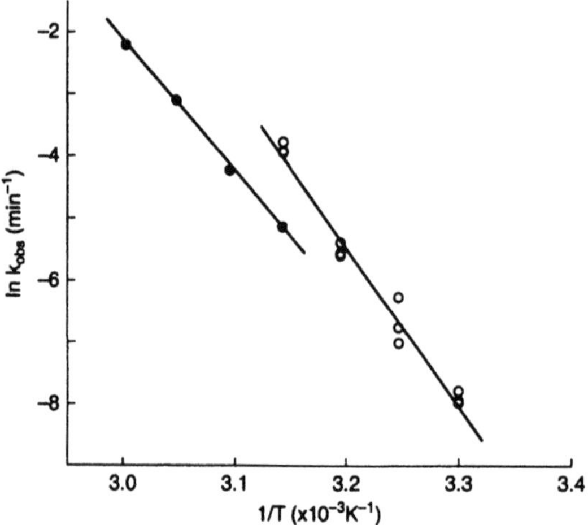

Figure 211. Arrhenius plots for the inactivation of α-chymotrypsin (○) and bromelain (●) at pH 7.4. (Reproduced from Ref. 879 with permission.)

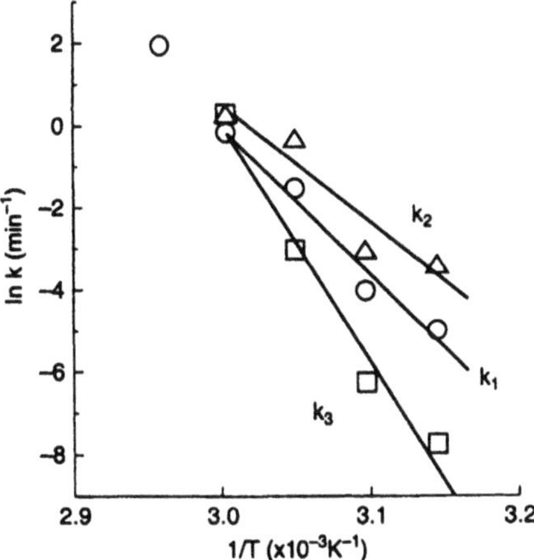

Figure 212. Arrhenius plots of k_1, k_2, and k_3, for kallikrein inactivation in aqueous solution at pH 7.4. Rate constants k_1, k_2, and k_3 were calculated according to the reaction scheme in Eq. (5.1). (Reproduced from Ref. 879 with permission.)

three peptides. When Arrhenius behavior is adhered to, accelerated-temperature testing of solutions or formulations is possible, whereas accelerated testing cannot be applied even to chemical degradations when the mechanism varies with temperature. This is the case for degradation of interleukin-1β, which exhibits deamidation as its primary degradation pathway below 30°C and oxidation at higher temperatures.[885]

Even when mechanisms and pathways are unknown, it is sometimes possible to use Arrhenius or empirical equations to determine apparent rate constants at other temperatures. Apparent first-order rate constants for α-chymotrypsin and bromelain inactivation exhibited linear Arrhenius plots (Fig. 211).[879] Apparent rate constants for kallikrein inactivation calculated according to the reaction scheme in Eq. (5.1) resulted in the temperature dependence seen in Fig. 212.[850] For inactivation of various peptide and protein pharmaceuticals

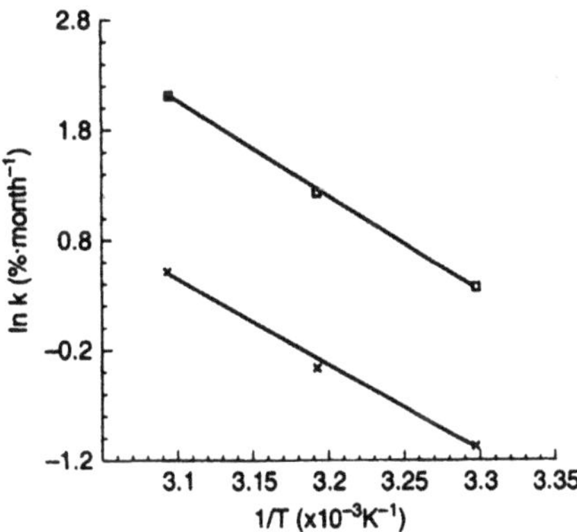

Figure 213. Arrhenius plots of the apparent zero-order rate constant for urokinase inactivation in the presence (□) and absence (×) of human serum albumin. (Reproduced from Ref. 838 with permission.)

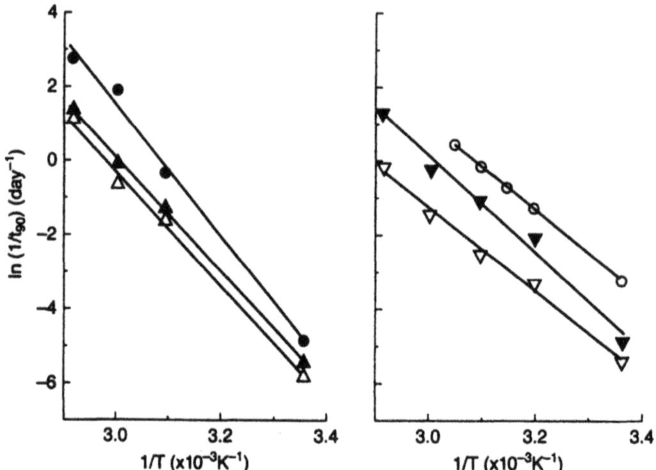

Figure 214. Arrhenius plots for the inactivation of marketed solid digestive enzyme formulations. ▲, α-Chymotrypsin troche; △, α-chymotrypsin tablet; ●, β-galactosidase powder; ▽, ▼ bromelain tablets; ○, kallikrein capsule. (Reproduced from Ref. 888 with permission.)

including interferon,[886] accelerated testing appears to be possible. Activation energies have been reported for inactivation of digestive enzymes (23–58 kcal/mol)[880,881] and horse serum cholinesterase (25 kcal/mol).[882]

Apparent activation energy values have also been reported for solid-state degradation of peptide and protein pharmaceuticals. The reported values include 1–2 kcal/mol for human interferon-β formulated with human serum albumin[887]; 15 kcal/mol for freeze-dried urokinase (Fig. 213)[838]; and 11–18 kcal/mol for α-chymotrypsin and bromelain tablets, kallikrein capsules, and β-galactosidase powder (Fig. 214).[888]

The kinetics of denaturation of peptide and protein drugs have not been extensively treated. Some studies on the temperature dependence of denaturation rates have been reported. A kinetic study of the denaturation of G actin in solution using DSC yielded linear Arrhenius plots, from which values for the activation energy and the frequency factor of 231 kJ/mol (55.2 kcal/mol) and 76.8 s^{-1}, respectively, were obtained (Fig. 215).[889] Linear

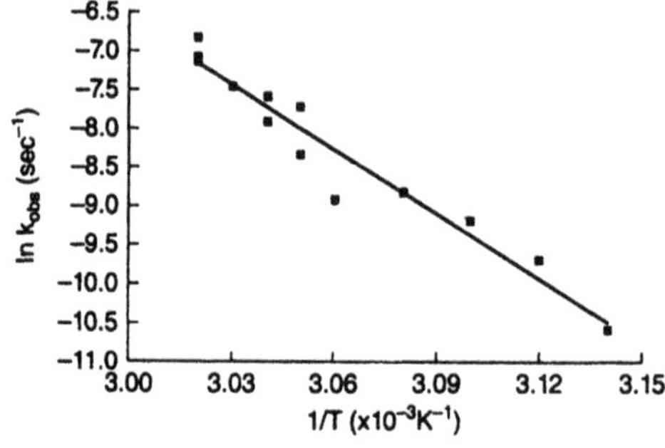

Figure 215. Arrhenius plot for the denaturation of G actin in aqueous solution at pH 8.0. (Reproduced from Ref. 889 with permission.)

Arrhenius plots were also observed for the denaturation of β-galactosidase in solution.[890] These linear Arrhenius plots suggest that prediction of denaturation rate by accelerated testing is possible if the denaturation mechanism does not change in the temperature range in question. Because peptide and protein drugs generally denature via complicated mechanisms (thermal denaturation, denaturation at interfaces, etc.), denaturation at lower temperature may occur via mechanisms different from those at higher temperature. This makes it difficult to predict the stability of peptide and protein drugs by accelerated testing. In cases where only thermal denaturation occurs, however, prediction of denaturation rate by accelerated testing may be possible.

Chapter 6

Regulations

The efficacy and safety of pharmaceuticals cannot be ensured unless the quality of the pharmaceuticals is maintained during their specified shelf lives. New drug applications need to submit scientific data that guarantee the stability of the product over a specified time period when maintained under specific storage conditions. The International Conference on Harmonisation of Technical Requirements for Registration of Pharmaceuticals for Human Use (ICH) was organized in order to harmonize stability testing requirements for new drug applications within the European Union (EU), the United States, and Japan. ICH Guidelines for Stability Testing of New Drug Substances and Products and for Photostability Testing of New Drug Substances and Products were officially adopted in October 1993 and November 1996, respectively. In this chapter, the ICH harmonized guidelines are introduced, and the major concerns raised by the EU, the United States, and Japan are briefly discussed.

6.1. ICH Harmonised Tripartite Guideline for Stability Testing of New Drug Substances and Products

Preamble

The following guideline sets out the stability testing requirement for a Registration Application within the three areas of the EC, Japan and the USA. It does not seek necessarily to cover the testing that may be required for registration in or export to other areas of the world.

The guideline seeks to exemplify the core stability data package required for new drug substances and products. It is not always necessary to follow this when there are scientifically justifiable reasons for using alternative approaches.

The guideline provides a general indication on the requirements for stability testing, but leaves sufficient flexibility to encompass the variety of different practical situations required for specific scientific situations and characteristics of the materials being evaluated.

The principle that information on stability generated in any one of the three areas of the EC, Japan and the USA would be mutually acceptable in both of the other two areas has been established, provided it meets the appropriate requirements of this guideline and the labelling is in accord with national/regional requirements.

Details of the specific requirements for sampling, test requirements for particular dosage forms/packaging etc., are not covered in this guideline.

Objective

The purpose of stability testing is to provide evidence on how the quality of a drug substance or drug product varies with time under the influence of a variety of environmental factors such as temperature, humidity and light, and enables recommended storage conditions, re-test periods and shelf lives to be established.

Scope

The guideline primarily addresses the information required in Registration Applications for new molecular entities and associated drug products.

This guideline does not currently seek to cover the information required for abbreviated or abridged applications, variations, clinical trial applications, etc.

The choice of test conditions defined in this guideline is based on an analysis of the effects of climatic conditions in the three areas of the EC, Japan and the USA. The mean kinetic temperature in any region of the world can be derived from climatic data (Grimm, W. *Drugs Made in Germany* **28**, 196–202, 1985 and 29, 39–47, 1986).

Drug Substance

General

Information on the stability of the drug substance is an integral part of the systematic approach to stability evaluation.

Stress Testing

Stress testing helps to determine the intrinsic stability of the molecule by establishing degradation pathways in order to identify the likely degradation products and to validate the stability indicating power of the analytical procedures used.

Formal Studies

Primary stability studies are intended to show that the drug substance will remain within specification during the re-test period if stored under recommended storage conditions.

Selection of Batches

Stability information from accelerated and long term testing is to be provided on at least three batches. The long term testing should cover a minimum of 12 months duration on at least three batches at the time of submission.

The batches manufactured to a minimum of Pilot plant scale should be by the same synthetic route and use a method of manufacture and procedure that simulates the final process to be used on a manufacturing scale.

The overall quality of the batches of drug substance placed on stability should be representative of both the quality of the material used in pre-clinical and clinical studies and the quality of material to be made on a manufacturing scale.

Supporting information may be provided using stability data on batches of drug substance made on a laboratory scale.

The first three production batches of drug substance manufactured post approval, if not submitted in the original Registration Application, should be placed on long term stability studies using the same stability protocol as in the approved drug application.

Test Procedures and Test Criteria

The testing should cover those features susceptible to change during storage and likely to influence quality, safety and/or efficacy. Stability information should cover as necessary the physical, chemical and microbiological test characteristics. Validated stability-indicating testing methods must be applied. The need for the extent of replication will depend on the results of validation studies.

Specification

Limits of acceptability should be derived from the profile of the material as used in the pre-clinical and clinical batches. It will need to include individual and total upper limits for impurities and degradation products, the justification for which should be influenced by the levels observed in material used in preclinical studies and clinical trials.

Storage Conditions

The length of the studies and the storage conditions should be sufficient to cover storage, shipment and subsequent use. Application of the same storage conditions as applied to the drug product will facilitate comparative review and assessment. Other storage conditions are allowable if justified. In particular, temperature sensitive drug substances should be stored under an alternative, lower temperature condition which will then become the designated long term testing storage temperature. The six months accelerated testing should then be carried out at a temperature at least 15°C above this designated long term storage temperature (together with the appropriate relative humidity conditions for that temperature). The designated long term testing conditions will be reflected in the labelling and re-test date.

	Conditions	Minimum Time Period at Submission
Long term testing	25°C ± 2°C/60% RH ± 5%	12 months
Accelerated testing	40°C ± 2°C/75% RH ± 5%	6 months

Where significant change occurs during six months storage under conditions of accelerated testing at 40°C ± 2°C/75 percent RH ± 5 percent, additional testing at an intermediate condition (such as 30°C ± 2°C/60 percent RH ± 5 percent) should be conducted for drug substances to be used in the manufacture of dosage forms tested long term at 25°C/60 percent

RH and this information included in the Registration Application. The initial Registration Application should include a minimum of 6 months data from a 12 months study.

'Significant change' at 40°C/75 percent RH or 30°C/60 percent RH is defined as failure to meet the specification.

The long term testing will be continued for a sufficient period of time beyond 12 months to cover all appropriate re-test periods, and the further accumulated data can be submitted to the Authorities during the assessment period of the Registration Application.

The data (from accelerated testing or from testing at an intermediate condition) may be used to evaluate the impact of short term excursions outside the label storage conditions such as might occur during shipping.

Testing Frequency

Frequency of testing should be sufficient to establish the stability characteristics of the drug substance. Testing under the defined long term conditions will normally be every three months, over the first year, every six months over the second year and then annually.

Packaging/Containers

The containers to be used in the long term, real time stability evaluation should be the same as or simulate the actual packaging used for storage and distribution.

Evaluation

The design of the stability study is to establish, based on testing a minimum of three batches of the drug substance and evaluating the stability information (covering as necessary the physical, chemical and microbiological test characteristics), a re-test period applicable to all future batches of the bulk drug substance manufactured under similar circumstances. The degree of variability of individual batches affects the confidence that a future production batch will remain within specification until the re-test date.

An acceptable approach for quantitative characteristics that are expected to decrease with time is to determine the time at which the 95% one-sided confidence limit for the mean degradation curve intersects the acceptable lower specification limit. If analysis shows that the batch to batch variability is small, it is advantageous to combine the data into one overall estimate and this can be done by first applying appropriate statistical tests (for example, p values for level of significance of rejection of more than 0.25) to the slopes of the regression lines and zero time intercepts for the individual batches. If it is inappropriate to combine data from several batches, the overall re-test period may depend on the minimum time a batch may be expected to remain within acceptable and justified limits.

The nature of any degradation relationship will determine the need for transformation of the data for linear regression analysis. Usually the relationship can be represented by a linear, quadratic or cubic function on an arithmetic or logarithmic scale. Statistical methods should be employed to test the goodness of fit of the data on all batches and combined batches (where appropriate) to the assumed degradation line or curve.

The data may show so little degradation and so little variability that it is apparent from looking at the data that the requested re-test period will be granted. Under the circumstances, it is normally unnecessary to go through the formal statistical analysis but merely to provide a full justification for the omission.

Limited extrapolation of the real time data beyond the observed range to extend expiration dating at approval time, particularly where the accelerated data supports this, may be undertaken. However, this assumes that the same degradation relationship will continue to apply beyond the observed data and hence the use of extrapolation must be justified in each application in terms of what is known about the mechanism of degradation, the goodness of fit of any mathematical model, batch size, existence of supportive data etc.

Any evaluation should cover not only the assay, but the levels of degradation products and other appropriate attributes.

Statements/Labelling

A storage temperature range may be used in accordance with relevant national/regional requirements. The range should be based on the stability evaluation of the drug substance. Where applicable, specific requirements should be stated, particularly for drug substances that cannot tolerate freezing. The use of terms such as 'ambient conditions' or 'room temperature' is unacceptable.

A re-test period should be derived from the stability information.

Drug Product

General

The design of the stability programme for the finished product should be based on the knowledge of the behavior and properties of the drug substance and the experience gained from clinical formulation studies and from the stability studies on the drug substance. The likely changes on storage and the rationale for the selection of product variables to include in the testing programme should be stated.

Selection of Batches

Stability information from accelerated and long term testing is to be provided on three batches of the same formulation and dosage form in the containers and closure proposed for marketing. Two of the three batches should be at least pilot scale. The third batch may be smaller (e.g., 25,000 to 50,000 tablets or capsules for solid oral dosage forms). The long term testing should cover at least 12 months duration at the time of submission. The manufacturing process to be used should meaningfully simulate that which would be applied to large scale batches for marketing. The process should provide product of the same quality intended for marketing, and meeting the same quality specification as to be applied for release of material. Where possible, batches of the finished product should be manufactured using identifiably different batches of drug substance.

Data on laboratory scale batches is not acceptable as primary stability information. Data on associated formulations or packaging may be submitted as supportive information. The first three production batches manufactured post approval, if not submitted in the original Registration Application, should be placed on accelerated and long term stability studies using the same stability protocols as in the approved drug application.

Test Procedures and Test Criteria

The testing should cover those features susceptible to change during storage and likely to influence quality, safety and/or efficacy. Analytical test procedures should be fully validated and the assays should be stability-indicating. The need for the extent of replication will depend on the results of validation studies.

The range of testing should cover not only chemical and biological stability but also loss of preservative, physical properties and characteristics, organoleptic properties and, where required, microbiological attributes. Preservative efficacy testing and assays on stored samples should be carried out to determine the content and efficacy of antimicrobial preservatives.

Specifications

Limits of acceptance should relate to the release limits (where applicable), to be derived from consideration of all the available stability information. The shelf life specification could allow acceptable and justifiable derivations from the release specification based on the stability evaluation and the changes observed on storage. It will need to include specific upper limits for degradation products, the justification for which should be influenced by the levels observed in material used in pre-clinical studies and clinical trials. The justification for the limits proposed for certain other tests such as particle size and/or dissolution rate will require reference to the results observed for batch(es) used in bioavailability and/or clinical studies. Any differences between the release and shelf life specifications for antimicrobial preservatives should be supported by preservative efficacy testing.

Storage Test Conditions

The length of the studies and the storage conditions should be sufficient to cover storage, shipment and subsequent use (e.g., reconstitution or dilution as recommended in the labelling).

See table below for accelerated and long term storage conditions and minimum times. An assurance that long term testing will continue to cover the expected shelf life should be provided.

Other storage conditions are allowable if justified. Heat sensitive drug products should be stored under an alternative lower temperature condition which will eventually become the designated long term storage temperature. Special consideration may need to be given to products which change physically or even chemically at lower storage conditions, e.g., suspensions or emulsions which may sediment or cream, oils and semi-solid preparations which may show an increased viscosity. Where a lower temperature condition is used, the six months accelerated testing should be carried out at a temperature at least 15°C above its designated long term storage temperature (together with appropriate relative humidity conditions for that temperature). For example, for a product to be stored long term under refrigerated conditions, accelerated testing should be conducted at 25°C ± 2°C/60 percent RH ± 5 percent RH. The designated long term testing conditions will be reflected in the labelling and expiration date.

Storage under conditions of high relative humidities applies particularly to solid dosage forms. For products such as solutions, suspensions etc., contained in packs designed to provide a permanent barrier to water loss, specific storage under conditions of high relative humidity is not necessary but the same range of temperatures should be applied. Low relative humidity (e.g., 10–20 percent RH) can adversely affect products packed in semi-permeable containers (e.g., solutions in plastic bags, nose drops in small plastic containers etc.) and consideration should be given to appropriate testing under such conditions.

	Conditions	Minimum Time Period at Submission
Long term testing	25°C ± 2°C/60% RH ± 5%	12 months
Accelerated testing	40°C ± 2°C/75% RH ± 5%	6 months

Where 'significant change' occurs due to accelerated testing, additional testing at an intermediate condition e.g., 30°C ± 2°C/60 percent ± 5 percent RH should be conducted. 'Significant change' at the accelerated condition is defined as:

1. A 5 percent potency loss from the initial assay value of a batch;
2. Any specified degradant exceeding its specification limit;
3. The product exceeding its pH limits;
4. Dissolution exceeding the specification limits for 12 capsules or tablets.
5. Failure to meet specifications for appearance and physical properties e.g., color, phase separation, resuspendibility, delivery per actuation, caking, hardness, etc.

Should significant change occur at 40°C/75 percent RH then the initial Registration Application should include a minimum of 6 months data from an ongoing one year study at 30°C/60 percent RH; the same significant change criteria shall then apply.

The long term testing will be continued for a sufficient time beyond 12 months to cover shelf life at appropriate test periods. The further accumulated data should be submitted to the authorities during the assessment period of the Registration Application.

The first three production batches manufactured post approval, if not submitted in the original Registration Application, should be placed on accelerated and long term stability studies using the same stability protocol as in the approved drug application.

The first three production batches manufactured post approval, if not submitted in the original Registration Application, should be placed on accelerated and long term stability studies using the same stability protocol as in the approved drug application.

Testing Frequency

Frequency of testing should be sufficient to establish the stability characteristics of the drug product. Testing will normally be every three months over the first year, every six months over the second year and then annually.

The use of matrixing or bracketing can be applied, if justified. *(See Glossary.)*

Packaging Materials

The testing should be carried out in the final packaging proposed for marketing. Additional testing of unprotected drug product can form a useful part of the stress testing and pack evaluation, as can studies carried out in other related packaging materials in supporting the definitive pack(s).

Evaluation

A systematic approach should be adopted in the presentation and evaluation of the stability information which should cover as necessary physical, chemical, biological, microbiological quality characteristics, including particular properties of the dosage form (for example, dissolution rate for oral solid dose forms).

The design of the stability study is to establish, based on testing a minimum of three batches of the drug product, a shelf-life and label storage instructions applicable to all future batches of the dosage form manufactured and packed under similar circumstances. The degree of variability of individual batches affects the confidence that a future production batch will remain within specification until the expiration date.

An acceptable approach for quantitative characteristics that are expected to decrease with time is to determine the time at which the 95% one-sided confidence limit for the mean degradation curve intersects the acceptable lower specification limit. If analysis shows that the batch to batch variability is small, it is advantageous to combine the data into one overall estimate and this can be done by first applying appropriate statistical tests (for example, p values for level of significance of rejection of more than 0.25) to the slopes of the regression lines and zero time intercepts for the individual batches. If it is inappropriate to combine data from several batches, the overall shelf-life may depend on the minimum time a batch may be expected to remain within acceptable and justified limits.

The nature of the degradation relationship will determine the need for transformation of the data for linear regression analysis. Usually the relationship can be represented by a linear, quadratic or cubic function on an arithmetic or logarithmic scale. Statistical methods should be employed to test the goodness of fit on all batches and combined batches (where appropriate) to the assumed degradation line or curve.

Where the data shows so little degradation and so little variability that it is apparent from looking at the data that the requested shelf-life will be granted, it is normally unnecessary to go through the formal statistical analysis but only to provide a justification for the omission.

Limited extrapolation of the real time data beyond the observed range to extend expiration dating at approval time, particularly where the accelerated data supports this, may be undertaken. However, this assumes that the same degradation relationship will continue to apply beyond the observed data and hence the use of extrapolation must be justified in each application in terms of what is known about the mechanisms of degradation, the goodness of fit of any mathematical model, batch size, existence of supportive data, etc.

Any evaluation should consider not only the assay but the levels of degradation products and appropriate attributes. Where appropriate, attention should be paid to reviewing the adequacy of the mass balance, different stability and degradation performance.

The stability of the drug products after reconstituting or diluting according to labelling should be addressed to provide appropriate and supportive information.

Statements/Labelling

A storage temperature range may be used in accordance with relevant national/regional requirements. The range should be based on the stability evaluation of the drug product. Where applicable, specific requirements should be stated particularly for drug products that cannot tolerate freezing.

The use of terms such as 'ambient conditions' or 'room temperature' is unacceptable.

There should be a direct linkage between the label statement and the demonstrated stability characteristics of the drug product.

Annex I. Glossary and Information

The following terms have been in general use and the following definitions are **provided** to facilitate interpretation of the guideline.

Accelerated Testing

Studies designed to increase the rate of chemical degradation or physical change of an active drug substance or drug product by using exaggerated storage conditions as part of the formal, definitive, storage programme.

These data, in addition to long term stability studies, may also be used to assets longer term chemical effects at non-accelerated conditions and to evaluate the impact of short term excursions outside the label storage conditions such as might occur during shipping. Results from accelerated testing studies are not always predictive of physical changes.

Active Substance; Active Ingredient; Drug Substance; Medicinal Substance

The unformulated drug substance which may be subsequently formulated with excipients to produce the drug product.

The design of a stability schedule so that at any time point only the samples on the extremes, for example of container size and/or dosage strengths, are tested. The design assumes that the stability of the intermediate condition samples are represented by those at the extremes.

Where a range of dosage strengths is to be tested, bracketing designs may be particularly applicable if the strengths are very closely related in composition (e.g., for a tablet range made with different compression weights of a similar basic granulation, or a capsule range made by filling different plug fill weights of the same basic composition into different size capsule shells).

Where a range of sizes of immediate containers are to be evaluated, bracketing designs may be applicable if the material of composition of the container and the type of closure are the same throughout the range.

Climatic Zones

The concept of dividing the world into four zones based on defining the prevalent annual climatic conditions.

Dosage Form; Preparation

A pharmaceutical product type, for example tablet, capsule, solution, cream etc. that contains a drug ingredient generally, but not necessarily, in association with excipients.

Drug; Finished Product

The dosage form in the final immediate packaging intended for marketing.

Excipient

Anything other than the drug substance in the dosage form.

Expiry/Expiration Date

The date placed on the container/labels of a drug product designating the time during which a batch of the product is expected to remain within the approved shelf-life specification if stored under defined conditions, and after which it must not be used.

Formal (Systematic) Studies

Formal studies are those undertaken to a pre-approval stability protocol that embraces the principles of these guidelines.

Long Term (Real Time) Testing

Stability evaluation of the physical, chemical, biological and microbiological characteristics of a drug product and a drug substance, covering the expected duration of the shelf life and re-test period, which are claimed in the submission and will appear on the labelling.

Mass Balance; Material Balance

The process of adding together the assay value and levels of degradation products to see how closely these add up to 100 per cent of the initial value, with due consideration of the margin of analytical precision.

This concept is a useful scientific guide for evaluating data, but it is not achievable in all circumstances. The focus may instead be on assuring the specificity of the assay, the completeness of the investigation of routes of degradation, and the use, if necessary, of identified degradants as indicators of the extent of degradation via particular mechanisms.

Matrixing

The statistical design of a stability schedule so that only a fraction of the total number of samples are tested at any specified sampling point. At a subsequent sampling point, different sets of samples of the total number would be tested. The design assumes that the stability of the samples tested represents the stability of all samples. The differences in the samples for the same drug product should be identified as, for example, covering different batches, different strengths, different sizes of the same container and closure and possibly in some cases different container/closure systems.

Matrixing can cover reduced testing when more than one variable is being evaluated. Thus the design of the matrix will be dictated by the factors needing to be covered and evaluated. This potential complexity precludes inclusion of specific details and examples, and it may be desirable to discuss design in advance with the Regulatory Authority, where this is possible. In every case it is essential that all batches are tested initially and at the end of the long term testing.

Mean Kinetic Temperature

When establishing the mean value of the temperature, the formula of J. D. Haynes (*J. Pharm. Sci.* **60**, 927–929, 1971) can be used to calculate the mean kinetic temperature. It is higher than the arithmetic mean temperature and takes into account the Arrhenius equation from which Haynes derived his formula.

New Molecular Entity; New Active Substance

A substance that has not previously been registered as a new drug substance with the national or regional authority concerned.

Pilot Plant Scale

The manufacture of either a drug substance or drug product by a procedure fully representative of and simulating that to be applied on a full manufacturing scale.

For oral solid dosage forms this is generally taken to be at a minimum scale of one-tenth that of full production or 100,000 tablets or capsules, whichever is the larger.

Primary Stability Data

Data on the drug substance stored in the proposed packaging under storage conditions that support the proposed re-test date.

Data on the drug product stored in the proposed container-closure for marketing under storage conditions that support the proposed shelf life.

Re-Test Date

The date when samples of the drug substance should be re-examined to ensure that material is still suitable for use.

Re-Test Period

The period of time during which the drug substance can be considered to remain within the specifications and is therefore acceptable for use in the manufacture of a given drug product, provided that it has been stored under the defined conditions; after this period, the batch should be re-tested for compliance with specifications and then used immediately.

Shelf-Life; Expiration Dating Period

The time interval that a drug product is expected to remain within the approved shelf-life specification provided that it is stored under the conditions defined on the label in the proposed containers and closure.

Specification—Release

The combination of physical, chemical, biological and microbiological test requirements that determine that a drug product is suitable for release at the time of its manufacture.

Specification—Check/Shelf-Life

The combination of physical, chemical, biological and microbiological test requirements that a drug substance must meet up to its re-test date or a drug product must meet throughout its shelf life.

Storage Conditions Tolerances

The acceptable variation in temperature and relative humidity of storage facilities.

The equipment must be capable of controlling temperature to a range of ±2°C and relative humidity to ±5%. The actual temperatures and humidities should be monitored during stability storage. Short term spikes due to opening of doors of the storage facility are accepted as unavoidable. The effect of excursions due to equipment failure should be addressed by the applicant and reported if judged to impact stability results. Excursions that exceed these ranges (i.e., ±2°C and/or ±5 percent RH) for more than 24 hours should be described in the study report and their impact assessed.

Stress Testing (Drug Substance)

These studies are undertaken to elucidate intrinsic stability characteristics. Such testing is part of the development strategy and is normally carried out under more severe conditions than those used for accelerated tests.

Stress testing is conducted to provide data on forced decomposition products and decomposition mechanisms for the drug substance. The severe conditions that may be encountered during distribution can be covered by stress testing of definitive batches of drug substance.

These studies should establish the inherent stability characteristics of the molecule, such as the degradation pathways, and lead to identification of degradation products and hence support the suitability of the proposed analytical procedures. The detailed nature of the studies will depend on the individual drug substance and type of drug product.

This testing is likely to be carried out on a single batch of material and to include the effect of temperatures in 10°C increments above the accelerated temperature test condition (e.g., 50°C, 60°C, etc.); humidity where appropriate (e.g., 75 per cent or greater); oxidation and photolysis on the drug substance plus its susceptibility to hydrolysis across a wide range of pH values when in solution or suspension.

Results from these studies will form an integral part of the information provided to regulatory authorities.

Light testing should be an integral part of stress testing. [The standard conditions for light testing are still under discussion and will be considered in a further ICH document.]

It is recognized that some degradation pathways can be complex and that under forcing conditions decomposition products may be observed which are unlikely to be formed under accelerated or long term testing. This information may be useful in developing and validating suitable analytical methods, but it may not always be necessary to examine specifically for all degradation products, if it has been demonstrated that in practice these are not formed.

Stress Testing (Drug Product)

Light testing should be an integral part of stress testing. Special test conditions for specific products (e.g., metered dose inhalations and creams and emulsions) may require additional stress studies.

Supporting Stability Data

Data other than primary stability data, such as stability data on early synthetic route batches of drug substance, small scale batches of materials, investigational formulations not proposed for marketing, related formulations, product presented in containers and/or closures other than those proposed for marketing, information regarding test results on containers, and other scientific rationale that support the analytical procedures, the proposed re-test period or shelf life and storage conditions.

6.2 ICH Harmonised Tripartite Guideline for Photostability Testing of New Drug Substances and Products

1. General

The ICH Harmonized Tripartite Guideline covering the Stability Testing of New Drug Substances and Products (hereafter referred to as the Parent Guideline) notes that light testing should be an integral part of stress testing. This document is an annex to the Parent Guideline and addresses the recommendations for photostability testing.

A. *Preamble*

The intrinsic photostability characteristics of new drug substances and products should be evaluated to demonstrate that, as appropriate, light exposure does not result in unacceptable change. Normally, photostability testing is carried out on a single batch of material selected as described under Selection of Batches in the Parent Guideline. Under some circumstances these studies should be repeated if certain variations and changes are made to the product (e.g., formulation, packaging). Whether these studies should be repeated depends on the photostability characteristics determined at the time of initial filing and the type of variation and/or change made.

The guideline primarily addresses the generation of photostability information for submission in Registration Applications for new molecular entities and associated drug products. The guideline does not cover the photostability of drugs after administration (i.e., under conditions of use) and those applications not covered by the Parent Guideline. Alternative approaches may be used if they are scientifically sound and justification is provided.

A systematic approach to photostability testing is recommended covering, as appropriate, studies such as:

(i) Tests on the drug substance;
(ii) Tests on the exposed drug product outside of the immediate pack; and if necessary;
(iii) Tests on the drug product in the immediate pack; and, if necessary;
(iv) Tests on the drug product in the marketing pack.

The extent of drug product testing should be established by assessing whether or not acceptable change has occurred at the end of the light exposure testing as described in the Decision Flow Chart for Photostability Testing of Drug Products (Fig. 216). Acceptable change is change within limits justified by the applicant.

The formal labeling requirements for photolabile drug substances and drug products are established by national/regional requirements.

B. Light Sources

The light sources described below may be used for photostability testing. The applicant should either maintain an appropriate control of temperature to minimize the effect of localized temperature changes or include a dark control in the same environment unless otherwise justified. For both options 1 and 2, a pharmaceutical manufacturer/applicant may rely on the spectral distribution specification of the light source manufacturer.

Option 1

Any light source that is designed to produce an output similar to the D65/ID65 emission standard such as an artificial daylight fluorescent lamp combining visible and ultraviolet (UV) outputs, xenon, or metal halide lamp. D65 is the internationally recognized standard for outdoor daylight as defined in ISO 10977 (1993). ID65 is the equivalent indoor indirect daylight standard. For a light source emitting significant radiation below 320 nm, an appropriate filter(s) may be fitted to eliminate such radiation.

Option 2

For option 2 the same sample should be exposed to both the cool white fluorescent and near ultraviolet lamp.

1. A cool white fluorescent lamp designed to produce an output similar to that specified in ISO 10977 (1993); and
2. A near UV fluorescent lamp having a spectral distribution from 320 nm to 400 nm with a maximum energy emission between 350 nm and 370 nm; a significant proportion of UV should be in both bands of 320 to 360 nm and 360 to 400 nm.

C. Procedure

For confirmatory studies, samples should be exposed to light providing an overall illumination of not less than 1.2 million lux hours and an integrated near ultraviolet energy of not less than 200 watt hours/square meter to allow direct comparisons to be made between the drug substance and drug product.

Samples may be exposed side-by-side with a validated chemical actinometric system to ensure the specified light exposure is obtained, or for the appropriate duration of time when conditions have been monitored using calibrated radiometers/lux meters. An example of an actinometric procedure is provided in the Annex.

If protected samples (e.g., wrapped in aluminum foil) are used as dark controls to evaluate the contribution of thermally induced change to the total observed change, these should be placed alongside the authentic sample.

2. Drug Substance

For drug substances, photostability testing should consist of two parts: forced degradation testing and confirmatory testing.

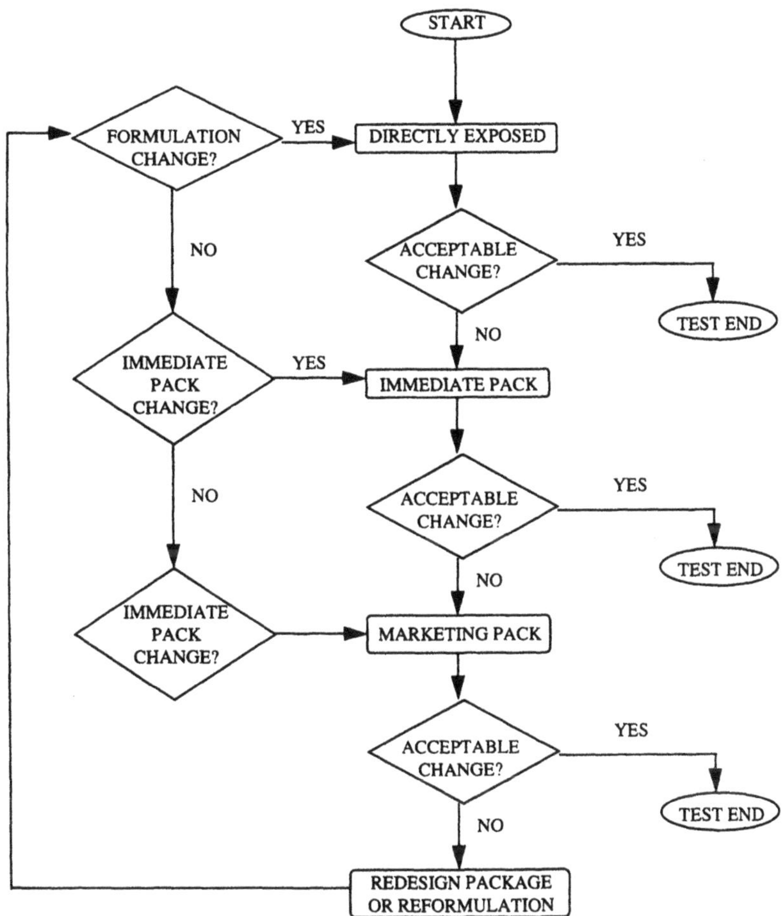

The purpose of forced degradation testing studies is to evaluate the overall photosensitivity of the material for method development purposes and/or degradation pathway elucidation. This testing may involve the drug substance alone and/or in simple solutions/suspensions to validate the analytical procedures. In these studies, the samples should be in chemically inert and transparent containers. In these forced degradation studies, a variety of exposure conditions may be used, depending on the photosensitivity of the drug substance involved and the intensity of the light sources used. For development and validation purposes it is appropriate to limit exposure and end the studies if extensive decomposition occurs. For photostable materials, studies may be terminated after an appropriate exposure level has been used. The design of these experiments is left to the applicant's discretion although the exposure levels used should be justified.

Under forcing conditions, decomposition products may be observed that are unlikely to be formed under the conditions used for confirmatory studies. This information may be useful in developing and validating suitable analytical methods. If in practice it has been demonstrated they are not formed in the confirmatory studies, these degradation products need not be further examined.

Confirmatory studies should then be undertaken to provide the information necessary for handling, packaging, and labeling (see section I.C., Procedure, and II.A., Presentation, for information on the design of these studies).

Normally, only one batch of drug substance is tested during the development phase, and then the photostability characteristics should be confirmed on a single batch selected as described in the Parent Guideline if the drug is clearly photostable or photolabile. If the results of the confirmatory study are equivocal, testing of up to two additional batches should be conducted. Samples should be selected as described in the Parent Guideline.

A. *Presentation of Samples*

Care should be taken to ensure that the physical characteristics of the samples under test are taken into account and efforts should be made, such as cooling and/or placing the samples in sealed containers, to ensure that the effects of the changes in physical states such as sublimation, evaporation or melting are minimized. All such precautions should be chosen to provide minimal interference with the exposure of samples under test. Possible interactions between the samples and any material used for containers or for general protection of the sample should also be considered and eliminated wherever not relevant to the test being carried out.

As a direct challenge for samples of solid drug substances, an appropriate amount of sample should be taken and placed in a suitable glass or plastic dish and protected with a suitable transparent cover if considered necessary. Solid drug substances should be spread across the container to give a thickness of typically not more than 3 millimeters. Drug substances that are liquids should be exposed in chemically inert and transparent containers.

B. *Analysis of Samples*

At the end of the exposure period, the samples should be examined for any changes in physical properties (e.g., appearance, clarity, or color of solution) and for assay and degradants by a method suitably validated for products likely to arise from photochemical degradation processes.

Where solid drug substance samples are involved, sampling should ensure that a representative portion is used in individual tests. Similar sampling considerations, such as homogenization of the entire sample, apply to other materials that may not be homogeneous after exposure. The analysis of the exposed sample should be performed concomitantly with that of any protected samples used as dark controls if these are used in the test.

C. *Judgment of Results*

The forced degradation studies should be designed to provide suitable information to develop and validate test methods for the confirmatory studies. These test methods should be capable of resolving and detecting photolytic degradants that appear during the confirmatory studies. When evaluating the results of these studies, it is important to recognize that they form part of the stress testing and are not therefore designed to establish qualitative or quantitative limits for change.

The confirmatory studies should identify precautionary measures needed in manufacturing or in formulation of the drug product, and if light resistant packaging is needed. When evaluating the results of confirmatory studies to determine whether change due to exposure to light is acceptable, it is important to consider the results from other formal stability studies in order to assure that the drug will be within justified limits at time of use (see the relevant ICH-Stability and Impurity Guidelines).

3. Drug Product

Normally, the studies on drug products should be carried out in a sequential manner starting with testing the fully exposed product then progressing as necessary to the product in the immediate pack and then in the marketing pack. Testing should progress until the results demonstrate that the drug product is adequately protected from exposure to light. The drug product should be exposed to the light conditions described under the procedure in section I.C.

Normally, only one batch of drug product is tested during the development phase, and then the photostability characteristics should be confirmed on a single batch selected as described in the Parent Guideline if the product is clearly photostable or photolabile. If the results of the confirmatory study are equivocal, testing of up to two additional batches should be conducted.

For some products where it has been demonstrated that the immediate pack is completely impenetrable to light, such as aluminium tubes or cans, testing should normally only be conducted on directly exposed drug product.

It may be appropriate to test certain products such as infusion liquids, dermal creams, etc., to support their photostability in-use. The extent of this testing should depend on and relate to the directions for use, and is left to the applicant's discretion.

The analytical procedures used should be suitably validated.

A. *Presentation of Samples*

Care should be taken to ensure that the physical characteristics of the samples under test are taken into account and efforts, such as cooling and/or placing the samples in sealed containers, should be made to ensure that the effects of the changes in physical states are minimized, such as sublimation, evaporation, or melting. All such precautions should be chosen to provide a minimal interference with the irradiation of samples under test. Possible interactions between the samples and any material used for containers or for general protection of the sample should also be considered and eliminated wherever not relevant to the test being carried out.

Where practicable when testing samples of the drug product outside of the primary pack, these should be presented in a way similar to the conditions mentioned for the drug substance. The samples should be positioned to provide maximum area of exposure to the light source. For example, tablets, capsules, etc. should be spread in a single layer.

If direct exposure is not practical (e.g., due to oxidation of a product), the sample should be placed in a suitable protective inert transparent container (e.g., quartz).

If testing of the drug product in the immediate container or as marketed is needed, the samples should be placed horizontally or transversely with respect to the light source, whichever provides for the most uniform exposure of the samples. Some adjustment of testing conditions may have to be made when testing large volume containers (e.g., dispensing packs).

B. Analysis of Samples

At the end of the exposure period, the samples should be examined for any changes in physical properties (e.g., appearance, clarity or color of solution, dissolution/disintegration for dosage forms such as capsules, etc.) and for assay and degradants by a method suitably validated for products likely to arise from photochemical degradation processes.

When powder samples are involved, sampling should ensure that a representative portion is used in individual tests. For solid oral dosage form products, testing should be conducted on an appropriately sized composite of, for example, 20 tablets or capsules. Similar sampling considerations, such as homogenization or solubilization of the entire sample, apply to other materials that may not be homogeneous after exposure (e.g., creams, ointments, suspensions, etc.). The analysis of the exposed sample should be performed concomitantly with that of any protected samples used as dark controls if these are used in the test.

C. Judgement of Results

Depending on the extent of change, special labeling or packaging may be needed to mitigate exposure to light. When evaluating the results of photostability studies to determine whether change due to exposure to light is acceptable, it is important to consider the results obtained from other formal stability studies in order to assure that the product will be within proposed specifications during the shelf life (see the relevant ICH Stability and Impurity Guidelines).

4. Annex

A. Quinine Chemical Actinometry

The following provides details of an actinometric procedure for monitoring exposure to a near UV fluorescent lamp (based on FDA/National Institute of Standards and Technology study). For other light sources/actinometric systems, the same approach may be used, but each actinometric system should be calibrated for the light source used.

Prepare a sufficient quantity of a 2 per cent weight/volume aqueous solution of quinine monohydrochloride dehydrate (if necessary, dissolve by heating).

Option 1

Put 10 milliliters (ml) of the solution into a 20 ml colorless ampoule, seal it hermetically, and use this as the sample. Separately, put 10 ml of the solution into a 20 ml colourless ampoule, seal it hermetically, wrap in aluminum foil to protect completely from light, and use this as the control. Expose the sample and control to the light source for an appropriate number of hours. After exposure determine the absorbances of the sample (A_T) and the control (A_o) at 400 nm using a 1 centimeter (cm) pathlength. Calculate the change in absorbance, $\Delta A = A_T - A_o$. The length of exposure should be sufficient to ensure a change in absorbance of at least 0.9.

Option 2

Fill a 1 cm quartz cell and use this as the sample. Separately fill a 1 cm quartz cell, wrap in aluminum foil to protect completely from light, and use this as the control. Expose the sample and control to the light source for an appropriate number of hours. After exposure determine the absorbances of the sample (A_T) and the control (A_0) at 400 nm. Calculate the change in absorbance, $\Delta A = A_T - A_0$. The length of exposure should be sufficient to ensure a change in absorbance of at least 0.5.

Alternative packaging configurations may be used if appropriately validated. Alternative validated chemical actinometers may be used.

5. Glossary

Immediate (primary) pack is that constituent of the packaging that is in direct contact with the drug substance or drug product, and includes any appropriate label.

Marketing pack is the combination of immediate pack and other secondary packaging such as a carton.

Forced degradation testing studies are those undertaken to degrade the sample deliberately. These studies, which may be undertaken in the development phase normally on the drug substances, are used to evaluate the overall photosensitivity of the material for method development purposes and/or degradation pathway elucidation.

Confirmatory studies are those undertaken to establish photostability characteristics under standardized conditions. These studies are used to identify precautionary measures needed in manufacturing or formulation and whether light resistant packaging and/or special labeling is needed to mitigate exposure to light. For the confirmatory studies, the batch(es) should be selected according to batch selection for long-term and accelerated testings which is described in the Parent Guideline.

6. References

Quinine Actinometry as a method for calibrating ultraviolet radiation intensity in light-stability testing of pharmaceuticals. Yoshioka S. et al., *Drug Development and Industrial Pharmacy*, **20** (13), 2049–2062 (1994).

6.3 Major Concerns Raised by the EU, the United States, and Japan at the International Conference on Harmonisation of Technical Requirements for Registration of Pharmaceuticals for Human Use

6.3.1. Storage Conditions for Stability Testing

The standard storage condition for long-term storage testing was harmonized at 25°C, 60% RH. This standard temperature was decided upon on the basis of mean kinetic temperature. Mean kinetic temperature calculated according to Eq. (4.12) was lower than 25°C for all areas of the EU and Japan and most areas of the United States. Therefore, long-term storage testing carried out at 25°C would allow one to evaluate the stability of the pharmaceuticals distributed in these geographical areas, taking into account the effect of temperature fluctuations throughout the year. It was pointed out that the mean kinetic temperature may be higher than 25°C for some parts of the United States. This problem was addressed by taking into account the relevant national/regional requirements for labeling. The storage condition for long-term storage testing for temperature-sensitive pharmaceuticals designated for refrigerated storage and for freezer storage is now under discussion and will be harmonized at 5 and –15°C, respectively, in 1999.

The standard temperature for accelerated testing for ordinary products is 40°C, a temperature at least 15°C higher than that for normal long-term testing. For products designated for refrigerated and freezer storage, 25 and 5°C, respectively, are recommended as accelerated temperatures. However, there are ongoing discussions on this issue, which will be harmonized in 1999.

The standard humidity condition for long-term storage testing at 25°C is 60% RH, although there has been continued discussion that higher humidity should be considered for humidity-sensitive pharmaceutical products, especially for those products to be distributed in areas where high humidity in summer is prevalent (e.g., some areas in Japan). However, it was concluded that accelerated testing at 40°C/75% RH could cover the effect of higher humidity. On the other hand, the effect of low humidity should be evaluated for liquid or suspension products packed in semipermeable containers. Low-humidity conditions of 40% RH and 20% RH have been proposed for long-term storage testing and for accelerated testing, respectively. These conditions are now under discussion and will be harmonized in 1999.

6.3.2. Photostability Testing

For photostability testing, the choice of light source for testing purposes was one of the matters debated at the ICH conference. Photodegradation of pharmaceuticals depends largely on the spectral distribution of the light source; thus, testing using different light sources might bring about differences in the evaluation of photostability. Therefore, the choice of light source for testing is a very important issue that determines under what conditions the photostability of pharmaceuticals should be evaluated. The photostability of pharmaceuticals has been evaluated in European countries using light sources that simulate sunlight, such as a xenon lamp,[891] whereas the effect of room lighting has been evaluated in Japan using white fluorescent lamps.

A consensus on the light sources for testing could not be achieved at the ICH conference, and two options for light sources (option 1: lamps having output similar to standard daylight; option 2: white fluorescent lamps and near-UV fluorescent lamps) were adopted in the ICH guidelines. To reduce the differences in evaluation between the two options, the minimum requirement for exposure level for both visible and near UV light was designated at 1.2 million lux hours and 200 watt hours per square meter, respectively.

Exposure levels of a product should be confirmed by monitoring that uses calibrated radiometers or lux meters. Validated chemical actinometry can also be used for this purpose. For near-UV light, quinine actinometry is noted in the appendix of the guideline. Quinine actinometry was adopted because its usefulness was confirmed by a collaborative study of the ICH working group[892] and by a study carried out by the U.S. Food and Drug Administration.[893] However, concern has been expressed about the usefulness of quinine actinometry, because the degradation mechanism has not been elucidated and the effect of temperature on the calibration is not clear.[894]

6.3.3. Bracketing and Matrixing

Pharmaceutical products containing new drug substances are usually marketed in more than one form. Also, products may be marketed simultaneously in different packaging (e.g., different sizes or different materials). Other minor variations in formulation are also common, such as changes in the ratio of drug substance to excipient without significant change in qualitative or quantitative composition. Large-scale stability testing is needed if stability evaluations are conducted separately for each of these forms/products. To reduce the total number of samples to be tested, the ICH guidelines introduced the concept of bracketing and matrixing.

Bracketing is a design of a stability testing schedule in which the experimental stability data are obtained only from the extremes of packaging size or strengths. The stability of the intermediate-condition samples is evaluated based on the data from the extremes. Matrixing is a statistical design for stability testing that requires experimental stability data to be obtained from all forms of the drug product but permits only a fraction of the total number of samples to be tested at any specified sampling point according to a specific sampling design.[895] Based on the data, a single shelf life applicable to all of these forms of drug product is derived. A prerequisite for the employment of the matrixing design is that no significant differences in stability exist among different forms. If stability shows significant variation due to different packaging or formulation, an unsuitable shelf life may be estimated by matrixing.[896]

A similar situation involves the evaluation of stability data from different batches. The ICH guideline says that if batch-to-batch variation is small, stability data from several batches can be combined into one overall estimate to determine the shelf life of a pharmaceutical product in which quantitative characteristics decrease with time. This can be performed if batch variation is not found to be significant according to appropriate statistical tests.

The details on the statistical method for assessing stability variations among batches, packagings, or formulations have not yet been harmonized. One method of assessment is an analysis of variance (ANOVA). However, the power of ANOVA depends largely on the assay error; it decreases markedly with increasing assay error.[896,897] Thus, stability variation is more easily overlooked when the stability data have a larger assay error. As another method for assessing stability variations, the assessment of shelf-life equivalence based on the range of shelf-life estimates (difference between the largest and smallest estimates) has been proposed. A simulation study showed that the effect of assay error on the power of this analysis was less than on ANOVA.[898] International harmonization on the application of bracketing and matrixing remains a matter of discussion.

References

1. R. I. Ellin and J. H. Wills, Oximes antagonistic to inhibitors of cholinesterase, Part I, *J. Pharm. Sci.* **53**, 995–1007 (1964).
2. T. D. Sokoloski, L. A. Mitscher, J. V. Juvarkar, and B. Hoener, Rate and proposed mechanism of anhydrotetracycline epimerization in acid solution, *J. Pharm. Sci.* **66**, 1159–1165 (1977).
3. B. A. Hoener, T. D. Sokoloski, L. A. Mitscher, and L. Malspeis, Kinetics of dehydration of epitetracycline in solution, *J. Pharm. Sci.* **63**, 1901–1904 (1974).
4. G. L. Amidon, Benzylpenicillin monograph, in *Chemical Stability of Pharmaceuticals: A Handbook for Pharmacists*, 2nd ed., (K. A. Connors, G. L. Amidon, and V. J. Stella, eds.), pp. 274–283, John Wiley & Sons, New York, 1986, and references therein.
5. G. Gosselin, J. L. Girardet, C. Perigaud, S. Benzaria, I. Lefebvre, N. Schlienger, A. Pompon, and J. L. Imbach, New insights regarding the potential of the pronucleotide approach in antiviral chemotherapy, *Acta Biochim. Pol.* **43**, 196–208 (1996).
6. J. P. Hou and J. W. Poole, The amino acid nature of ampicillin and related penicillins, *J. Pharm. Sci.* **58**, 1510–1515 (1969).
7. M. S. Gogate, Erythromycin monograph, in *Chemical Stability of Pharmaceuticals: A Handbook for Pharmacists*, 2nd ed., (K. A. Connors, G. L. Amidon, and V. J. Stella, eds.), pp. 457–463, John Wiley & Sons, New York, 1986, and references therein.
8. B. D. Anderson, M. B. Wygant, T. X. Xiang, W. A. Waugh, and V. J. Stella, Preformulation solubility and kinetic studies of 2′,3′-dideoxypurine nucleosides: Potential anti-AIDS agents, *Int. J. Pharm.* **45**, 27–37 (1988).
9. T. Higuchi, A. Havinga, and L. W. Busse, The kinetics of the hydrolysis of procaine, *J. Am. Pharm. Assoc., Sci. Ed.* **39**, 405–410 (1950).
10. A. D. Marcus and S. Baron, A comparison of the kinetics of the acid catalyzed hydrolyses of procainamide, procaine, and benzocaine, *J. Am. Pharm. Assoc., Sci. Ed.* **48**, 85–90 (1959).
11. L. J. Edwards, The hydrolysis of aspirin: A determination of the thermodynamic dissociation constant and a study of the reaction kinetics by ultraviolet spectrophotometry, *Trans. Faraday Soc.* **46**, 723–735 (1950).
12. E. R. Garrett, The kinetics of solvolysis of acyl esters of salicylic acid, *J. Am. Chem. Soc.* **79**, 3401–3408 (1957).
13. T. Higuchi and C. D. Bias, The kinetics of degradation of chloramphenicol in solution I. A study of the rate of formation of chloride ion in aqueous media, *J. Am. Pharm. Assoc., Sci. Ed.* **42**, 707–714 (1953).
14. T. Higuchi, A. D. Marcus, and C. D. Bias, The kinetics of degradation of chloramphenicol in solution II. Overall disappearance rate from buffered solutions, *J. Am. Pharm. Assoc., Sci. Ed.* **43**, 129–134 (1954).
15. T. Higuchi and A. D. Marcus, The kinetics of degradation of chloramphenicol in solution III. The nature, specific hydrogen ion catalysis, and temperature dependencies of the degradative reactions, *J. Am. Pharm. Assoc., Sci. Ed.* **43**, 530–535 (1954).
16. A. A. Kondritzer and P. Zvirblis, Stability of atropine in aqueous solution, *J. Am. Pharm. Assoc., Sci. Ed.* **46**, 531–535 (1957).
17. P. Zvirblis, I. Socholitsky, and A. A. Kondritzer, The kinetics of the hydrolysis of atropine, *J. Am. Pharm. Assoc., Sci. Ed.* **45**, 450–454 (1956).
18. J. L. Patel and A. P. Lemberger, The kinetics of the hydrolysis of homatropine methylbromide and atropine methylbromide, *J. Am. Pharm. Assoc., Sci. Ed.* **48**, 106–109 (1959).
19. S. Siegel, L. Lachman, and L. Malspeis, A kinetic study of the hydrolysis of methyl DL-alpha-phenyl-2-piperidylacetate, *J. Am. Pharm. Assoc., Sci. Ed.* **48**, 431–439 (1959).
20. R. J. Washkuhn, V. K. Patel, and J. R. Robinson, Linear free energy models for ester solvolysis with a critical examination of the alcohol and phenol dissociation model, *J. Pharm. Sci.* **60**, 736–744 (1971).

21. M. L. Maniar, D. S. Kalonia, and A. P. Simonelli, Alkaline hydrolysis of oligomers of tartrate esters: Effect of a neighboring carboxyl on the reactivity of ester groups, *J. Pharm. Sci.* **81**, 705–709 (1992).
22. S. M. Blaug and D. E. Grant, Kinetics of degradation of the parabens, *J. Soc. Cosmet. Chem.* **25**, 495–506 (1974).
23. I. A. Hamid and E. L. Parrott, Effect of temperature on solubilization and hydrolytic degradation of solubilized benzocaine and homatropine, *J. Pharm. Sci.* **60**, 901–906 (1971).
24. I. K. Winterborn, B. J. Meakin, and D. J. G. Davies, The effect of cetyltrimethylammonium bromide on the hydrolysis of *p*-substituted ethylbenzoates, *J. Pharm. Pharmacol.* **24**, 113P (1972).
25. I. Oh, S. Chi, B. R. Vishnuvajjala, and B. D. Anderson, Stability and solubilization of oxathiin carboxanilide, a novel anti-HIV agent, *Int. J. Pharm.* **73**, 23–31 (1991).
26. W. Lund and T. Waaler, The kinetics of atropine and apoatropine in aqueous solutions, *Acta Chem. Scand.* **22**, 3085–3097 (1968).
27. J. J. Windheuser, J. L. Sutter, and A. Sarrif, Analysis of scopolamine and its degradation products by GLC and liquid partition chromatography, *J. Pharm. Sci.* **61**, 1311–1313 (1972).
28. R. M. Patel, T. Chin, and J. L. Lach, Kinetic study of the acid hydrolysis of meperidine hydrochloride, *Am. J. Hosp. Pharm.* **25**, 256–261 (1968).
29. E. R. Garrett, Prediction of stability in pharmaceutical preparations X. Alkaline hydrolysis of hydrocortisone hemisuccinate, *J. Pharm. Sci.* **51**, 445–450 (1962).
30. J. W. Mauger, A. N. Paruta, and R. J. Gerraughty, Consecutive first-order kinetic consideration of hydrocortisone hemisuccinate, *J. Pharm. Sci.* **58**, 574–578 (1969).
31. B. D. Anderson and V. Taphouse, Initial rate studies of hydrolysis and acyl migration in methylpredinisolone 21-hemisuccinate and 17-hemisuccinate, *J. Pharm. Sci.* **70**, 181–186 (1981).
32. T. Suzuki, Studies on decomposition and stabilization of drugs in solution. X. Chemical kinetic studies on aqueous solution of succinylcholine chloride. 2. Overall velocity constants for succinylcholine chloride hydrolysis as a function of pH, *Chem. Pharm. Bull.* **10**, 912–921 (1962).
33. J. J. Boehm and R. I. Poust, Hydrolysis of succinylcholine chloride in pH range 3.0 to 4.5, *Chem. Pharm. Bull.* **32**, 1113–1119 (1984).
34. E. R. Garrett and K. Seyda, Prediction of stability in pharmaceutical preparations XX: Stability evaluation and bioanalysis of cocaine and benzoylecgonine by high-performance liquid chromatography, *J. Pharm. Sci.* **72**, 258–271 (1983).
35. E. R. Garrett and D. Maruhn, Prediction of stability in pharmaceutical preparations. XXI: The analysis and kinetics of hydrolysis of a cocaine degradation product, ecgonine methyl ester, plus the pharmacokinetics of cocaine in the dog, *J. Pharm. Sci.* **83**, 269–272 (1994).
36. P. Chung, T. Chin, and J. L. Lach, Kinetics of the hydrolysis of pilocarpine in aqueous solution, *J. Pharm. Sci.* **59**, 1300–1306 (1970).
37. M. A. Nunes and E. Brochmann-Hanssen, Hydrolysis and epimerization kinetics of pilocarpine in aqueous solution, *J. Pharm. Sci.* **63**, 716–721 (1974).
38. H. Bundgaard and S. H. Hansen, Hydrolysis and epimerization kinetics of pilocarpine in basic aqueous solution as determined by HPLC, *Int. J. Pharm.* **10**, 281–289 (1982).
39. C. M. Won, Epimerization and hydrolysis of dalvastatin, a new hydroxymethylglutaryl coenzyme A (HMG-CoA) reductase inhibitor, *Pharm. Res.* **11**, 165–170 (1994).
40. E. R. Garrett, B. C. Lippold, and J. B. Mielck, Kinetics and mechanisms of lactonization of coumarinic acids and hydrolysis of coumarins I, *J. Pharm. Sci.* **60**, 396–405 (1971).
41. B. C. Lippold and E. R. Garrett, Kinetics and mechanisms of lactonization of coumarinic acids and hydrolysis of coumarins II, *J. Pharm. Sci.* **60**, 1019–1027 (1971).
42. J. Fassberg and V. J. Stella, A kinetic and mechanistic study of the hydrolysis of camptothecin and some analogues, *J. Pharm. Sci.* **81**, 676–684 (1992).
43. M. Hara, H. Hayashi, and T. Yoshida, Kinetics and mechanism of degradation of chlorphenesin carbamate in strongly acidic aqueous solutions, *Chem. Pharm. Bull.* **34**, 3481–3484 (1986).
44. V. J. Stella, W. K. Anderson, A. Benedetti, W. A. Waugh, and R. B. Killion, Jr., Stability of carmethizole hydrochloride (NSC-602668), an experimental cytotoxic agent, *Int. J. Pharm.* **71**, 157–165 (1991).
45. K. Umprayn, W. Waugh, and V. J. Stella, Mechanism of degradation of the investigational cytotoxic agent, cyclodisone (NSC-348948), *Int. J. Pharm.* **66**, 253–262 (1990).

46. M. Paborji, W. Waugh, and V. J. Stella, Mechanistic investigation of the degradation of sulfamic acid 1,7-heptanediyl ester, an experimental cytotoxic agent, in water and ^{18}oxygen-enriched water, *J. Pharm. Sci.* **76**, 161–165 (1987).
47. A. D. Marcus, The hydrolysis of hydrocortisone phosphate in essentially neutral solutions, *J. Am. Pharm. Assoc., Sci. Ed.* **49**, 383–385 (1960).
48. G. L. Flynn and D. J. Lamb, Factors influencing solvolysis of corticosteroid-21-phosphate esters, *J. Pharm. Sci.* **59**, 1433–1438 (1970).
49. A. Hussain, P. Schurman, V. Peter, and G. Milosovich, Kinetics and mechanism of degradation of echothiophate iodide in aqueous solution, *J. Pharm. Sci.* **57**, 411–418 (1968).
50. M. C. Crew and F. J. Dicarlo, Identification and assay of isomeric 14C-glyceryl nitrates, *J. Chromatogr.* **35**, 506–512 (1968).
51. H. Nagai, M. Kikuchi, H. Nagano, and M. Shiba, The stability of nicorandil in aqueous solution. I. Kinetics and mechanism of decomposition of *N*-(2-hydroxyethyl)nicotinamide nitrate (ester) in aqueous solution, *Chem. Pharm. Bull.* **32**, 1063–1070 (1984).
52. C. J. Herman and M. J. Groves, Hydrolysis kinetics of phospholipids in thermally stressed intravenous lipid emulsion formulations, *J. Pharm. Pharmacol.* **44**, 539–542 (1991).
53. M. Grit, N. J. Zuidam, W. J. M. Underberg, and D. J. A. Crommelin, Hydrolysis of partially saturated egg phosphatidylcholine in aqueous liposome dispersions and the effect of cholesterol incorporation on hydrolysis kinetics, *J. Pharm. Pharmacol.* **45**, 490–495 (1992).
54. K. T. Koshy and J. L. Lach, Stability of aqueous solutions of *N*-acetyl-*p*-aminophenol, *J. Pharm. Sci.* **50**, 113–118 (1961).
55. A. A. Forist, L. W. Brown, and M. E. Royer, Acid stability of lincomycin, *J. Pharm. Sci.* **54**, 476–477 (1965).
56. S. Goto, W. Tseng, M. Kai, S. Aizawa, and S. Iguchi, Kinetics of degradation of indomethacin in aqueous solution [in Japanese], *Yakuzaigaku* **29**, 118–124 (1969).
57. S. Goto, N. Murao, and S. Iguchi, Kinetics of hydrolysis of compounds related to indomethacin in alkaline region [in Japanese], *Yakuzaigaku* **33**, 139–145 (1973).
58. B. R. Hajratwala and J. E. Dawson, Kinetics of indomethacin degradation I: Presence of alkali, *J. Pharm. Sci.* **66**, 27–29 (1977).
59. E. Pawelczyk, B. Knitter, and W. Alejska, Drug decomposition kinetics. LVI. Mechanism and kinetics of hydrolysis of indomethacin in the acid medium, *Acta Pol. Pharm.* **36**, 181–188 (1979).
60. B. J. Meakin, I. P. Tansey, and D. J. G. Davies, The effect of heat, pH and some buffer materials on the hydrolytic degradation of sulphacetamide in aqueous solution, *J. Pharm. Pharmacol.* **23**, 252–261 (1971).
61. S. P. King, K. W. Sigvardson, J. Dudzinski, and G. Torosian, Degradation kinetics and mechanisms of moricizine hydrochloride in acidic medium, *J. Pharm. Sci.* **81**, 586–591 (1992).
62. W. D. Korte and M. L. Shih, Degradation of three related bis(pyridinium)aldoximes in aqueous solutions at high concentrations: Examples of unexpectedly rapid amide group hydrolysis, *J. Pharm. Sci.* **82**, 782–786 (1993).
63. P. Finholt, G. Jurgensen, and H. Kristiansen, Catalytic effect of buffers on degradation of penicillin G in aqueous solution, *J. Pharm. Sci.* **54**, 387–393 (1965).
64. J. M. Blaha, A. M. Knevel, D. P. Kessler, J. W. Mincy, and S. L. Hem, Kinetic analysis of penicillin degradation in acidic media, *J. Pharm. Sci.* **65**, 1165–1170 (1976).
65. J. P. Hou and J. W. Poole, Kinetics and mechanism of degradation of ampicillin in solution, *J. Pharm. Sci.* **58**, 447–454 (1969).
66. H. Bundgaard, Polymerization of penicillins. II. Kinetics and mechanism of dimerization and self-catalyzed hydrolysis of amoxycillin in aqueous solution, *Acta Pharm. Suec.* **14**, 47–66 (1977).
67. H. Zia, M. Tehrani, and R. Zargarbashi, Kinetics of carbenicillin degradation in aqueous solutions, *Can. J. Pharm. Sci.* **9**, 112–117 (1974).
68. M. A. Schwartz, A. P. Granatek, and F. H. Buckwalter, Stability of potassium phenethicillin. I. Kinetics of degradation in aqueous solution, *J. Pharm. Sci.* **51**, 523–526 (1962).
69. M. A. Schwartz, E. Bara, I. Rubycz, and A. P. Granatek, Stability of methicillin, *J. Pharm. Sci.* **54**, 149–150 (1965).
70. T. Yamana and A. Tsuji, Comparative stability of cephalosporins in aqueous solution: Kinetics and mechanisms of degradation, *J. Pharm. Sci.* **65**, 1563–1574 (1976).

71. A. Tsuji, E. Nakashima, Y. Deguchi, K. Nishide, T. Shimizu, S. Horiuchi, K. Ishikawa, and T. Yamana, Degradation kinetics and mechanism of aminocephalosporins in aqueous solution: Cefadroxil, *J. Pharm. Sci.* **70**, 1120–1128 (1981).
72. A. Tsuji, E. Nakashima, K. Nishide, Y. Deguchi, S. Hamano, and T. Yamana, Physicochemical properties of amphoteric β-lactam antibiotics. III. Stability, solubility, and dissolution behavior of cefatrizine and cefadroxil as a function of pH, *Chem. Pharm. Bull.* **31**, 4057–4069 (1983).
73. S. M. Berge, N. L. Henderson, and M. J. Frank, Kinetics and mechanism of degradation of cefotaxime sodium in aqueous solution, *J. Pharm. Sci.* **72**, 59–63 (1983).
74. H. Fabre, N. H. Eddine, and G. Berge, Degradation kinetics in aqueous solution of cefotaxime sodium, a third-generation cephalosporin, *J. Pharm. Sci.* **73**, 611–618 (1984).
75. V. D. Gupta, Stability of cefotaxime sodium as determined by high-performance liquid chromatography, *J. Pharm. Sci.* **73**, 565–567 (1984).
76. S. W. Baertschi, D. E. Dorman, J. L. Occolowitz, L. A. Spangle, M. W. Collins, M. E. Wildfeuer, and L. J. Lorenz, Isolation and structure elucidation of a novel product of the acidic degradation of cefaclor, *J. Pharm. Sci.* **82**, 622–626 (1993).
77. M. J. Skibic, K. W.Taylor, J. L. Occolowitz, M. W. Collins, J. W. Paschal, L. J. Lorenz, L. A. Spangle, D. E. Dorman, and S. W. Baertschi, Aqueous acidic degradation of the carbacephalosporin loracarbef, *J. Pharm. Sci.* **82**, 1010–1017 (1993).
78. H. Bundgaard, Polymerization of penicillins: Kinetics and mechanism of di- and polymerization of ampicillin in aqueous solution, *Acta Pharm. Suec.* **13**, 9–26 (1976).
79. L. Malspeis and D. Gold, Stability of cycloserine in buffered aqueous solutions, *J. Pharm. Sci.* **53**, 1173–1180 (1964).
80. E. R. Garrett, J. T. Bojarski, and G. J. Yakatan, Kinetics of hydrolysis of barbituric acid derivatives, *J. Pharm. Sci.* **60**, 1145–1154 (1971).
81. L. A. Gardner and J. E. Goyan, Mechanism of phenobarbital degradation, *J. Pharm. Sci.* **62**, 1026–1027 (1973).
82. J. T. Carstensen, E. G. Serenson, and J. J. Vance, Use of Hammett graphs in stability programs, *J. Pharm. Sci.* **53**, 1547–1548 (1964).
83. Z. R. Zaidi, F. J. Sena, and C. P. Basilio, Stability assay of allantoin in lotions and creams by high-pressure liquid chromatography, *J. Pharm. Sci.* **71**, 997–999 (1982).
84. B. R. Vishnuvajjala and J. C. Cradock, Tricyclo[4.2.2.0]dec-9-ene-3,4,7,8-tetracarboxylic acid diimide: Formulation and stability studies, *J. Pharm. Sci.* **75**, 301–303 (1986).
85. J. M. Sisco and V. J. Stella, Is ICRF-187 [(+)-1,2-bis(3,5-dioxopiperazinyl-1-yl)propane] unusually reactive for an imide? *Pharm. Res.* **9**, 1076–1082 (1992).
86. B. B. Hasinoff, An HPLC and spectrophotometric study of the hydrolysis of ICRF-187 (dextrazoxane, (+)-1,2-bis(3,5-dioxopiperazinyl-1-yl)propane) and its one-ring opened intermediates, *Int. J. Pharm.* **107**, 67–76 (1994).
87. W. W. Han, G. J. Yakatan, and D. D. Maness, Kinetics and mechanisms of hydrolysis of 1,4-benzodiazepines II: Oxazepam and diazepam, *J. Pharm. Sci.* **66**, 573–577 (1977).
88. A. G. Davidson, S.-M. Chee, F. M. Millar, and D. Watson, Isolation and identification of an alkali-catalyzed hydrolysis product of nitrazepam, *Int. J. Pharm.* **63**, 29–34 (1990).
89. A. G. Davidson and G. A. Smail, Isolation and characterisation of an acid-catalyzed intermediate hydrolysis product of nitrazepam, *Int. J. Pharm.* **69**, 1–3 (1991).
90. H. V. Maulding, J. P. Nazareno, J. E. Pearson, and A. F. Michaelis, Practical kinetics III: Benzodiazepine hydrolysis, *J. Pharm. Sci.* **64**, 278–284 (1975).
91. W. W. Han, G. J. Yakatan, and D. D. Maness, Kinetics and mechanisms of hydrolysis of 1,4-benzodiazepines I: Chlordiazepoxide and demoxepam, *J. Pharm. Sci.* **65**, 1198–1204 (1976).
92. M. Konishi, K. Hirai, and Y. Mori, Kinetics and mechanism of the equilibrium reaction of triazolam in aqueous solution, *J. Pharm. Sci.* **71**, 1328–1334 (1982).
93. M. Ikeda and T. Nagai, Kinetics of hydrolysis of oxazolam in aqueous solution, *Chem. Pharm. Bull.* **32**, 1080–1090 (1984).
94. T. Kuwayama, Y. Kurono, T. Muramatsu, T. Yashiro, and K. Ikeda, The behavior of 1,4-benzodiazepine drugs in acidic media. V. Kinetics of hydrolysis of flutazolam and haloxazolam in aqueous solution, *Chem. Pharm. Bull.* **34**, 320–326 (1986).

95. Y. Kurono, T. Kuwayama, K. Kamiya, T. Yashiro, and K. Ikeda, The behavior of 1,4-benzodiazepine drugs in acidic media. II. Kinetics and mechanism of the acid–base equilibrium reaction of oxazolam, *Chem. Pharm. Bull.* **33**, 1633–1640 (1985).
96. Y. Kurono, H. Tani, T. Kuwayama, K. Hatano, T. Yashiro, and K. Ikeda, Kinetics and mechanism of the acid–base equilibrium and the hydrolysis of benzodiazepinooxazines, *Chem. Pharm. Bull.* **39**, 3323–3326 (1991).
97. Y. Kurono, H. Tamaki, T. Kuwayama, T. Ichii, H. Sawabe, T. Yashiro, K. Ikeda, and H. Bundgaard, Kinetics and mechanism of the acid–base equilibrium of mexazolam analogues: A modification of the mechanism proposed previously for mexazolam, *Int. J. Pharm.* **80**, 135–143 (1992).
98. S. Yoshioka, H. Ogata, T. Shibazaki, and T. Inoue, Stability of sulpyrine. I. Reversible hydrolysis in alkaline solution, *Chem. Pharm. Bull.* **25**, 475–483 (1977).
99. S. Yoshioka, H. Ogata, T. Shibazaki, and T. Inoue, Stability of sulpyrine. II. Hydrolysis in acid solution, *Chem. Pharm. Bull.* **25**, 484–490 (1977).
100. J. E. Cruz, D. D. Maness, and G. J. Yakatan, Kinetics and mechanism of hydrolysis of furosemide, *Int. J. Pharm.* **2**, 275–281 (1979).
101. A. G. Ghanekar, V. D. Gupta, and C. W. Gibbs, Stability of furosemide in aqueous systems, *J. Pharm. Sci.* **67**, 808–811 (1978).
102. J. J. Windheuser and T. Higuchi, Kinetics of thiamine hydrolysis, *J. Pharm. Sci.* **51**, 354–364 (1962).
103. S. M. Walters, Influence of pH on hydrolytic decomposition of diethylpropion hydrochloride: Stability studies on drug substance and tablets using high-performance liquid chromatography, *J. Pharm. Sci.* **69**, 1206–1209 (1980).
104. J. H. Beijnen and W. J. M. Underberg, Degradation of mitomycin C in acidic solution, *Int. J. Pharm.* **24**, 219–229 (1985).
105. W. J. M. Underberg and H. Lingeman, Aspects of the chemical stability of mitomycin and porfiromycin in acidic solution, *J. Pharm. Sci.* **72**, 549–553 (1983).
106. F. J. Alvarez and R. T. Slade, Kinetics and mechanism of degradation of zileuton, a potent 5-lipoxygenase inhibitor, *Pharm. Res.* **9**, 1465–1473 (1992).
107. A. J. Ross, M. C. Go, D. L. Casey, and D. J. Palling, Kinetics and mechanism of the hydrolysis of a 2-substituted imidazoline, cibenzoline (cifenline), *J. Pharm. Sci.* **76**, 306–309 (1987).
108. G. P. Moloney, D. J. Craik, and M. N. Iskander, Qualitative analysis of the stability of the oxazine ring of various benzoxazine and pyridooxazine derivatives with proton nuclear magnetic resonance spectroscopy, *J. Pharm. Sci.* **81**, 692–697 (1992).
109. N. Inotsume and M. Nakano, Reversible azomethine bond cleavage of nitrofurantoin in acidic solutions at body temperature, *Int. J. Pharm.* **8**, 111–119 (1981).
110. G. Ertan, Y. Karasulu, and T. Guneri, Degradation and gastrointestinal stability of nitrofurantoin in acidic and alkaline media, *Int. J. Pharm.* **96**, 243–248 (1993).
111. J. K. Seydel, Physicochemical studies on rifampicin, *Antibiot. Chemother.* **16**, 380–391 (1970).
112. T. Yamana, Y. Mizukami, and A. Tsuji, Studies on the stability of drugs. XVII. Stability of benzothiadiazines. (8). Kinetics of acid–base catalysed hydrolysis of chlorothiazide, *Chem. Pharm. Bull.* **16**, 396–404 (1968).
113. J. A. Mollica, C. R. Rehm, and J. B. Smith, Hydrolysis of hydrochlorothiazide, *J. Pharm. Sci.* **58**, 635–636 (1969).
114. J. A. Mollica, C. R. Rehm, J. B. Smith, and H. K. Govan, Hydrolysis of benzothiadiazines, *J. Pharm. Sci.* **60**, 1380–1384 (1971).
115. E. R. Garrett and G. J. Yakatan, Solvolysis of 5-halouridines and related nucleosides, *J. Pharm. Sci.* **57**, 1478–1487 (1968).
116. R. E. Notari and J. L. DeYoung, Kinetics and mechanisms of degradation of the antileukemic agent 5-azacytidine in aqueous solution, *J. Pharm. Sci.* **64**, 1148–1157 (1975).
117. K. K. Chan, D. D. Giannini, J. A. Staroscik, and W. Sadee, 5-Azacytidine hydrolysis kinetics measured by high-pressure liquid chromatography and 13C-NMR spectroscopy, *J. Pharm. Sci.* **68**, 807–812 (1979).
118. R. E. Notari, M. L. Chin, and R. Wittebort, Arabinosylcytosine stability in aqueous solutions: pH profile and shelflife predictions, *J. Pharm. Sci.* **61**, 1189–1196 (1972).
119. R. E. Notari, M. L. Chin, and A. Cardoni, Intermolecular and intramolecular catalysis in deamination of cytosine nucleosides, *J. Pharm. Sci.* **59**, 28–32 (1970).
120. H. Ehrsson, S. Eksborg, I. Wllin, and S. Nilsson, Degradation of chlorambucil in aqueous solution, *J. Pharm. Sci.* **69**, 1091–1094 (1980).

121. D. C. Chatterji, R. L. Yeager, and J. F. Gallelli, Kinetics of chlorambucil hydrolysis using high-pressure liquid chromatography, *J. Pharm. Sci.* **71**, 50–54 (1982).
122. K. P. Flora, J. C. Cradock, and J. A. Kelley, The hydrolysis of spirohydantoin mustard, *J. Pharm. Sci.* **71**, 1206–1211 (1982).
123. T. O. Oesterling, Aqueous stability of clindamycin, *J. Pharm. Sci.* **59**, 63–67 (1970).
124. A. F. Fell and S. M. Plag, Stability-indicating assay for azathioprine and 6-mercaptopurine by reversed-phase high-performance liquid chromatography, *J. Chromatogr.* **186**, 691–704 (1979).
125. M. J. Reader and C. B. Lines, Decomposition of thimerosal in aqueous solution and its determination by high-performance liquid chromatography, *J. Pharm. Sci.* **72**, 1406–1409 (1983).
126. I. Caraballo, A. M. Rabasco, and M. Fernadez-Arevalo, Study of thimerosal degradation mechanism, *Int. J. Pharm.* **89**, 213–221 (1993).
127. M. A. Allsopp, G. J. Sewell, C. G. Rowland, C. M. Riley, and R. L. Schowen, The degradation of carboplatin in aqueous solutions containing chloride or other selected nucleophiles, *Int. J. Pharm.* **69**, 197–210 (1991).
128. S. A. Khalil and S. El-Masry, Instability of digoxin in acid medium using a nonisotopic method, *J. Pharm. Sci.* **67**, 1358–1360 (1978).
129. L. A. Sternson and R. D. Shaffer, Kinetics of digoxin stability in aqueous solution, *J. Pharm. Sci.* **67**, 327–330 (1978).
130. N. T. Nguyen and R. E. Notari, Substituent effects on degradation rates and pathways of cytosine nucleosides, *J. Pharm. Sci.* **78**, 802–806 (1989).
131. L. J. Ravin, C. A. Simpson, A. F. Zappala, and J. J. Gulesich, Hydrolysis of idoxuridine, *J. Pharm. Sci.* **53**, 1064–1066 (1964).
132. A. V. Schepdael, N. Ossembe, L. P. Herdewijn, E. Roets, and J. Hoogmartens, Kinetics of the hydrolysis of 2,3-dideoxyguanosine: A potent anti-HIV agent, *Int. J. Pharm.* **73**, 105–110 (1991).
133. M. Brandl, R. Strickley, T. Bregante, L. Gu, and T. W. Chan, Degradation of 4′-azidothymidine in aqueous solution, *Int. J. Pharm.* **93** 75–83 (1993).
134. M. Safadi, D. S. Bindra, T. Williams, R. C. Moschel, and V. J. Stella, Kinetics and mechanism of the acid-catalyzed hydrolysis of O^6-benzylguanine, *Int. J. Pharm.* **90**, 239–246 (1993).
135. M. L. Wolfrom, R. D. Schuetz, and L. F. Cavalieri, Chemical interaction of amino compounds and sugars. III. The conversion of D-glucose to 5-(hydroxymethyl)-2-furaldehyde, *J. Am. Chem. Soc.* **70**, 514–517 (1948).
136. R. B. Taylor, B. M. Jappy, and J. M. Neil, Kinetics of dextrose degradation under autoclaving conditions, *J. Pharm. Pharmacol.* **24**, 121–129 (1972).
137. R. B. Taylor and V. C. Sood, An h.p.l.c. study of the initial stages of dextrose decomposition in neutral solution, *J. Pharm. Pharmacol.* **30**, 510–511 (1978).
138. C. A. Brownley and L. Lachman, Browning of spray-processed lactose, *J. Pharm. Sci.* **53**, 452–454 (1964).
139. K. T. Koshy, R. N. Duvall, A. E. Troup, and J. W. Pyles, Factors involved in the browning of spray-dried lactose, *J. Pharm. Sci.* **54**, 549–554 (1965).
140. P. Kurath, P. H. Jones, R. S. Egan, and T. J. Perun, Acid degradation of erythromycin A and erythromycin B, *Experientia* **27**, 362–363 (1971).
141. P. J. Atkins, T. O. Herbert, and N. B. Jones, Kinetic studies on the decomposition of erythromycin A in aqueous acidic and neutral buffers, *Int. J. Pharm.* **30**, 199–207 (1986).
142. D. C. Monkhouse, L. V. Campen, and A. J. Aguiar, Kinetics of dehydration and isomerization of prostaglandins E_1 and E_2, *J. Pharm. Sci.* **62**, 576–580 (1973).
143. R. G. Stehle and T. O. Oesterling, Stability of prostaglandin E_1 and dinoprostone (prostaglandin E_2) under strongly acidic and basic conditions, *J. Pharm. Sci.* **66**, 1590–1595 (1977).
144. H. Lee, H. J. Lambert, and R. L. Schowen, On the mechanism of dehydration of a beta-hydroxycyclopentanone analogue of prostaglandin E_1, *J. Pharm. Sci.* **73**, 306–310 (1984).
145. T. Kawaguchi and Y. Suzuki, Dehydration and epimerization of 7-thiaprostaglandin E_1 analogues, *J. Pharm. Sci.* **75**, 992–994 (1986).
146. M. N. Nassar, C. A. House, and S. N. Agharkar, Stability of batanopride hydrochloride in aqueous solutions, *J. Pharm. Sci.* **81**, 1088–1091 (1992).
147. R. E. Notari and S. M. Caiola, Catalysis of streptovitacin A dehydration: Kinetics and mechanisms, *J. Pharm. Sci.* **58**, 1203–1208 (1969).
148. J. M. T. Hamilton-Miller, The effect of pH and of temperature on the stability and bioactivity of nystatin and amphotericin B, *J. Pharm. Pharmacol.* **25**, 401–407 (1973).

149. G. J. Friis and H. Bundgaard, Kinetics of degradation of cyclosporin A in acidic aqueous solution and its implication in its oral absorption, *Int. J. Pharm.* **82**, 79–83 (1992).
150. B. G. Snider, T. A. Runge, P. E. Fagerness, R. H. Robins, and B. D. Kaluzny, Identification of degradation products occurring in acidic solutions of a 21-aminosteroid (tirilazad), *Int. J. Pharm.* **66**, 63–70 (1990).
151. J. R. D. McCormick, S. M. Fox, L. L. Smith, B. A. Bitler, J. Reichenthal, V. E. Origoni, W. H. Muller, R. Winterbottom, and A. P. Doerschuk, Studies of the reversible epimerization occurring in the tetracycline family. The preparation, properties and proof of structure of some 4-epi-tetracyclines, *J. Am. Chem. Soc.* **79**, 2849–2858 (1957).
152. W. Naidong, S. Hua, E. Roets, R. Busson, and J. Hoogmartens, Investigation of keto–enol tautomerism and ionization of doxycycline in aqueous solutions, *Int. J. Pharm.* **96**, 13–21 (1993).
153. D. W. Hughes, W. L. Wilson, A. G. Butterfield, and N. J. Pound, Stability of rolitetracycline in aqueous solution, *J. Pharm. Pharmacol.* **26**, 79–80 (1974).
154. A. G. Butterfield, D. W. Hughes, W. L. Wilson, and N. J. Pound, Simultaneous high-speed liquid chromatographic determination of tetracycline and rolitetracycline in rolitetracycline formulations, *J. Pharm. Sci.* **64**, 316–320 (1975).
155. H. Ott, A. Hofmann, and A. J. Frey, Acid-catalyzed isomerization in the peptide part of ergot alkaloids, *J. Am. Chem. Soc.* **88**, 1251–1256 (1966).
156. J. H. Beijnen, J. J. M. Holthuis, H. G. Kerkdijk, O. A. G. J. van der Houwen, A. C. A. Paalman, A. Bult, and W. J. M. Underberg, Degradation kinetics of etoposide in aqueous solution, *Int. J. Pharm.* **41**, 169–178 (1988).
157. Y. Aso, Y. Hayashi, S. Yoshioka, Y. Takeda, Y. Kita, Y. Nishimura, and Y. Arata, Epimerization and hydrolysis of etoposide analogues in aqueous solution, *Chem. Pharm. Bull.* **37**, 422–424 (1989).
158. L. C. Schroeter and T. Higuchi, A kinetic study of acid-catalyzed racemization of epinephrine, *J. Am. Pharm. Assoc.* **47**, 426–430 (1958).
159. Y. Aso, S. Yoshioka, T. Shibazaki, and M. Uchiyama, The kinetics of the racemization of oxazepam in aqueous solution, *Chem. Pharm. Bull.* **36**, 1834–1840 (1988).
160. N. Hashimoto, T. Tasaki, and H. Tanaka, Degradation and epimerization kinetics of moxalactam in aqueous solution, *J. Pharm. Sci.* **73**, 369–373 (1984).
161. N. Hashimoto and H. Tanaka, Epimerization kinetics of moxalactam, its derivatives, and carbenicillin in aqueous solution, *J. Pharm. Sci.* **74**, 68–71 (1985).
162. T. Fujita, H. Kimoto, and A. Koshiro, Stability of disodium latamoxef in aqueous solution and in the admixtures of 5% glucose solution, *Yakugaku Zasshi* **106**, 68–74 (1986).
163. A. Tsuji, Y. Itatani, and T. Yamana, Hydrolysis and epimerization kinetics of hetacillin in aqueous solution, *J. Pharm. Sci.* **66**, 1004–1009 (1977).
164. P. A. Twomey, High-performance liquid chromatographic analysis of carbenicillin and its degradation products, *J. Pharm. Sci.* **70**, 824–826 (1981).
165. Y. Aso, S. Yoshioka, and Y. Takeda, Epimerization and racemization of some chiral drugs in the presence of human serum albumin, *Chem. Pharm. Bull.* **38**, 180–184 (1990).
166. T. Fujita and A. Koshiro, Kinetics and mechanism of the degradation and epimerization of sodium cefsulodin in aqueous solution, *Chem. Pharm. Bull.* **32**, 3651–3661 (1984).
167. B. Vilanova, F. Munoz, J. Donoso, J. Frau, and F. G. Blanco, Alkaline hydrolysis of cefotaxime. A HPLC and ^{1}H NMR study, *J. Pharm. Sci.* **83**, 322–327 (1994).
168. N. Tanaka and M. Nakagaki, Physicochemical studies on the decomposition of *p*-aminosalicylic acid and its salts. I. Effect of pH, and addition of propylene glycol, polyvinylpyrrolidone, and some surface-active agents on the decarboxylation of *p*-aminosalicylic acid in the solution [in Japanese], *Yakugaku Zasshi* **81**, 591–596 (1961).
169. H. Bundgaard and N. Mork, Kinetics of the decarboxylation of foscarnet in acidic aqueous solution and its implication in its oral absorption, *Int. J. Pharm.* **63**, 213–218 (1990).
170. Y. J. Lee, J. Padula, and H. Lee, Kinetics and mechanisms of etodolac degradation in aqueous solutions, *J. Pharm. Sci.* **77**, 81–86 (1988).
171. C. Jackson, T. A. Crabb, M. Gibson, R. Godgrey, R. Saunders, and D. E. Thurston, Studies on the stability of trimelamol, a carbinolamine-containing antitumor drug, *J. Pharm. Sci.* **80**, 245–251 (1991).
172. C. M. Won, Kinetics of degradation of levothyroxine in aqueous solution and in solid state, *Pharm. Res.* **9**, 131–137 (1992).

173. M. A. F. Hamelijnck, P. J. Stevenson, P. K. Kadaba, and L. A. Damani, Triazolines. XXI: Preformulation degradation kinetics and chemical stability of a novel triazoline anticonvulsant, *J. Pharm. Sci.* **81**, 392–396 (1992).
174. T. Martens, D. Langevin-Bermond, and M. B. Fleury, Ditiocarb: Decomposition in aqueous solution and effect of the volatile product on its pharmacological use, *J. Pharm. Sci.* **82**, 379–383 (1993).
175. A. O. Dekker and R. G. Dickinson, Oxidation of ascorbic acid by oxygen with cupric ion as catalyst, *J. Am. Chem. Soc.* **62**, 2165–2171 (1940).
176. A. Weissberger, J. E. LuValle, and D. S. Thomas, Oxidation processes. XVI. The autoxidation of ascorbic acid, *J. Am. Chem. Soc.* **65**, 1934–1939 (1943).
177. M. M. T. Khan and A. E. Martell, Metal ion and metal chelate catalyzed oxidation of ascorbic acid by molecular oxygen. I. Cupric and ferric ion catalyzed oxidation, *J. Am. Chem. Soc.* **89**, 4176–4185 (1967).
178. M. M. T. Khan and A. E. Martell, Metal ion and metal chelate catalyzed oxidation of ascorbic acid by molecular oxygen. II. Cupric and ferric chelate catalyzed oxidation, *J. Am. Chem. Soc.* **89**, 7104–7111 (1967).
179. S. M. Blaug and B. Hajratwala, Kinetics of aerobic oxidation of ascorbic acid, *J. Pharm. Sci.* **61**, 556–562 (1972).
180. R. J. Sassetti and H. H. Fudenberg, Alpha-methyldopa melanin. Synthesis and stabilization in vitro, *Biochem. Pharmacol.* **20**, 57–66 (1971).
181. T. D. Sokoloski and T. Higuchi, Kinetics of degradation in solution of epinephrine by molecular oxygen, *J. Pharm. Sci.* **51**, 172–177 (1962).
182. R. K. Palsmeier, D. M. Radzik, and C. E. Lunte, Investigation of the degradation mechanism of 5-aminosalicylic acid in aqueous solution, *Pharm. Res.* **9**, 933–938 (1992).
183. J. Jensen, C. Cornett, C. E. Olsen, J. Tjornelund, and S. H. Hansen, Identification of major degradation products of 5-aminosalicylic acid formed in aqueous solutions and in pharmaceuticals, *Int. J. Pharm.* **88**, 177–187 (1992).
184. T. M. Chen and L. Chafetz, Kinetics of procaterol auto-oxidation in buffered acid solutions, *J. Pharm. Sci.* **76**, 703–706 (1987).
185. I. L. Doerr, I. Wempen, D. A. Clarke, and J. J. Fox, Thiation of nucleosides. III. Oxidation of 6-mercaptopurines, *J. Org. Chem.* **26**, 3401–3409 (1961).
186. P. Timmins, I. M. Jackson, and Y. J. Wang, Factors affecting captopril stability in aqueous solution, *Int. J. Pharm.* **11**, 329–336 (1982).
187. R. G. Strickley and B. D. Anderson, Solubilization and stabilization of an anti-HIV thiocarbamate, NSC 629243, for parenteral delivery, using extemporaneous emulsions, *Pharm. Res.* **10**, 1076–1082 (1993).
188. W. J. M. Underberg, Oxidative degradation of pharmaceutically important phenothiazines I: Isolation and identification of oxidation products of promethazine, *J. Pharm. Sci.* **67**, 1128–1131 (1978).
189. W. J. M. Underberg, Oxidative degradation of pharmaceutically important phenothiazines III: Kinetics and mechanism of promethazine oxidation, *J. Pharm. Sci.* **67**, 1133–1138 (1978).
190. J. K. Guillory and T. Higuchi, Solid state stability of some crystalline vitamin A compounds, *J. Pharm. Sci.* **51**, 100–105 (1962).
191. B. A. Stewart, S. L. Midland, and S. R. Byrn, Degradation of crystalline ergocalciferol [Vitamin D_2 (3,5*Z*,22*E*)-9,10-secoergosta-5,7,10(19),22-tetraen-3-ol], *J. Pharm. Sci.* **73**, 1322–1323 (1984).
192. L. T. Grady and K. D. Thakker, Stability of solid drugs: Degradation of ergocalciferol (vitamin D_2) and cholecalciferol (vitamin D_3) at high humidities and elevated temperatures, *J. Pharm. Sci.* **69**, 1099–1102 (1980).
193. E. R. Garrett, Studies on the stability of fumagillin III. Thermal degradation in the presence and absence of air, *J. Am. Pharm. Assoc., Sci. Ed.* **43**, 539–543 (1954).
194. J. E. Tingstad and E. R. Garrett, Studies on the stability of filipin I. Thermal degradation in the presence of air, *J. Am. Pharm. Assoc., Sci. Ed.* **49**, 352–355 (1960).
195. G. B. Whitfield, T. D. Brock, A. Ammann, D. Gottlieb, and H. E. Carter, Filipin, an antifungal antibiotic: Isolation and properties, *J. Am. Chem. Soc.* **77**, 4799–4801 (1955).
196. E. Pawelczyk and R. Wachowiak, Chemical characteristics of decay products of drugs. IV. Chemical mechanisms of the decay of phenylbutazone in pharmaceutical preparations, *Acta Pol. Pharm.* **26**, 433–438 (1969).
197. D. V. C. Awang, A. Vincent, and F. Matsui, Pattern of phenylbutazone degradation, *J. Pharm. Sci.* **62**, 1673–1676 (1973).

198. H. Fabre, B. Mandrou, and H. Eddine, Quality control of phenylbutazone II: Analysis of phenylbutazone and its decomposition products in drugs by high-pressure liquid chromatography, *J. Pharm. Sci.* **71**, 120–122 (1982).
199. H. Fabre, N. Hussam-Eddine, D. Lerner, and B. Mandrou, Autoxidation and hydrolysis kinetics of the sodium salt of phenylbutazone in aqueous solution, *J. Pharm. Sci.* **73**, 1709–1713 (1984).
200. S. Yoshioka, H. Ogata, T. Shibazaki, and A. Ejima, Stability of sulpyrine. III. Copper(II) ion catalyzed oxidation by molecular oxygen, *Chem. Pharm. Bull.* **26**, 2723–2728 (1978).
201. S. Yoshioka, H. Ogata, T. Shibazaki, and A. Ejima, Oxidative degradation of sulpyrine by molecular oxygen, *Chem. Pharm. Bull.* **312**, 81–86 (1979).
202. S. Yeh and J. L. Lach, Stability of morphine in aqueous solution III. Kinetics of morphine degradation in aqueous solution, *J. Pharm. Sci.* **50**, 35–42 (1961).
203. G. Boccardi, C. Deleuze, M. Gachon, G. Palmisano, and J. P. Vergnaud, Autoxidation of tetrazepam in tablets: Prediction of degradation impurities from the oxidative behavior in solution, *J. Pharm. Sci.* **81**, 183–185 (1992).
204. I. H. Pitman, T. Higuchi, M. Alton, and R. Wiley, Deuterium isotope effects on degradation of hydrocortisone in aqueous solution, *J. Pharm. Sci.* **61**, 818–820 (1972).
205. D. E. Guttman and P. D. Meister, The kinetics of the base-catalyzed degradation of prednisolone, *J. Am. Pharm. Assoc.* **47**, 773–778 (1958).
206. M. Ogata, R. Shimizu, and H. Abe, New degradation product of spiradoline mesylate in aqueous solution: Formation of an imidazolidine ring, *J. Pharm. Sci.* **82**, 91–94 (1993).
207. K. Nord, J. Karlsen, and H. H. Tonnesen, Photochemical stability of biologically active compounds. IV. Photochemical degradation of chloroquine, *Int. J. Pharm.* **42**, 11–18 (1991).
208. S. Kristensen, A. Grislingaas, J. V. Greenhill, T. Skjetne, J. Karlsen, and H. H. Tonnesen, Photochemical stability of biologically active compounds: V. Photochemical degradation of primaquine in an aqueous medium, *Int. J. Pharm.* **100**, 15–23 (1993).
209. E. R. Garrett and T. E. Eble, Studies on the stability of fumagillin I. Photolytic degradation in alcohol solution, *J. Am. Pharm. Assoc., Sci. Ed.* **43**, 385–390 (1954).
210. T. E. Eble and E. R. Garrett, Studies on the stability of fumagillin II. Photolytic degradation of crystalline fumagillin, *J. Am. Pharm. Assoc., Sci. Ed.* **43**, 536–538 (1954).
211. L. J. Ravin, L. Kennon, and J. V. Swintosky, A note on the photosensitivity of phenothiazine derivatives, *J. Am. Pharm. Assoc., Sci. Ed.* **47**, 760 (1958).
212. I. A. Majeed, W. J. Murray, D. W. Newton, S. Othman, and W. A. Al-Turk, Spectrophotometric study of the photodecomposition kinetics of nifedipine, *J. Pharm. Pharmacol.* **39**, 1044–1046 (1987).
213. K. Thoma and R. Klimek, Untersuchungen zur Photoinstabilitat von Nifedipin 2. Mitt.: Einfluss von Milieubedingungen, *Pharm. Ind.* **47**, 319–327 (1985).
214. G. S. Sadana and A. B. Ghogare, Mechanistic studies on photolytic degradation of nifedipine by use of H-NMR and C-NMR spectroscopy, *Int. J. Pharm.* **70**, 195–199 (1991).
215. G. E. Wright and T. Y. Tang, Photooxidation of reserpine, *J. Pharm. Sci.* **61**, 299–300 (1972).
216. M. C. Bonferoni, G. Mellerio, P. Giunchedi, C. Caramella, and U. Conte, Photostability evaluation of nicardipine HCl solutions, *Int. J. Pharm.* **80**, 109–117 (1992).
217. A. L. Zanocco, L. Diaz, L. J. Nunez-Vergara, and J. A. Squella, Polarographic study of the photodecomposition of nimodipine, *J. Pharm. Sci.* **81**, 920–924 (1992).
218. M. Schiavi, S. Serafini, A. Italia, M. Villa, G. Fronza, and A. Selva, Identification of the major degradation products of 4-methoxy-2-(3-phenyl-2-propynyl)phenol formed by exposure to air and light, *J. Pharm. Sci.* **81**, 812–814 (1992).
219. F. Bosca, M. A. Miranda, and F. Vargas, Photochemistry of tiaprofenic acid, a nonsteroidal anti-inflammatory drug with phototoxic side effects, *J. Pharm. Sci.* **81**, 181–182 (1992).
220. M. Hanamori, T. Mizuno, K. Akimoto, and H. Nakagawa, Photo-stability of the new antiplatelet agent, KBT-3022 (ethyl 2-[4,5-bis(4-methoxyphenyl)thiazole-2-yl]pyrrol-1-ylacetate) in aqueous solutions containing acetonitrile, *Chem. Pharm. Bull.* **40**, 3048–3051 (1992).
221. T. Mizuno, M. Hanamori, K. Akimoto, H. Nakagawa, and K. Arakawa, Wavelength dependency and mechanism of the photodegradation of ethyl 2-[4,5-bis(4-methoxyphenyl)thiazole-2-yl]pyrrol-1-ylacetate (KBT-3022) in solution, *Chem. Pharm. Bull.* **42**, 160–162 (1994).

222. P. J. G. Cornelissen, and G.M.J.B. Henegouwen, Photochemical decomposition of 1,4-benzodiazepines. Quantitative analyses of decomposed solutions of chlordiazepoxide and diazepam, *Pharm. Weekbl. Sci. Ed.* **2**, 547–556 (1980).

223. H. H. Tonnesen and D. E. Moore, Photochemical stability of biologically active compounds. III. Mefloquine as a photosensitizer, *Int. J. Pharm.* **70**, 95–101 (1991).

224. N. Yagi, H. Kenmotsu, H. Sekikawa, and M. Takada, Studies on the photolysis and hydrolysis of furosemide in aqueous solution, *Chem. Pharm. Bull.* **39**, 454–457 (1991).

225. G. L. Mosher and J. McBee, Photoreactivity of LY277359 maleate, a 5-hydroxytryptamine3 (5-HT3) receptor antagonist, in solution, *Pharm. Res.* **8**, 1215–1222 (1991).

226. J. Philip and D. H. Szulczewski, Photolytic decomposition of *N*-(2,6-dichloro-*m*-tolyl)anthranilic acid (meclofenamic acid), *J. Pharm. Sci.* **62**, 1479–1482 (1973).

227. C. Lin, P. Perrier, G. G. Clay, P. A. Sutton, and S. R. Byrn, Solid-state photooxidation of 21-cortisol *tert*-butylacetate to 21-cortisone *tert*-butylacetate, *J. Org. Chem.* **47**, 2978–2981 (1982).

228. H. Okada, V. Stella, J. Haslam, and N. Yata, Photolytic degradation of α-[(dibutylamino)methyl]-6,8-dichloro-2-(30′,4′-dichlorophenyl)-4-quinoline methanol: An experimental antimalarial, *J. Pharm. Sci.* **64**, 1665–1667 (1975).

229. F. Vargas, C. Rivas, R. Machado, and Z. Sarabia, Photodegradation of benzydamine: Phototoxicity of an isolated photoproduct on erythrocytes, *J. Pharm. Sci.* **82**, 371–372 (1993).

230. J. C. Vire, G. J. Patriarche, and G. D. Christian, Electrochemical study of the degradation of vitamins of the K group, *Pharmazie* **35**, 209–212 (1980).

231. T. Higuchi and L. C. Schroeter, Kinetics and mechanism of formation of sulfonate from epinephrine and bisulfite, *J. Am. Chem. Soc.* **82**, 1904–1907 (1960).

232. G. B. Smith, L. M. Weinstock, F. E. Roberts, G. S. Brenner, A. M. Hoinowski, B. H. Arison, and A. W. Douglas, Kinetics of equilibration of bisulfite and dexamethasone-21-phosphate in aqueous solution, *J. Pharm. Sci.* **61**, 708–716 (1972).

233. R. N. Duvall, K. T. Koshy, and J. W. Pyles, Comparison of reactivity of amphetamine, methamphetamine, and dimethylamphetamine with lactose and related compounds, *J. Pharm. Sci.* **54**, 607–611 (1965).

234. W. Wu, T. Chin, and J. L. Lach, Interaction of isoniazid with magnesium oxide and lactose, *J. Pharm. Sci.* **59**, 1234–1242 (1970).

235. S. M. Blaug and W. Huang, Interaction of dextroamphetamine sulfate with spray-dried lactose, *J. Pharm. Sci.* **61**, 1770–1775 (1972).

236. S. M. Blaug and W. Huang, Browning of dextrates in solid-solid mixtures containing dextroamphetamine sulfate, *J. Pharm. Sci.* **63**, 1415–1418 (1974).

237. K. Kigazawa, N. Ikari, K. Ohkubo, H. Iimura, and S. Haga, Decomposition and stabilization of drugs. III. Reaction products of norphenylephrine with degradation products of sugars [in Japanese], *Yakugaku Zasshi* **93**, 925–927 (1973).

238. S. Yoshioka, H. Ogata, T. Shibazaki, and A. Ejima, Stability of sulpyrine. IV. Reaction with glucose in aqueous solution, *Chem. Pharm. Bull.* **27**, 907–915 (1979).

239. A. L. Jacobs, A. E. Dilatush, S. Weinstein, and J. J. Windheuser, Formation of acetylcodeine from aspirin and codeine, *J. Pharm. Sci.* **55**, 893–895 (1966).

240. L. Liu and E. L. Parrott, Solid-state reaction between sulfadiazine and acetylsalicylic acid, *J. Pharm. Sci.* **80**, 564–566 (1991).

241. K. T. Koshy, A. E. Troup, R. N. Duvall, R. C. Conwell, and L. L. Shankle, Acetylation of acetaminophen in tablet formulations containing aspirin, *J. Pharm. Sci.* **56**, 1117–1121 (1967).

242. H. W. Jun, C. W. Whitworth, and L. A. Luzzi, Decomposition of aspirin in polyethylene glycols, *J. Pharm. Sci.* **61**, 1160–1162 (1972).

243. C. W. Whitworth, H. W. Jun, and L. A. Luzzi, Stability of aspirin in liquid and semisolid bases I: Substituted and nonsubstituted polyethylene glycols, *J. Pharm. Sci.* **62**, 1184–1185 (1973).

244. V. Kumar and G. S. Banker, Incompatibility of polyvinyl acetate phthalate with benzocaine: Isolation and characterization of 4-phthalimidobenzoic acid ethyl ester, *Int. J. Pharm.* **79**, 61–65 (1992).

245. K. A. Connors, *Chemical Kinetics: The Study of Reaction Rates in Solution*, 1st ed., VCH Publishers, New York, 1990.

246. W. N. Charman, L. E. McCrossin, N. L. Pochopin, and S. L. A. Munro, Study of the relative lactonization rates of pilocarpic acid and isopilocarpic acid in acidic media, *Int. J. Pharm.* **88**, 397–404 (1992).

247. A. S. Kearney, S. C. Mehta, and G. W. Radebaugh, Aqueous stability and solubility of CI-988, a novel "dipeptoid" cholecystokinin-B receptor antagonist, *Pharm. Res.* **9**, 1092–1095 (1992).
248. A. S. Kearney, S. C. Mehta, and G. W. Radebaugh, The effect of structural changes on the intramolecular degradation of the "dipeptoid" CI-988, *Int. J. Pharm.* **92**, 63–70 (1993).
249. S. M. Blaug and J. W. Wesolowski, The stability of acetylsalicylic acid in suspension, *J. Am. Pharm. Assoc., Sci. Ed.* **48**, 691–694 (1959).
250. M. Konishi, K. Hirai, and Y. Mori, Kinetics and mechanism of the equilibrium reaction of triazolam in aqueous solution, *J. Pharm. Sci.* **71**, 1328–1334 (1982).
251. M. B. Devani, C. J. Shishoo, K. J. Doshi, and H. B. Patel, Kinetic studies of the interaction between isoniazid and reducing sugars, *J. Pharm. Sci.* **74**, 427–432 (1985).
252. T. Yamana, Y. Mizukami, A. Tsuji, and F. Ichimura, Studies on the stability of drugs. XX. Stability of benzothiadiazines.(11) Studies on the hydrolysis of hydrochlorothiazide in the various pH solutions [in Japanese], *Yakugaku Zasshi* **89**, 740–744 (1969).
253. J. A. Mollica, C. R. Rehm, J. B. Smith, and H. K. Govan, Hydrolysis of benzothiadiazines, *J. Pharm. Sci.* **60**, 1380–1384 (1971).
254. S. Yoshioka, Y. Aso, T. Shibazaki, and M. Uchiyama, Stability of pilocarpine ophthalmic formulation, *Chem. Pharm. Bull.* **34**, 4280–4286 (1986).
255. Y. Namiki, T. Tanabe, T. Kobayashi, J. Tanabe, Y. Okimura, S. Koda, and Y. Morimoto, Degradation kinetics and mechanisms of a new cephalosporin, cefixime, in aqueous solution, *J. Pharm. Sci.* **76**, 208–214 (1987).
256. M. Hara, H. Hayashi, T. Yoshida, and H. Murayama, Studies on the kinetics and mechanism of degradation of chlorphenesin carbamate in strongly alkaline aqueous solutions, *Chem. Pharm. Bull.* **34**, 1764–1769 (1986).
257. Y. Matsuhisa, K. Umemoto, K. Itakura, and Y. Usui, Degradation kinetics of carumonam in aqueous solution, *Chem. Pharm. Bull.* **34**, 3779–3790 (1986).
258. T. Yamana, Y. Mizukami, F. Ichimura, and H. Koike, Studies on the stability of drugs. XII. Stability of benzothiadiazines. (4). A simplified method for the calculation of rate constants of complicated reactions [in Japanese], *Yakugaku Zasshi* **84**, 974–976 (1964).
259. R. Oliyai, L. Yuan, T. C. Dahl, S. Swaminathan, K. Wang, and W. A. Lee, Biexponential decomposition of a neuraminidase inhibitor prodrug (GS-4104) in aqueous solution, *Pharm. Res.* **15**, 1300–1304 (1998).
260. H. Bundgaard, Polymerization of penicillins: Kinetics and mechanism of di- and polymerization of ampicillin in aqueous solution, *Acta Pharm. Suec.* **13**, 9–26 (1976).
261. H. Bundgaard, Polymerization of penicillins. II. Kinetics and mechanism of dimerization and self-catalyzed hydrolysis of amoxycillin in aqueous solution, *Acta Pharm. Suec.* **14**, 47–66 (1977).
262. H. Bundgaard, Polymerization of penicillins. III. Structural effects influencing rate of dimerization of amino-penicillins in aqueous solution, *Acta Pharm. Suec.* **14**, 67–80 (1977).
263. E. Nakashima, A. Tsuji, M. Nakamura, and T. Yamana, Physicochemical properties of amphoteric β-lactam antibiotics. IV. First- and second-order degradations of cefaclor and cefatrizine in aqueous solution and kinetic interpretation of the intestinal absorption and degradation of the concentrated antibiotics, *Chem. Pharm. Bull.* **33**, 2098–2106 (1985).
264. K. A. Connors and J. A. Mollica, Theoretical analysis of comparative studies of complex formation, *J. Pharm. Sci.* **55**, 772–780 (1966).
265. F. Sendo, C. M. Riley, and V. J. Stella, Kinetics of hydrolysis of fetindomide (NSC-373965), bis-*N,N'*-phenylalanyloxymethyl prodrug of mitindomide (NSC-284356); an unexpected catalytic effect of generated formaldehyde, *Int. J. Pharm.* **45**, 207–216 (1988).
266. W. Jander, Reaktionen im festen Zustande bei hoheren Temperaturen, I. Reaktionsgeschwindingkeiten endotherm verlaufender Umsetzungen, *Z. Anorg. Allg. Chem.* **163**, 1–30 (1927).
267. V. W. Jander and E. Hoffmann, Reaktionen im festen Zustande bei hoheren Temperaturen, Die Bestimmung von Reaktionsgeschwindigkeiten beiumsetzungen, die mit einer Gasabgave verbunden sind, *Z. Anorg. Allg. Chem.* **200**, 245–256 (1931).
268. E. G. Prout and F. C. Tompkins, The thermal decomposition of mercuric oxalate, *Trans. Faraday Soc.* **43**, 148–157 (1947).
269. K. Okazaki, R. Nishigaki, and M. Hanano, Equation and accelerating factor of degradation in cocarboxylase free ester lyophilizate [in Japanese], *Yakuzaigaku* **49**, 141–147 (1989).

270. K. Okazaki, R. Nishigaki, and M. Hanano, Relationship between degradation rate and relative humidity, temperature, or physical pretreatments of cocarboxylase free ester lyophilizate [in Japanese], *Yakuzaigaku* **50**, 141–148 (1990).

271. M. Horioka, T. Aoyama, K. Takata, T. Maeda, and K. Shirahama, Degradation of propantheline bromide and dried aluminum hydroxide gel in powdered preparations, *Yakuzaigaku* **34**, 16–21 (1974).

272. S. Yoshioka, T. Shibazaki, and A. Ejima, Stability of solid dosage forms. I. Hydrolysis of meclofenoxate hydrochloride in the solid state, *Chem. Pharm. Bull.* **30**, 3734–3741 (1982).

273. S. Yoshioka, T. Shibazaki, and A. Ejima, Stability of solid dosage forms. II. Hydrolysis of meclofenoxate hydrochloride in commercial tablets, *Chem. Pharm. Bull.* **31**, 2513–2517 (1983).

274. S. Yoshioka and M. Uchiyama, Kinetics and mechanism of the solid-state decomposition of propantheline bromide, *J. Pharm. Sci.* **75**, 92–96 (1986).

275. J. Hasegawa, M. Hanano, and S. Awazu, Decomposition of acetylsalicylic acid and its derivatives in solid state, *Chem. Pharm. Bull.* **23**, 86–97 (1975).

276. E. G. Prout and F. C. Tompkins, The thermal decomposition of potassium permanganate, *Trans. Faraday Soc.* **40**, 188–198 (1944).

277. J. T. Carstensen and F. Attarchi, Decomposition of aspirin in the solid state in the presence of limited amounts of moisture. III. Effect of temperature and a possible mechanism, *J. Pharm. Sci.* **77**, 318–321 (1988).

278. K. Kawakita, The kinetics of the reduction of ferric oxide, *Rev. Phys. Chem. Jpn.* **14**, 79–85 (1940).

279. M. Avrami, Kinetics of phase change. I. General theory, *J. Chem. Phys.* **7**, 1103–1112 (1939).

280. M. Avrami, Kinetics of phase change. II. Transformation–time relations for random distribution of nuclei, *J. Chem. Phys.* **8**, 212–224 (1940).

281. J. T. Carstensen and M. N. Musa, Decomposition of benzoic acid derivatives in solid state, *J. Pharm. Sci.* **61**, 1112–1118 (1972).

282. J. T. Carstensen and R. C. Kothari, Decarboxylation kinetics of 5-(tetradecyloxy)-2-furoic acid, *J. Pharm. Sci.* **69**, 123–124 (1980).

283. J. T. Carstensen and R. Kothari, Solid-state decomposition of alkoxyfuroic acids, *J. Pharm. Sci.* **70**, 1095–1100 (1981).

284. J. T. Carstensen and R. C. Kothari, Solid-state decomposition of alkoxyfuroic acids in the presence of microcrystalline cellulose, *J. Pharm. Sci.* **72**, 1149–1154 (1983).

285. L. J. Leeson and A. M. Mattocks, Decomposition of aspirin in the solid state, *J. Am. Pharm. Assoc., Sci. Ed.* **47**, 329–333 (1958).

286. W. Yang and D. Brooke, Rate equation for solid state decomposition of aspirin in the presence of moisture, *Int. J. Pharm.* **11**, 271–276 (1982).

287. J. T. Carstensen, F. Attarchi, and X. Hou, Decomposition of aspirin in the solid state in the presence of limited amounts of moisture, *J. Pharm. Sci.* **74**, 741–745 (1985).

288. J. T. Carstensen and F. Attarchi, Decomposition of aspirin in the solid state in the presence of limited amounts of moisture. II. Kinetics and salting-in of aspirin in aqueous acetic acid solutions, *J. Pharm. Sci.* **77**, 314–317 (1988).

289. S. Yoshioka, H. Ogata, T. Shibazaki, and A. Ejima, Stability of sulpyrine. V. Oxidation with molecular oxygen in the solid state, *Chem. Pharm. Bull.* **27**, 2363–2371 (1979).

290. S. S. Komblum and B. J. Sciarrone, Decarboxylation of *p*-aminosalicylic acid in the solid state, *J. Pharm. Sci.* **53**, 935–941 (1964).

291. J. T. Carstensen and P. Pothisiri, Decomposition of *p*-aminosalicylic acid in the solid state, *J. Pharm. Sci.* **64**, 37–44 (1975).

292. N. Okusa, Prediction of stability of drugs. III. Application of Weibull probability paper to prediction of stability, *Chem. Pharm. Bull.* **23**, 794–802 (1975).

293. H. Seki, T. Hayashi, and N. Okusa, An application of weighted least-squares analysis to Weibull probability paper for prediction of stability of drug products [in Japanese], *Yakuzaigaku* **40**, 201–207 (1980).

294. K. C. Yeh, Kinetic parameter estimation by numerical algorithms and multiple liner regression: Theoretical, *J. Pharm. Sci.* **66**, 1688–1691 (1977).

295. S. A. Sande and J. Karlsen, Curve fitting of stability data by personal computer. Software in pharmaceutics II, *Int. J. Pharm.* **73**, 147–156 (1991).

296. D. S. Kalonia and A. P. Simonelli, Analysis of consecutive pseudo-first-order reactions II: Calculation of the rate constants from the co-product or co-reactant data, *J. Pharm. Sci.* **78**, 78–84 (1989).

297. K. Yamaoka, Y. Tanigawara, T. Nakagawa, and T. Uno, A pharmacokinetic analysis program (MULTI) for microcomputer, *J. Pharmacobio-Dyn.* **4**, 879–885 (1981).
298. S. Arrhenius, Uber die Reactionsgeschwindigkeit bei der Inversion von Rohrzucker durch Sauren, *Z. Phys. Chem.* **4**, 226–248 (1889).
299. H. Eyring, The activated complex in chemical reactions, *J. Chem. Phys.* **3**, 107–115 (1935).
300. R. Brodersen, Inactivation of penicillin G by acids as a function of temperature, *Acta Chem. Scand.* **1**, 403–414 (1947).
301. E. R. Garrett and R. F. Carper, Prediction of stability in pharmaceutical preparations I. Color stability in a liquid multisulfa preparations, *J. Am. Pharm. Assoc., Sci. Ed.* **44**, 515–518 (1955).
302. E. R. Garrett, Prediction of stability in pharmaceutical preparations II. Vitamin stability in liquid multivitamin preparations, *J. Am. Pharm. Assoc., Sci. Ed.* **45**, 171–178 (1956).
303. E. R. Garrett, Prediction of stability in pharmaceutical preparations III. Comparison of vitamin stabilities in different multivitamin preparations, *J. Am. Pharm. Assoc., Sci. Ed.* **45**, 470–473 (1956).
304. J. T. Carstensen, Stability patterns of vitamin A in various pharmaceutical dosage forms, *J. Pharm. Sci.* **53**, 839–840 (1964).
305. J. T. Carstensen, J. B. Johnson, W. Valentine, and J. J. Vance, Extrapolation of appearance of tablets and powders from accelerated storage tests, *J. Pharm. Sci.* **53**, 1050–1054 (1964).
306. I. Horikoshi and S. Hata, Physicochemical studies on water contained in solid medicaments. VI. On the evaluating of the coloring rate of solid medicaments [in Japanese], *Yakugaku Zasshi* **86**, 356–366 (1966).
307. R. Tardif, Reliability of accelerated storage tests to predict stability of vitamins (A, B_1, C) in tablets, *J. Pharm. Sci.* **54**, 281–284 (1965).
308. F. Matsui, D. L. Robertson, P. Lafontaine, H. Kolasinski, and E. G. Lovering, Stability studies of phenylbutazone and phenylbutazone–antacid oral formulations, *J. Pharm. Sci.* **67**, 646–650 (1978).
309. M. Grit, W. J. M. Underberg, and D. J. A. Crommelin, Hydrolysis of saturated soybean phosphatidylcholine in aqueous liposome dispersions, *J. Pharm. Sci.* **82**, 362–366 (1993).
310. J. T. Carstensen and T. Morris, Chemical stability of indomethacin in the solid amorphous and molten states, *J. Pharm. Sci.* **82**, 657–659 (1993).
311. A. K. Kishore and J. B. Nagwekar, Influence of temperature and hydrophobic group-associated icebergs on the activation energy of drug decomposition and its implication in drug shelf-life prediction, *Pharm. Res.* **7**, 730–735 (1990).
312. B. D. Anderson and R. T. Darrington, Letters to the editor, *Pharm. Res.* **8**, 661 (1991).
313. J. B. Nagwekar, Letters to the editor, *Pharm. Res.* **8**, 661–662 (1991).
314. J. G. Slater, H. A. Stone, B. T. Palermo, and R. N. Duvall, Reliability of Arrhenius equation in predicting vitamin A stability in multivitamin tablets, *J. Pharm. Sci.* **68**, 49–52 (1979).
315. O. L. Davies and D. A. Budgett, Accelerated storage tests on pharmaceutical products: Effect of error structure of assay and errors in recorded temperature, *J. Pharm. Pharmacol.* **32**, 155–159 (1980).
316. K. D. Ertel and J. T. Carstensen, Examination of a modified Arrhenius relationship for pharmaceutical stability prediction, *Int. J. Pharm.* **61**, 9–14 (1990).
317. D. L. Bentley, Statistical techniques in predicting thermal stability, *J. Pharm. Sci.* **59**, 464–468 (1970).
318. J. T. Carstensen and K. S. E. Su, Statistical aspects of Arrhenius plotting, *Bull. Parenteral Drug Assoc.* **25**, 287–302 (1971).
319. S. P. King, M. Kung, and H. Fung, Statistical prediction of drug stability based on nonlinear parameter estimation, *J. Pharm. Sci.* **73**, 657–662 (1984).
320. H. J. Borchardt and F. Daniels, The application of differential thermal analysis to the study of reaction kinetics, *J. Am. Chem. Soc.* **79**, 41–46 (1957).
321. R. E. Davis, Temperature as a variable during a kinetic experiment, *J. Phys. Chem.* **63**, 307–309 (1959).
322. A. R. Rogers, An accelerated storage test with programmed temperature rise, *J. Pharm. Pharmacol.* **15**, 101T–105T (1963).
323. S. P. Eriksen and H. Stelmach, Single step stability studies, *J. Pharm. Sci.* **54**, 1029–1034 (1965).
324. B. R. Cole and L. Leadbeater, A critical assessment of an accelerated storage test, *J. Pharm. Pharmacol.* **18**, 101–111 (1966).
325. J. T. Carstensen, A. Koff, and S. H. Rubin, Programmed kinetic studies, *J. Pharm. Pharmacol.* **20**, 485–486 (1968).
326. M. A. Zoglio, J. J. Windheuser, R. Vatti, H. V. Maulding, S. S. Kornblum, A. Jacobs, and H. Hamot, Linear nonisothermal stability studies, *J. Pharm. Sci.* **57**, 2080–2085 (1968).

327. H. V. Maulding and M. A. Zoglio, Flexible nonisothermal stability studies, *J. Pharm. Sci.* **59**, 333–337 (1970).
328. M. A. Zoglio, H. V. Maulding, W. H. Streng, and W. C. Vincek, Nonisothermal kinetic studies III: Rapid nonisothermal–isothermal method for stability prediction, *J. Pharm. Sci.* **64**, 1381–1383 (1975).
329. A. I. Kay and T. H. Simon, Use of an analog computer to simulate and interpret data obtained from linear nonisothermal stability studies, *J. Pharm. Sci.* **60**, 205–208 (1971).
330. B. Edel and M. O. Baltzer, Nonisothermal kinetics with programmed temperature steps, *J. Pharm. Sci.* **69**, 287–290 (1980).
331. B. W. Madsen, R. A. Anderson, D. Herbison-Evans, and W. Sneddon, Integral approach to nonisothermal estimation of activation energies, *J. Pharm. Sci.* **63**, 777–781 (1974).
332. I. G. Tucker and W. R. Owen, Estimation of all parameters from nonisothermal kinetic data, *J. Pharm. Sci.* **71**, 969–974 (1982).
333. J. M. Hempenstall, W. J. Irwin, A. Li Wan Po, and A. H. Andrews, Nonisothermal kinetics using a microcomputer: A derivative approach to the prediction of the stability of penicillin formulations, *J. Pharm. Sci.* **72**, 668–673 (1983).
334. S. Yoshioka, Y. Aso, and M. Uchiyama, Statistical evaluation of nonisothermal prediction of drug stability, *J. Pharm. Sci.* **76**, 794–798 (1987).
335. N. Okusa and K. Kinuno, Kinetic studies on prediction of stability of pharmaceuticals. 1. Kinetic analysis by a nonisothermal method [in Japanese], *Yakuzaigaku* **28**, 17–20 (1968).
336. N. Okusa and K. Kinuno, Kinetic studies on prediction of stability of pharmaceuticals. 2. Prediction of stability of adenosine-triphosphate disodium powders by a nonisothermal method [in Japanese], *Yakuzaigaku* **28**, 23–26 (1968).
337. I. Utsumi, I. Sugimoto, M. Kobayashi, and M. Ohtsuka, Estimation of the stability of pharmaceutical by nonisothermal method [in Japanese], *Yakuzaigaku* **34**, 173–179 (1974).
338. N. Okusa, Prediction of stability of drugs. IV. Prediction of stability by multilevel nonisothermal method, *Chem. Pharm. Bull.* **23**, 803–809 (1975).
339. J. Waltersson and P. Lundgren, Nonisothermal kinetics applied to drugs in pharmaceutical suspensions, *Acta Pharm. Suec.* **20**, 145–154 (1983).
340. J. Waltersson and P. Lundgren, Nonisothermal kinetics applied to pharmaceuticals, *Acta Pharm. Suec.* **20**, 145–154 (1983).
341. A. Li Wan Po, A. N. Elias, and W. J. Irwin, Non-isothermal and non-isopH kinetics in formulation studies, *Acta Pharm. Suec.* **20**, 277–286 (1983).
342. D. Kersten and B. Goeber, Auswertung nichtisothermer Stabilitatstests am Beispiel der Rohrzuckerinversion, *Pharmazie* **39**, 393–395 (1984).
343. D. Kersten and B. Goeber, Zur Auswertung nichtisothermer Kurzzeittests von Arzneistoffloesungen, *Pharmazie* **39**, 541–546 (1984).
344. J. E. Kipp, M. M. Jensen, K. Kronholm, and M. McHalsky, Automated liquid chromatography for non-isothermal kinetic studies, *Int. J. Pharm.* **34**, 1–8 (1986).
345. K. Inaba and J. Mizutani, Prediction of stability for prostaglandin E_1–α-cyclodextrin complex [in Japanese], *Yakuzaigaku* **46**, 58–62 (1986).
346. K. Ellstrom and H. Nyqvist, Non-isothermal stability testing of drug substances in the solid state, *Acta Pharm. Suec.* **24**, 115–122 (1987).
347. J. E. Kipp and J. J. Hlavaty, Nonisothermal stability assessment of stable pharmaceuticals: Testing of a clinamycin phosphate formulation, *Pharm. Res.* **8**, 570–575 (1991).
348. S. Yoshioka, Y. Aso, and M. Uchiyama, Statistical evaluation of non-isothermal prediction of drug stability. II. Experimental design for practical drug products, *Int. J. Pharm.* **46**, 121–132 (1988).
349. S. Yoshioka, Y. Aso, and Y. Takeda, Isothermal and nonisothermal kinetics in the stability prediction of vitamin A preparations, *Pharm. Res.* **7**, 388–391 (1990).
350. N. Hashimoto, T. Ichihashi, E. Yamamoto, K. Hirano, M. Inoue, H. Tanaka, and H. Yamada, Epimerization kinetics of moxalactam in frozen solution, *Pharm. Res.* **5**, 266–271 (1988).
351. R. Shija, V. B. Sunderland, and C. McDonald, Alkaline hydrolysis of methyl, ethyl and *n*-propyl 4-hydroxybenzoate esters in the liquid and frozen states, *Int. J. Pharm.* **80**, 203–211 (1992).
352. C. McDonald, V. B. Sunderland, H. Lau, and R. Shija, The stability of amoxycillin sodium in normal saline and glucose (5%) solutions in the liquid and frozen states, *J. Clin. Pharm. Ther.* **14**, 45–52 (1989).
353. E. Hayakawa, K. Sugiyama, K. Furuya, and T. Kuroda, Stability of mitomycin C and adriamycin in frozen state [in Japanese], *Yakuzaigaku* **49**, 289–296 (1989).

354. R. Brodersen, Inactivation of penicillin G by acids—a reaction-kinetic investigation, *Trans. Faraday Soc.* **43**, 351–355 (1947).
355. O. A. G. J. van der Houwen, J. H. Beijnen, A. Bult, and W. J. M. Underberg, A general approach to the interpretation of pH degradation profiles, *Int. J. Pharm.* **45**, 181–188 (1988).
356. O. A. G. J. van der Houwen, O. Bekers, J. H. Beijnen, A. Bult, and W. J. M. Underberg, A general approach to the interpretation of pH degradation profiles in the presence of ligands, *Int. J. Pharm.* **67**, 155–162 (1991).
357. T. Loftsson, B. J. Olafsdottir, and J. Baldvinsdottir, Estramustine: Hydrolysis, solubilization, and stabilization in aqueous solutions, *Int. J. Pharm.* **79**, 107–112 (1992).
358. C. M. Won and A. K. Iula, Kinetics of hydrolysis of diltiazem, *Int. J. Pharm.* **79**, 183–190 (1992).
359. D. M. Johnson, W. F. Taylor, G. F. Thompson, and R. A. Pritchard, Degradation of fenprostalene in aqueous solution, *J. Pharm. Sci.* **72**, 946–948 (1983).
360. J. D. de Vries, J. Winkelhorst, W. J. M. Underberg, R. E. C. Henrar, and J. H. Beijnen, A systematic study on the chemical stability of the novel indoloquinone antitumour agent EO9, *Int. J. Pharm.* **100**, 181–188 (1993).
361. O. A. G. J. van der Houwen, J. H. Beijnen, A. Bult, O. Bekers, R. M. J. Goossen, and W. J. M. Underberg, Degradation kinetics of the antitumour drug nimustine (ACNU) in aqueous solution, *Int. J. Pharm.* **99**, 73–78 (1993).
362. C. E. Pasini and J. M. Indelicato, Pharmaceutical properties of loracarbef: The remarkable solution stability of an oral 1-carba-1-dethiacephalosporin antibiotic, *Pharm. Res.* **9**, 250–254 (1992).
363. T. Yamana, A. Tsuji, and Y. Mizukami, Kinetic approach to the development in β-lactam antibiotics. I. Comparative stability of semisynthetic penicillins and 6-aminopenicillanic acid in aqueous solution, *Chem. Pharm. Bull.* **22**, 1186–1197 (1974).
364. L. J. Edwards, The hydrolysis of aspirin, *Trans. Faraday Soc.* **48**, 696–699 (1952).
365. A. R. Fersht and A. J. Kirby, The hydrolysis of aspirin. Intramolecular general base catalysis of ester hydrolysis, *J. Am. Chem. Soc.* **89**, 4857–4863 (1967).
366. C. M. Won, J. J. Zalipsky, D. M. Patel, and N. H. Reavey-Cantwell, Kinetics of hydrolysis of fenclorac, *J. Pharm. Sci.* **66**, 73–77 (1977).
367. M. Brandl and J. A. Kennedy, Degradation of the antiarthritic prodrug, 3-carboxy-5-methyl-*N*-[4-(trifluoromethoxy)phenyl]-4-isoxazolecarboxamide, in aqueous solution, *Pharm. Res.* **11**, 345–348 (1994).
368. T. Fujita and A. Koshiro, Stability of flomoxef in aqueous solution and in intravenous admixture of 5% glucose infusion, [in Japanese], *Yakugaku Zasshi* **108**, 777–787 (1988).
369. N. Muhammad, G. Adams, and H. Lee, Kinetics and mechanisms of vinpocetine degradation in aqueous solutions, *J. Pharm. Sci.* **77**, 126–131 (1988).
370. C. F. Carney, Solution stability of ciclosidomine, *J. Pharm. Sci.* **76**, 393–397 (1987).
371. A. K. Mitra and M. M. Narurkar, Kinetics of azathioprine degradation in aqueous solution, *Int. J. Pharm.* **35**, 165–171 (1986).
372. F. Ichimura, O. Nakano, K. Shimazaki, M. Kimura, M. Suda, K. Wakiya, and T. Yamana, Stability kinetics of cefotiam dihydrochloride in aqueous solution [in Japanese], *Yakuzaigaku* **43**, 169–173 (1983).
373. S. G. Jivani and V. J. Stella, Mechanism of decarboxylation of *p*-aminosalicylic acid, *J. Pharm. Sci.* **74**, 1274–1282 (1985).
374. A. S. Kearney and V. J. Stella, Hydrolysis of pharmaceutically relevant phosphate monoester monoanions: Correlation to an established structure–reactivity relationship, *J. Pharm. Sci.* **82**, 69–72 (1993).
375. B. V. Shetty, R. L. Schowen, M. Slavik, and C. M. Riley, Degradation of dacarbazine in aqueous solution, *J. Pharm. Biomed. Anal.* **10**, 675–683 (1992).
376. J. M. Sisco and V. J. Stella, An unexpected hydrolysis pH–rate profile, at pH values less than 7, of the labile imide, ICRF-187:(+)-1,2-bis-(3,5-dioxopiperazin-1-yl)propane, *Pharm. Res.* **9**, 1209–1214 (1992).
377. C. M. Won, Kinetics and mechanism of hydrolysis of tyrphostins, *Int. J. Pharm.* **104**, 29–40 (1994).
378. M. A. Schwarts, Catalysis of penicillin hydrolysis by catechol and 3,6-bis(dimethylaminomethyl)catechol, *J. Pharm. Sci.* **53**, 1433 (1964).
379. R. J. Prankerd, J. M. Walters, and J. H. Parnes, Kinetics for degradation of rifampicin, an azomethine-containing drug which exhibits reversible hydrolysis in acidic solutions, *Int. J. Pharm.* **78**, 59–67 (1992).
380. R. G. Strickley, G. C. Visor, L. Lin, and L. Gu, An unexpected pH effect on the stability of moexipril lyophilized powder, *Pharm. Res.* **6**, 971–975 (1989).
381. P. Finholt, G. Jurgensen, and H. Kristiansen, Catalytic effect of buffers on degradation of penicillin G in aqueous solution, *J. Pharm. Sci.* **54**, 387–393 (1965).

382. H. Zia, M. Teharan, and R. Zargarbashi, Kinetics of carbenicillin degradation in aqueous solutions, *Can. J. Pharm. Sci.* **9**, 112–117 (1974).
383. M. F. Powell, Enhanced stability of codeine sulfate: Effect of pH, buffer, and temperature on the degradation of codeine in aqueous solution, *J. Pharm. Sci.* **75**, 901–903 (1986).
384. Y. Pramar and V. D. Gupta, Preformulation studies of spironolactone: Effect of pH, two buffer species, ionic strength, and temperature on stability, *J. Pharm. Sci.* **80**, 551–553 (1991).
385. G. K. Poochikian, J. C. Cradock, and J. P. Davignon, Heroin: Stability and formulation approaches, *Int. J. Pharm.* **13**, 219–226 (1983).
386. A. R. Fersht and A. J. Kirby, Intramolecular nucleophilic catalysis of ester hydrolysis by the ionized carboxyl group. The hydrolysis of 3,5-dinitroaspirin anion, *J. Am. Chem. Soc.* **90**, 5818–5826 (1968).
387. R. B. Killion and V. J. Stella, The nucleophilicity of dextrose, sucrose, sorbitol, and mannitol with *p*-nitrophenyl esters in aqueous solution, *Int. J. Pharm.* **66**, 149–155 (1990).
388. T. Fujita, Y. Harima, and A. Koshiro, Kinetic study of the degradation of cefotiam dihydrochloride in aqueous solution and of the reaction with aminoglycosides, *Chem. Pharm. Bull.* **31**, 2103–2109 (1983).
389. M. Zajac and A. Jelinska, Effect of amino acids, amines and polyhydroxy compounds on the stability of latamoxef, *Pharmazie* **46**, 115–117 (1991).
390. J. N. Bronsted, Zur Theorie der chemischen Reaktionsgeschwindigkeit, *Z. Phys. Chem.* **102**, 169–207 (1922).
391. N. Bjerrum, Zur Theorie der chemischen Reaktionsgeschwindigkeit, *Z. Phys. Chem.*, **108**, 82–100 (1924).
392. J. N. Bronsted, Zur Theorie der chemischen Reaktionsgeschwindigkeit. II., *Z. Phys. Chem.* **115**, 337–364 (1925).
393. N. Bjerrum, Zur Theorie der chemischen Reaktionsgeschwindigkeit. II., *Z. Phys. Chem.* **118**, 251–254 (1925).
394. J. T. Carstensen, Kinetic salt effect in pharmaceutical investigations, *J. Pharm. Sci.* **59**, 1140–1143 (1970).
395. G. N. Singh, R. P. Gupta, and P. Prakash. Effect of ionic strength on the stability of norfloxacin, *Pharmazie* **43**, 134 (1988).
396. E. S. Amis, Coulomb's law and the quantitative interpretation of reaction rates, II, *J. Chem. Educ.* **30**, 351–353 (1953).
397. A. D. Marcus and A. J. Taraszka, A kinetic study of the specific hydrogen ion catalyzed solvolysis of chloramphenicol in water–propylene glycol systems, *J. Am. Pharm. Assoc., Sci. Ed.* **48**, 77–84 (1959).
398. S. Singh and R. L. Gupta, Dielectric constant effects on degradation of azathioprine in solution, *Int. J. Pharm.* **46**, 267–270 (1988).
399. K. Ikeda, Studies on decomposition and stabilization of drugs in solution. IV. Effect of dielectric constant on the stabilization of barbiturate in alcohol–water mixtures, *Chem. Pharm. Bull.* **8**, 504–509 (1960).
400. K. Thoma and M. Strive, Stabilisierung von Barbitursaure-Derivaten in wassrigen Loesungen, 2. Mitt.: Beziehungen zwishen der Hydrolysegeschwindigkeit von Barbitursaure-Derivaten und der Dielektrizitatskonstante der Losungsmittel, *Pharm. Ind.* **47**, 1190–1194 (1985).
401. W. P. Adams and H. B. Kostenbauder, Phenoxybenzamine stability in aqueous ethanolic solutions. II. Solvent effects upon kinetics, *Int. J. Pharm.* **25**, 313–327 (1985).
402. K. Akimoto, H. Nakagawa, and I. Sugimoto, Photo-stability of cianidanol in aqueous solution, *Drug Dev. Ind. Pharm.* **11**, 865–889 (1985).
403. K. Akimoto, K. Kurosaka, H. Nakagawa, and I. Sugimoto, A new approach to evaluating photo-stability of nifedipine and its derivatives in solution by actinometry, *Chem. Pharm. Bull.* **36**, 1483–1490 (1988).
404. I. Sugimoto, K. Tohgo, K. Sasaki, H. Nakagawa, Y. Matsuda, and R. Masahara, Wavelength dependency of the photodegradation of nifedipine tablets [in Japanese], *Yakugaku Zasshi* **101**, 1149–1153 (1981).
405. Y. Matsuda and Y. Minamida, Stability of solid dosage forms. II. Coloration and photolytic degradation of sulfisomidine tablets by exaggerated ultraviolet irradiation, *Chem. Pharm. Bull.* **24**, 2229–2236 (1976).
406. Y. Matsuda and Y. Minamida, Stability of solid dosage forms. I. Fading of several colored tablets by exaggerated light [in Japanese], *Yakugaku Zasshi* **96**, 425–433 (1976).
407. R. Teraoka and Y. Matsuda, Stabilization-oriented preformulation study of photolabile menatetrenone (vitamin K2), *Int. J. Pharm.* **93**, 85–90 (1993).
408. Y. Matsuda and R. Masahara, Comparative evaluation of coloration of photosensitive solid drugs under various light sources, *Yakugaku Zasshi*, **100**, 953–957 (1980).
409. M. Okamura, M. Hanano, and S. Awazu, Relationship between the morphological character of 5-nitroacetylsalicylic acid crystals and the decomposition rate in a humid environment, *Chem. Pharm. Bull.* **28**, 578–584 (1980).

410. F. U. Krahn and J. B. Mielck, Effect of type and extent of crystalline order on chemical and physical stability of carbamazepine, *Int. J. Pharm.* **53**, 25–34 (1989).
411. Y. Matsuda, R. Akazawa, R. Teraoka, and M. Totsuka, Pharmaceutical evaluation of carbamazepine modifications: Comparative study for photostability of carbamazepine polymorphs by using Fourier-transformed reflection–absorption infrared spectroscopy and colorimetric measurement, *J. Pharm. Pharmacol.* **46**, 162–167 (1994).
412. M. M. DeVilliers, J. G. van der Watt, and A. P. Lotter, Kinetic study of the solid-state photolytic degradation of two polymorphic forms of furosemide, *Int. J. Pharm.* **88**, 275–283 (1992).
413. M. J. Pikal, A. L. Lukes, J. E. Lang, and K. Gaines, Quantitative crystallinity determinations for β-lactam antibiotics by solution calorimetry: Correlations with stability, *J. Pharm. Sci.* **67**, 767–773 (1978).
414. J. Waltersson and P. Lundgren, The effect of mechanical comminution on drug stability, *Acta Pharm. Suec.* **22**, 291–300 (1985).
415. H. Weng and E. L. Parrott, Solid-solid reaction between sulfacetamide and phthalic anhydride, *J. Pharm. Sci.* **73**, 1059–1063 (1984).
416. R. Yamamoto and S. Kawai, Studies on the stability of dry preparations. VII. Relation between atmospheric humidity and the stability of ascorbic acid, sodium ascorbate, and their diluted preparations [in Japanese], *Yakuzaigaku* **19**, 35–39 (1959).
417. R. Yamamoto, Studies on the stability of dry preparations. VIII. Relation between the moisture content and stability of ascorbic acid, sodium ascorbate and their diluted preparations [in Japanese], *Yakuzaigaku* **19**, 39–43 (1959).
418. R. Yamamoto and K. Inazu, Studies on the stability of dry preparations. IX. Relation between atmospheric humidity or the moisture content and stability of diluted preparations of various thiamine salts [in Japanese], *Yakuzaigaku* **19**, 113–117 (1959).
419. R. Yamamoto and T. Takahashi, Studies on the stability of dry preparations. VI. Relationship between the appearance and stability of thiamine salts in powders [in Japanese], *Yakugaku Zasshi* **79**, 419–422 (1959).
420. R. Yamamoto and K. Inazu, Studies on the stability of dry preparations. X. Relation between atmospheric humidity and stability of diluted preparations of acetylsalicylic acid [in Japanese], *Yakuzaigaku* **19**, 117–119 (1959).
421. J. T. Carstensen, E. S. Aron, D. C. Spera, and J. J. Vance, Moisture stress tests in stability programs, *J. Pharm. Sci.* **55**, 561–563 (1966).
422. R. Teraoka, M. Otsuka, and Y. Matsuda, Effects of temperature and relative humidity on the solid-state chemical stability of ranitidine hydrochloride, *J. Pharm. Sci.* **82**, 601–604 (1993).
423. I. Horikoshi and I. Himuro, Physicochemical studies on water contained in solid medicaments. V. Relationship between the character of contained water and the coloring of solid medicaments [in Japanese], *Yakugaku Zasshi* **86**, 353–356 (1966).
424. H. Nakagawa, Y. Takahashi, and I. Sugimoto, The effects of grinding and drying on the solid state stability of sodium prasterone sulfate, *Chem. Pharm. Bull.* **30**, 242–248 (1982).
425. Y. Takahashi, K. Nakashima, H. Nakagawa, and I. Sugimoto, Effects of grinding and drying on the solid-state stability of ampicillin trihydrate, *Chem. Pharm. Bull.* **32**, 4963–4984 (1984).
426. S. Kitamura, A. Miyamae, S. Koda, and Y. Morimoto, Effect of grinding on the solid-state stability of cefixime trihydrate, *Int. J. Pharm.* **56**, 125–134 (1989).
427. S. Kitamura, S. Koda, A. Miyamae, T. Yasuda, and Y. Morimoto, Dehydration effect on the stability of cefixime trihydrate, *Int. J. Pharm.* **59**, 217–224 (1990).
428. D. Genton and U. W. Kesselring, Effect of temperature and relative humidity on nitrazepam stability in solid state, *J. Pharm. Sci.* **66**, 676–680 (1977).
429. M. Zajac, Stability of crystalline mecillinam sodium salt in the solid phase, *Acta Pol. Pharm.* **41**, 659–664 (1984).
430. Z. Plotkowiak, The effect of the chemical character of certain penicillins on the stability of the β-lactam group in their molecules, Part 7: Effect of humidity in the solid state, *Pharmazie* **44**, 837–839 (1989).
431. M. Aruga, S. Awazu, and M. Hanano, Kinetic studies on decomposition of glutathione. I. Decomposition in solid state, *Chem. Pharm. Bull.* **26**, 2081–2091 (1978).
432. S. Yoshioka and M. Uchiyama, Nonlinear estimation of kinetic parameters for solid-state hydrolysis of water-soluble drugs, *J. Pharm. Sci.* **75**, 459–462 (1986).

433. S. Yoshioka and J. T. Carstensen, Nonlinear estimation of kinetic parameters for solid-state hydrolysis of water-soluble drugs. II. Rational presentation mode below the critical moisture content, *J. Pharm. Sci.* **79**, 799–801 (1990).
434. F. J. Bandelin and J. V. Tuschhoff, The stability of ascorbic acid in various liquid media, *J. Am. Pharm. Assoc., Sci. Ed.* **44**, 241–244 (1955).
435. R. Yamamoto and T. Takahashi, Studies on the stability of dry preparations. III. Decomposition of thiamine by adsorbents. (1) [in Japanese], *Yakugaku Zasshi* **78**, 1242–1245 (1958).
436. R. Yamamoto and T. Takahashi, Studies on the stability of dry preparations. IV. Decomposition of thiamine by absorbents. (2) [in Japanese], *Yakugaku Zasshi* **78**, 1246–1249 (1958).
437. R. A. Castello and A. M. Mattocks, Discoloration of tablets containing amines and lactose, *J. Pharm. Sci.* **51**, 106–108 (1962).
438. G. Gold and J. A. Campbell, Effects of selected U.S.P. talcs on acetylsalicylic acid stability in tablets, *J. Pharm. Sci.* **53**, 52–54 (1964).
439. H. V. Maulding, M. A. Zoglio, and E. J. Johnston, Pharmaceutical heterogeneous systems II. Study of hydrolysis of aspirin in combination with fatty acid tablet lubricants, *J. Pharm. Sci.* **57**, 1873–1876 (1968).
440. M. A. Zoglio, H. V. Maulding, R. M. Haller, and S. Briggen, Pharmaceutical heterogeneous systems III. Inhibition of stearate lubricant induced degradation of aspirin by the use of certain organic acids, *J. Pharm. Sci.* **57**, 1877–1880 (1968).
441. M. R. Nazareth and C. L. Huyck, Effect of calcium succinate and calcium carbonate on the stability of aspirin tablets, *J. Pharm. Sci.* **50**, 608–611 (1961).
442. R. J. Manning and C. Washington, Chemical stability of total parenteral nutrition mixtures, *Int. J. Pharm.* **81**, 1–20 (1992).
443. D. A. Williams and J. Lokich, A review of the stability and compatibility of antineoplastic drugs for multiple-drug infusions, *Cancer Chemother. Pharmacol.* **31**, 171–181 (1992).
444. *Handbook of Pharmaceutical Excipients*, American Pharmaceutical Association and Pharmaceutical Society of Great Britain, 1986.
445. K. Yamaoka, Y. Nakajima, S. Okinaga, H. Ohata, T. Miyake, and Y. Takei, Effect of trace minerals on stability of vitamin in hyperalimentation fluids [in Japanese], *Jpn. J. Hosp. Pharm.* **17**, 373–380 (1991).
446. H. M. El-Banna, N. A. Daabis, and S. A. El-Fattah, Aspirin stability in solid dispersion binary systems, *J. Pharm. Sci.* **67**, 1631–1633 (1978).
447. S. Lee, H. G. DeKay, and G. S. Banker, Effect of water vapor pressure on moisture sorption and the stability of aspirin and ascorbic acid in tablet matrices, *J. Pharm. Sci.* **54**, 1153–1158 (1965).
448. K. H. Fromming and R. Hosemann, Die Stabilitat von Harnstoff-Einschlussverbindungen unter besonderer Berucksichtingung von Hilfsstoffzusatzen, *Arch. Pharm.* **301**, 548–554 (1968).
449. T. Nara, T. Komatsu, and T. Okano, Effect of water content and diluent on the stability of cysteine derivatives in powdered drugs [in Japanese], *Yakuzaigaku* **35**, 176–182 (1975).
450. N. Katori, S. Yoshioka, and M. Uchiyama, Stability of pharmaceuticals of cephalexine [in Japanese], *Eishi-houkoku* **104**, 30–35 (1986).
451. E. De Ritter, L. Magid, M. Osadca, and S. H. Rubin, Effect of silica gel on stability and biological availability of ascorbic acid, *J. Pharm. Sci.* **59**, 229–232 (1970).
452. A. Y. Gore and G. S. Banker, Surface chemistry of colloidal silica and a possible application to stabilize aspirin in solid matrixes, *J. Pharm. Sci.* **68**, 197–202 (1979).
453. J. T. Carstensen, M. Osadca, and S. H. Rubin, Degradation mechanisms for water-soluble drugs in solid dosage forms, *J. Pharm. Sci.* **58**, 549–553 (1969).
454. T. Aoyama, T. Maeda, and M. Horioka, Effect of water on degradation of propantheline bromide mixed with dried aluminum hydroxide gel, *Chem. Pharm. Bull.* **25**, 3376–3380 (1977).
455. P. R. Perrier and U. W. Kesselring, Quantitative assessment of the effect of some excipients on nitrazepam stability in binary powder, *J. Pharm. Sci.* **72**, 1072–1074 (1983).
456. C. Ahlneck and G. Alderborn, Solid state stability of acetylsalicylic acid in binary mixtures with microcrystalline and microfine cellulose, *Acta Pharm. Suec.* **25**, 41–52 (1988).
457. K. E. Fielden, J. M. Newton, P. O'Brien, and R. C. Rowe, Thermal studies on the interaction of water and microcrystalline cellulose, *J. Pharm. Pharmacol.* **40**, 674–678 (1988).
458. D. R. Heidemann and P. J. Jarosz, Preformulation studies involving moisture uptake in solid dosage forms, *Pharm. Res.* **8**, 292–297 (1991).

459. W. J. Irwin and M. Iqbal, Solid-state stability: The effect of grinding solvated excipients, *Int. J. Pharm.* **75**, 211–218 (1991).
460. T. Otsuka, S. Yoshioka, Y. Aso, and S. Kojima, Water mobility in aqueous solutions of macromoleular pharmaceutical excipients measured by oxygen-17 nuclear magnetic resonance, *Chem. Pharm. Bull.* **43**, 1221–1223 (1995).
461. S. Yoshioka, Y. Aso, T. Otsuka, and S. Kojima, Water mobility in polyethylene glycol–, poly(vinylpyrrolidone)–, and gelatin–water systems, as measured by spin–lattice relaxation time, dielectric relaxation time, and water activity, *J. Pharm. Sci.* **84**, 1072–1077 (1995).
462. S. Yoshioka, Y. Aso, and T. Terao, Effect of water mobility on drug hydrolysis rates in gelatin gels, *Pharm. Res.* **9**, 607–612 (1992).
463. C. W. Spancake, A. K. Mitra, and D. O. Kildsig, Kinetics of aspirin hydrolysis in aqueous solutions and gels of poloxamines (Tetronic 1508)—Influence of microenvironment, *Int. J. Pharm.* **75**, 231–239 (1991).
464. Y. Aso, S. Tang, S. Yoshioka, and S. Kojima, Amount of mobile water estimated from ^{2}H spin–lattice relaxation time, and its effects on the stability of cephalothin in mixtures with pharmaceutical excipients, *Drug Stability* **1**(4), 237–242 (1997).
465. T. T. Kararli and T. Catalano, Stabilization of misoprostol with hydroxypropyl methylcellulose (HPMC) against degradation by water, *Pharm. Res.* **7**, 1186–1189 (1990).
466. T. Takahashi and R. Yamamoto, Studies on the stability of vitamin D_2 powder preparations. I. Relationship between surface acidity of excipients and isomerization of vitamin D_2 (1) [in Japanese], *Yakugaku Zasshi* **89**, 909–913 (1969).
467. S. Benita, J. P. Benoit, F. Puisieux, and C. Thies, Characterization of drug-loaded poly(*d,l*-lactide)microspheres, *J. Pharm. Sci.* **73**, 1721–1724 (1984).
468. Y. Aso, S. Yoshioka, and T. Terao, Release characteristics stability of poly(l-lactide) microspheres and chemical stability of drugs in the microspheres, *Proc. Int. Symp. Control. Release Bioact. Mater.* **19**, 310 (1992).
469. M. Ikeda, F. Usui, and T. Nagai, Effect of microcrystalline cellulose on the stability of oxazolam in solid state [in Japanese], *Yakuzaigaku* **47**, 204–210 (1987).
470. M. Ikeda, F. Usui, K. Itoh, and T. Nagai, The characteristics of microcrystalline cellulose promoting the decomposition of oxazolam [in Japanese], *Yakuzaigaku* **52**, 296–302 (1992).
471. P. V. Mroso, A. Li Wan Po, and W. J. Irwin, Solid-state stability of aspirin in the presence of excipients: Kinetic interpretation, modeling, and prediction, *J. Pharm. Sci.* **71**, 1096–1101 (1982).
472. R. Baugh, R. T. Calvert, and J. T. Fell, Stability of phenylbutazone in presence of pharmaceutical colors, *J. Pharm. Sci.* **66**, 733–735 (1977).
473. D. S. Sidhu and J. K. Sugden, Effect of food dyes on the photostability of aqueous solutions of L-ascorbic acid, *Int. J. Pharm.* **83**, 263–266 (1992).
474. A. B. Thakur, K. Morris, J. A. Grosso, K. Himes, J. K. Thottathil, R. L. Jerzewski, D. A. Wadke, and J. T. Carstensen, Mechanism and kinetics of metal ion-mediated degradation of fosinopril sodium, *Pharm. Res.* **10**, 800–809 (1993).
475. S. Riegelman, The effect of surfactants on drug stability I, *J. Am. Pharm. Assoc.* **49**, 339–343 (1960).
476. B. J. Meakin, I. K. Winterborn, and D. J. G. Davies, The modifying effects of a cationic surfactant on the rates of base catalysed hydrolysis of esters of different structures, *J. Pharm. Pharmacol.* **23**, 25S–32S (1971).
477. M. Nakagaki and S. Yokoyama, Base-catalyzed hydrolysis of acetylcholine chloride in the presence of cationic and nonionic surfactants, *Bull. Chem. Soc. Jpn.* **59**, 19–23 (1986).
478. M. S. M. Zarina, E. B. Anwar, and R. Nighat, The effect of changes in the polarity of the reaction solvent (the medium) upon the reaction of a cationic micelle with benzocaine, *Pharm. Acta Helv.* **61**, 59–64 (1986).
479. J. E. Dawson, B. R. Hajratwala, and H. Taylor, Kinetics of indomethacin degradation II: Presence of alkali plus surfactant, *J. Pharm. Sci.* **66**, 1259–1263 (1977).
480. A. Cipiciani, C. Ebert, R. Germani, P. Linda, M. Lovrecich, F. Rubessa, and G. Savelli, Micellar effects on the basic hydrolysis of indomethacin and related compounds, *J. Pharm. Sci.* **74**, 1184–1187 (1985).
481. J. M. Patel and D. E. Wurster, Effect of hydrocarbon chain length on the hydrolysis of several naphthyl esters in the presence of *o*-isobenzoic acid and CTAB micelles, *Int. J. Pharm.* **86**, 43–48 (1992).
482. J. M. Patel and D. E. Wurster, Catalysis of carbaryl hydrolysis in micellar solutions of cetyltrimethylammonium bromide, *Pharm. Res.* **8**, 1155–1158 (1991).
483. T. J. Broxton, T. Ryan, and S. R. Morrison, Micellar catalysis of organic reactions. XIV. Hydrolysis of some 1,4-benzodiazepin-2-one drugs in acidic solution, *Aust. J. Chem.* **37**, 1895–1902 (1984).

484. M. E. Moro, J. N. Fertrell, M. M. Velazquez, and L. J. Rodriguez, Kinetics of the acid hydrolysis of diazepam, bromazepam, and flunitrazepam in aqueous and micellar systems, *J. Pharm. Sci.* **80**, 459–468 (1991).
485. A. Tsuji, E. Miyamoto, M. Matsuda, K. Nishimura, and T. Yamana, Effects of surfactants on the aqueous stability and solubility of β-lactam antibiotics, *J. Pharm. Sci.* **71**, 1313–1318 (1982).
486. A. G. Oliveira, I. M. Cuccovia, and H. Chaimovich, Micellar modification of drug stability: Analysis of the effect of hexadecyltrimethylammonium halides on the rate of degradation of cephaclor, *J. Pharm. Sci.* **79**, 37–42 (1990).
487. T. Yotsuyanagi, T. Hamada, H. Tomida, and K. Ikeda, Hydrolysis of procaine in liposomal suspension, *Acta Pharm. Suec.* **16**, 271–280 (1979).
488. T. Yotsuyanagi and K. Ikeda, Kinetic characterization of liposomes, *J. Pharm. Sci.* **69**, 745–746 (1980).
489. M. J. Habib and J. A. Rogers, Kinetics of hydrolysis and stabilization of acetylsalicylic acid in liposome formulations, *Int. J. Pharm.* **44**, 235–241 (1988).
490. M. J. Habib and J. A. Rogers, Stabilization of local anesthetics in liposomes, *Drug Dev. Ind. Pharm.* **13**, 1947–1971 (1987).
491. M. J. Habib and J. A. Rogers, Hydrolysis and stability of acetylsalicylic acid in stearylamine-containing liposomes, *J. Pharm. Pharmacol.* **45**, 496–499 (1992).
492. M. L. Bianconi and S. Schreier, ESR study of membrane partitioning, orientation, and membrane-modulated alkaline hydrolysis of a spin-labeled benzoic acid ester, *J. Phys. Chem.* **95**, 2483–2486 (1991).
493. T. Yotsuyanagi, T. Hamada, H. Tomida, and K. Ikeda, Hydrolysis of 2-diethylaminoethyl *p*-nitrobenzoate in liposomal suspension, *Acta Pharm. Suec.* **16**, 325–332 (1979).
494. D. Toledo-Velasquez, H. T. Gaud, and K. A. Connors, Misoprostol dehydration kinetics in aqueous solution in the presence of hydroxypropyl methylcellulose, *J. Pharm. Sci.* **81**, 145–148 (1992).
495. W. J. Irwin and M. Iqbal, Thermal stability of bropirimine, *Int. J. Pharm.* **41**, 41–48 (1988).
496. Y. Nakai, K. Yamamoto, K. Terada, T. Oguchi, and S. Izumikawa, Interactions between crystalline medicinals and porous clay, *Chem. Pharm. Bull.* **34**, 4760–4766 (1986).
497. R. A. Kenley, M. O. Lee, L. Sukumar, and M. F. Powell, Temperature and pH dependence of fluocinolone acetonide degradation in a topical cream formulation, *Pharm. Res.* **4**, 342–347 (1987).
498. G. P. Jacobs and P. A. Wills, Recent developments in the radiation sterilization of pharmaceuticals, *Radiat. Phys. Chem.* **31**, 685–691 (1988).
499. M. A. Abuirjeie, A. A. Abdel-Aziz, and M. E. Abdel-Humid, Feasibility studies on radiation sterilization of cephradine, *Drug Dev. Ind. Pharm.* **16**, 1661–1673 (1990).
500. P. J. M. Salemink, J. C. Kolkman-Roodbeen, T. C. J. Gribnau, P. S. L. Janssen, and A. J. van der Veen, The influence of gamma-irradiation upon the chemical and biological properties of insulin, *Pharm. Weekbl., Sci. Ed.* **9**, 172–178 (1987).
501. J. P. Jacobs, A review: Radiation sterilization of pharmaceuticals, *Radiat. Phys. Chem.* **26**, 133–142 (1985).
502. W. Bogl, Radiation sterilization of pharmaceuticals—chemical changes and consequences, *Radiat. Phys. Chem.* **25**, 425–435 (1985).
503. M. P. Kane and K. Tsuji, Radiolytic degradation scheme for ^{60}Co-irradiated corticosteroids, *J. Pharm. Sci.* **72**, 30–35 (1983).
504. I. A. Werner, H. Altorfer, and X. Perlia, The effectiveness of various scavengers on the ganma-irradiated, methanolic solution of medazepam, *Int. J. Pharm.* **63**, 155–166 (1990).
505. T. Miyazaki, T. Kaneko, T. Yoshimura, A.-S. Crucq, and B. Tilquin, Electron spin resonance study of radiosterilization of antibiotics: Ceftazidime, *J. Pharm. Sci.* **83**, 68–71 (1994).
506. S. C. Seth, L. V. Allen, and P. Pinnamaraju, Stability of hydrocortisone salts during iontophoresis, *Int. J. Pharm.* **106**, 7–14 (1994).
507. A. D'Emanuele and J. N. Staniforth, An electrically modulated drug delivery device. III. Factors affecting drug stability during electrophoresis, *Pharm. Res.* **9**, 312–315 (1992).
508. Y. Nakagawa, S. Itai, T. Yoshida, and T. Nagai, Physicochemical properties and stability in the acidic solution of a new macrolide antibiotic, clarithromycin, in comparison with erythromycin, *Chem. Pharm. Bull.* **40**, 725–728 (1992).
509. S. A. Varia, S. Schuller, and V. J. Stella, Phenytoin prodrugs IV: Hydrolysis of various 3-(hydroxymethyl)phenytoin esters, *J. Pharm. Sci.* **73**, 1074–1080 (1984).
510. H. Bundgaard, E. Jensen, and E. Falch, Water-soluble, solution-stable, and biolabile N-substituted (aminomethyl)benzoate ester prodrugs of acyclovir, *Pharm. Res.* **8**, 1087–1093 (1991).

511. V. H. Naringrekar and V. J. Stella, Mechanism of hydrolysis and structure–stability relationship of enaminones as potential prodrugs of model primary amines, *J. Pharm. Sci.* **79**, 138–146 (1990).
512. F. Haviv, T. D. Fitzpatrick, C. J. Nichols, R. E. Swenson, E. N. Bush, G. Diaz, A. Nguyen, H. N. Nellans, D. J. Hoffman, H. Ghanbari, E. S. Johnson, S. Love, V. Cybulski, and J. Greer, Stabilization of the N-terminal residues of luteinizing hormone-releasing hormone agonists and the effect on pharmacokinetics, *J. Med. Chem.* **35**, 3890–3894 (1992).
513. T. Higuchi and L. Lachman, Inhibition of hydrolysis of esters in solution by formation of complexes I. Stabilization of benzocaine with caffeine, *J. Am. Pharm. Assoc.* **44**, 521–526 (1955).
514. L. Lachman, L. J. Ravin, and T. Higuchi, Inhibition of hydrolysis of esters in solution by formation of complexes II. Stabilization of procaine with caffeine, *J. Am. Pharm. Assoc.* **45**, 290–295 (1956).
515. L. Lachman and T. Higuchi, Inhibition of hydrolysis of esters in solution by formation of complexes III. Stabilization of tetracaine with caffeine, *J. Am. Pharm. Assoc.* **46**, 32–36 (1957).
516. D. E. Guttman, Complex formation influence on reaction rate I. Effect of caffeine on riboflavin base-catalyzed degradation, *J. Pharm. Sci.* **51**, 1162–1166 (1962).
517. H. Fujiwara, S. Kawashima, and M. Ohhashi, Stabilization of ampicillin analogs in aqueous solution. I. Assay of ampicillin in solutions containing benzaldehyde by iodine colorimetry and the effect of benzaldehyde on the stability of ampicillin, *Chem. Pharm. Bull.* **30**, 1430–1436 (1982).
518. H. Fujiwara, S. Kawashima, and M. Ohhashi, Stabilization of ampicillin analogs in aqueous solution. II. Kinetic analysis of the mechanism of degradation of ampicillin with benzaldehyde in aqueous solution, *Chem. Pharm. Bull.* **30**, 2181–2188 (1982).
519. H. Fujiwara, S. Kawashima, Y. Yamada, and K. Yabu, Stabilization of ampicillin analogs in aqueous solution. III. Kinetics of degradation of ampicillin with furfural in aqueous solution and physical properties of ampicillin–aldehyde adducts, *Chem. Pharm. Bull.* **30**, 3310–3318 (1982).
520. H. Fujiwara, S. Kawashima, and Y. Yamada, Stabilization of ampicillin analogs in aqueous solution. IV. Effect of addition of furfural on the degradation of ampicillin in the presence of various carbohydrates in aqueous solution, *Chem. Pharm. Bull.* **30**, 4153–4158 (1982).
521. H. Fujiwara, S. Kawashima, and M. Ohhashi, Stabilization of ampicillin analogs in aqueous solution. V. Kinetic study of the effects of aldehydes on the degradation of cephalexin in aqueous solution, *Chem. Pharm. Bull.* **31**, 1335–1344 (1983).
522. H. Fujiwara and S. Kawashima, Stabilization of ampicillin analogs in aqueous solution. VI. Kinetic analysis of the stabilization mechanism of bacampicillin with benzaldehyde in aqueous solution, *Chem. Pharm. Bull.* **33**, 1202–1213 (1985).
523. T. Loftsson and H. Fridriksdottir, Stabilizing effect of tris(hydroxymethyl)aminomethane on *N*-nitrosoureas in aqueous solutions, *J. Pharm. Sci.* **81**, 197–198 (1992).
524. K. Uekama, F. Hirayama, T. Wakuda, and M. Otagiri, Effects of cyclodextrins on the hydrolysis of prostacyclin and its methyl ester in aqueous solution, *Chem. Pharm. Bull.* **29**, 213–219 (1981).
525. F. Hirayama, M. Kurihara, and K. Uekama, Improvement of chemical instability of prostacyclin in aqueous solution by complexation with methylated cyclodextrins, *Int. J. Pharm.* **35**, 193–199 (1987).
526. H. Adachi, T. Irie, F. Hirayama, and K. Uekama, Stabilization of prostaglandin E_1 in fatty alcohol propylene glycol ointment by acidic cyclodextrin derivative, *O*-carboxymethyl-*O*-ethyl-β-cyclodextrin, *Chem. Pharm. Bull.* **40**, 1586–1591 (1992).
527. F. Hirayama, M. Kurihara, and K. Uekama, Improving the aqueous stability of prostaglandin E_2 and prostaglandin A_2 by inclusion complexation with methylated-β-cyclodextrins, *Chem. Pharm. Bull.* **32**, 4237–4240 (1984).
528. I. J. Oh, H. M. Song, and K. C. Lee, Effect of 2-hydroxypropyl-β-cyclodextrin on the stability of prostaglandin E_2 in solution, *Int. J. Pharm.* **106**, 135–140 (1994).
529. K. Uekama, F. Hirayama, Y. Yamada, K. Inaba, and K. Ikeda, Improvements of dissolution characteristics and chemical stability of 16,16-dimethyl-*trans*-Δ^2-prostaglandin E_1 methyl ester by cyclodextrin complexation, *J. Pharm. Sci.* **68**, 1059–1060 (1979).
530. M. Yamamoto, F. Hirayama, and K. Uekama, Improvement of stability and dissolution of prostaglandin E_1 by maltosyl-β-cyclodextrin in lyophilized formulation, *Chem. Pharm. Bull.* **40**, 747–751 (1992).
531. K. Fujioka, Y. Kurosaki, S. Sato, T. Noguchi, T. Noguchi, and Y. Yamahira, Biopharmaceutical study of inclusion complexes. I. Pharmaceutical advantages of cyclodextrin complexes of bencyclane fumarate, *Chem. Pharm. Bull.* **31**, 2416–2423 (1983).

532. E. Pop, T. Loftsson, and N. Bodor, Solubilization and stabilization of a benzylpenicillin chemical delivery system by 2-hydroxypropyl-β-cyclodextrin, *Pharm. Res.* **8**, 1044–1049 (1991).
533. M. E. Brewster, T. Loftsson, K. S. Estes, J. Lin, H. Fridriksdottir, and N. Bodor, Effect of various cyclodextrins on solution stability and dissolution rate of doxorubicin hydrochloride, *Int. J. Pharm.* **79**, 289–299 (1992).
534. O. A. G. J. Houwen, J. Teeuwsen, O. Bekers, J. H. Beijnen, A. Bult, and W. J. M. Underberg, Degradation kinetics of 7-*N*-(*p*-hydroxyphenyl)mitomycin C (M-83) in aqueous solution in the presence of γ-cyclodextrin, *Int. J. Pharm.* **105**, 249–254 (1994).
535. R. J. Prankerd, H. W. Stone, K. B. Sloan, and J. H. Perrin, Degradation of aspartame in acidic aqueous media and its stabilization by complexation with cyclodextrins or modified cyclodextrins, *Int. J. Pharm.* **88**, 189–199 (1992).
536. A. Suing, O. Beakers, J. H. Beijnen, W. J. M. Underberg, T. Tanimoto, K. Koizumi, and M. Otagiri, Stabilization of daunorubicin and 4-demethoxydaunorubicin on complexation with octakis(2,6-di-*O*-methyl)-γ-cyclodextrin in acidic aqueous solution, *Int. J. Pharm.* **82**, 29–37 (1992).
537. T. Loftsson, B. J. Olafsdottir, and J. Baldvinsdottir, Estramustine: Hydrolysis, solubilization, and stabilization in aqueous solutions, *Int. J. Pharm.* **79**, 107–112 (1992).
538. D. G. Musson, D. P. Evitts, A. M. Bidgood, and O. Olejnik, Stabilization of aqueous thymoxamine using dimethyl-β-cyclodextrin, *Int. J. Pharm.* **99**, 85–92 (1993).
539. M. Krenn, M. P. Gamcsik, G. B. Vogelsang, O. M. Colvin, and K. W. Leong, Improvements in solubility and stability of thalidomide upon complexation with hydroxypropyl-β-cyclodextrin, *J. Pharm. Sci.* **81**, 685–689 (1992).
540. T. Loftsson and J. Baldvinsdottir, Degradation of tauromustine (TCNU) in aqueous solutions, *Acta Pharm. Nord.* **4**, 129–132 (1992).
541. T. Loftsson, J. Baldvinsdottir, and A. M. Sigurdardottir, The effect of cyclodextrins on the solubility and stability of medroxyprogesterone acetate and megestrol acetate in aqueous solution, *Int. J. Pharm.* **98**, 225–230 (1993).
542. T. Takahashi, I. Kagami, K. Kitamura, Y. Nakanishi, and Y. Imasato, Stabilization of AD-1590, a non-steroidal antiinflammatory agent, in suppository bases by β-cyclodextrin complexation, *Chem. Pharm. Bull.* **34**, 1770–1774 (1986).
543. D. Q. Ma, R. A. Rajewski, D. Vander Velde, and V. J. Stella, Comparative effects of $(SBE)7_m$-β-CD and HP-β-CD on the stability of two anti-neoplastic agents, melphalan and carmustine, *J. Pharm. Sci.* **89**, 275–287 (2000).
544. J. L. Lach and T. Chin, Schardinger dextrin interaction IV. Inhibition of hydrolysis by means of molecular complex formation, *J. Pharm. Sci.* **53**, 924–927 (1964).
545. T. Chin, P. Chung, and J. L. Lach, Influence of cyclodextrins on ester hydrolysis, *J. Pharm. Sci.* **57**, 44–48 (1968).
546. S. Choudhury and A. K. Mitra, Kinetics of aspirin hydrolysis and stabilization in the presence of 2-hydroxypropyl-β-cyclodextrin, *Pharm. Res.* **10**, 156–159 (1993).
547. O. Bekers, J. H. Beijnen, B. J. Vis, A. Suenaga, M. Otagiri, A. Bult, and W. J. M. Underberg, Effect of cyclodextrin complexation on the chemical stability of doxorubicin and daunorubicin in aqueous solutions, *Int. J. Pharm.* **72**, 123–130 (1991).
548. O. Bekers, J. H. Beijnen, Y. A. G. Kempers, A. Bult, and W. J. M. Underberg, Effects of cyclodextrins on *N*-trifluoroacetyldoxorubicin-14-valerate (AD-32) stability and solubility in aqueous media, *Int. J. Pharm.* **68**, 271–276 (1991).
549. Y. V. Pathak, A. Rubinstein, and S. Benita, Enhanced stability of physostigmine salicylate in submicron o/w emulsion, *Int. J. Pharm.* **65**, 169–175 (1990).
550. K. Yamamura, M. Nakao, K. Yano, K. Miyamoto, and T. Yotsuyanagi, Stability of forskolin in lipid emulsions and oil/water partition coefficients, *Chem. Pharm. Bull.* **39**, 1032–1034 (1991).
551. A. D. Vos, L. Vervoort, and R. Kinglet, Solubilization and stability of indomethacin in a transparent oil–water gel, *Int. J. Pharm.* **92**, 191–196 (1993).
552. C. B. Grissom, A. M. Chagovets, and Z. Wang, Use of viscosigens to stabilize vitamin B_{12} solutions against photolysis, *J. Pharm. Sci.* **82**, 641–643 (1993).
553. M. J. Kaufman, Applications of oxygen polarography to drug stability testing and formulation development: Solution-phase oxidation of hydroxymethylglutaryl coenzyme A (HMG-CoA) reductase inhibitors, *Pharm. Res.* **7**, 289–292 (1990).

554. J. Sawicka, The influence of excipients and technological process on cholecalciferol stability and its liberation from tablets, *Pharmazie* **46**, 519–521 (1991).
555. D. E. Moore, Antioxidant efficiency of polyhydric phenols in photooxidation of benzaldehyde, *J. Pharm. Sci.* **65**, 1447–1451 (1976).
556. C. L. Mayers and D. R. Jenke, Stabilization of oxygen-sensitive formulations via a secondary oxygen scavenger, *Pharm. Res.* **10**, 445–448 (1993).
557. Y. Matsuda, H. Inouye, and R. Nakanishi, Stabilization of sulfisomidine tablets by use of film coating containing UV absorber: Protection of coloration and photolytic degradation from exaggerated light, *J. Pharm. Sci.* **67**, 196–201 (1978).
558. Y. Matsuda, T. Itooka, and Y. Mitsuhashi, Photostability of indomethacin in model gelatin capsules: Effects of film thickness and concentration of titanium dioxide on the coloration and photolytic degradation, *Chem. Pharm. Bull.* **28**, 2665–2671 (1980).
559. S. R. Bechard, O. Quraishi, and E. Kwong, Film coating: Effect of titanium dioxide concentration and film thickness on the photostability of nifedipine, *Int. J. Pharm.* **87**, 133–139 (1992).
560. D. S. Desai, M. A. Abdelnasser, B. A. Rubitski, and S. A. Varia, Photostabilization of uncoated tablets of sorivudine and nifedipine by incorporation of synthetic iron oxides, *Int. J. Pharm.* **103**, 69–76 (1994).
561. K. Thoma and R. Klimek, Photostabilization of drugs in dosage forms without protection from packaging materials, *Int. J. Pharm.* **67**, 169–175 (1991).
562. H. Takeuchi, H. Sasaki, T. Niwa, T. Hino, Y. Kawashima, K. Uesugi, and H. Ozawa, Improvement of photostability of ubidecarenone in the formulation of a novel powdered dosage form termed redispersible dry emulsion, *Int. J. Pharm.* **86**, 25–33 (1992).
563. R. Yamamoto and T. Takahashi, Studies on hygroscopicity of medicine. IV [in Japanese], *Yakugaku Zasshi* **76**, 7–10 (1956).
564. T. Muraoka and A. Kanayama, Moisture proof packaging for pharmaceutical preparations. IV. Relationship between water vapor permeability of pharmaceuticals in the container [in Japanese], *Yakuzaigaku* **29**, 189–195 (1969).
565. H. Nogami, Y. Nakai, and E. Nakajima, Studies on powder preparations(I). Hygroscopicity of powder drugs and method of its protection [in Japanese], *Yakuzaigaku* **18**, 167–170 (1969).
566. I. Sugimoto, A. Kuchiki, and H. Nakagawa, Stability of nifedipine–polyvinylpyrrolidone coprecipitate, *Chem. Pharm. Bull.* **29**, 1715–1723 (1981).
567. K. Uekama, K. Ikegami, Z. Wang, Y. Horiuchi, and F. Hirayama, Inhibitory effect of 2-hydroxypropyl-β-cyclodextrin on crystal-growth of nifedipine during storage: Superior dissolution and oral bioavailability compared with polyvinylpyrrolidone K-30, *J. Pharm. Pharmacol.* **44**, 73–78 (1992).
568. Y. Matsuda and S. Kawaguchi, Physicochemical characterization of oxyphenbutazone and solid-state stability of its amorphous form under various temperature and humidity conditions, *Chem. Pharm. Bull.* **34**, 1289–1298 (1986).
569. C. Torricelli, A. Martini, L. Muggetti, and R. De Ponti, Stability studies on a steroidal drug/β-cyclodextrin coground mixture, *Int. J. Pharm.* **71**, 19–24 (1991).
570. F. Nakamura, M. Fujitani, Y. Machida, and T. Nagai, Physicochemical stability of halopredone acetate ground mixtures with biocompatible polymers [in Japanese], *Yakuzaigaku* **53**, 161–168 (1993).
571. Y. Matsuda, M. Totsuka, M. Onoe, and E. Tatsumi, Amorphism and physicochemical stability of spray-dried frusemide, *J. Pharm. Pharmacol.* **44**, 627–633 (1992).
572. T. Yamaguchi, M. Nishimura, R. Okamoto, T. Takeuchi, and K. Yamamoto, Glass formation of 4″-*O*-(4-methoxyphenyl)acetyltylosin and physicochemical stability of the amorphous solid, *Int. J. Pharm.* **85**, 87–96 (1992).
573. T. Yamaguchi, M. Nishimura, R. Okamoto, T. Takeuchi, and K. Yamamoto, Physicoechemical stability of glassy 16-membered macrolide compounds, *Chem. Pharm. Bull.* **41**, 1812–1816 (1993).
574. J. T. Carstensen and K. Van Scoik, Amorphous-to-crystalline transformation of sucrose, *Pharm. Res.* **7**, 1278–1281 (1990).
575. K. Van Scoik and J. T. Carstensen, Nucleation phenomena in amorphous sucrose systems, *Int. J. Pharm.* **58**, 185–196 (1990).
576. A. Saleki-Gerhardt and G. Zografi, Non-isothermal crystallization of sucrose from the amorphous state, *Pharm. Res.* **11**, 1166–1173 (1994).
577. M. P. te Booy, R. A. de Ruiter, and A. L. de Meere, Evaluation of the physical stability of freeze-dried sucrose-containing formulations by differential scanning calorimetry, *Pharm. Res.* **9**, 109–114 (1992).

578. T. Yokoyama, N. Ohnishi, T. Umeda, T. Kuroda, Y. Kita, K. Kuroda, and Y. Matsuda, Kinetic study on the isothermal transition of benoxaprofen polymorphs in the solid state at high temperature, *Chem. Pharm. Bull.* **34**, 917–921 (1986).

579. N. Ohnishi, T. Yokoyama, Y. Kiyohara, Y. Kita, and K. Kuroda, Kinetic study on the isothermal transition of bromovalerylurea polymorphs in the solid state at high temperature, *Chem. Pharm. Bull.* **35**, 1207–1213 (1987).

580. T. Durig and A. R. Fassihi, Preformulation study of moisture effect on the physical stability of pyridoxal hydrochloride, *Int. J. Pharm.* **77**, 315–319 (1991).

581. T. Miyata, H. Nakagawa, K. Akimoto, I. Sugimoto, E. Maarti, and O. Heiber, Crystal forms of cianidanol [in Japanese], *Yakugaku Zasshi* **105**, 59–64 (1985).

582. H. Nyqvist and T. Wadsten, Can ageing effects in drugs be measured and interpreted by TA-techniques? *Thermochim. Acta* **93**, 125–127 (1985).

583. M. Totsuka, N. Kaneniwa, K. Kawakami, and O. Umezawa, Effect of surface characteristics of theophylline anhydrate powder on hygroscopic stability, *J. Pharm. Pharmacol.* **42**, 606–610 (1990).

584. M. Totsuka, N. Kaneniwa, K. Otsuka, K. Kawakami, O. Umezawa, and Y. Matsuda, Effect of geometric factors on hydration kinetics of theophylline anhydrate, *J. Pharm. Sci.* **81**, 1189–1193 (1992).

585. M. Totsuka, R. Teraoka, and Y. Matsuda, Physicochemical stability of nitrofurantoin anhydrate and monohydrate under various temperature and humidity conditions, *Pharm. Res.* **8**, 1066–1068 (1991).

586. K. Sekiguchi, K. Shirotani, O. Sakata, and E. Suzuki, Kinetic study of the dehydration of sulfaguanidine under isothermal conditions, *Chem. Pharm. Bull.* **32**, 1558–1567 (1984).

587. M. Totsuka, M. Onoe, and Y. Matsuda, Physicochemical stability of phenobarbital polymorphs at various levels of humidity and temperature, *Pharm. Res.* **10**, 577–582 (1993).

588. H. Murayama, M. Takahashi, M. Asano, M. Washitake, H. Yuasa, and K. Asahina, Growth behaviors of whiskers on ethenzamide tablet [in Japanese], *Yakuzaigaku* **41**, 113–118 (1981).

589. H. Yuasa, K. Miyata, T. Ando, Y. Kanaya, K. Asahina, and H. Murayama, Quantitative measurement and mechanism of whisker growth [in Japanese], *Yakuzaigaku* **41**, 161–171 (1981).

590. H. Yuasa, Y. Kanaya, and K. Asahina, Studies on whisker growth on the tablet surface. III. Mechanism of whisker growth on aspirin tablet and its effect on the mechanical strength of the tablet, *Chem. Pharm. Bull.* **34**, 850–857 (1986).

591. H. Yuasa, J. Yamashita, and Y. Kanaya, Studies on whisker growth in solid preparations. VI. Whisker growth in mixtures composed of aspirin and porous glass powder, *Chem. Pharm. Bull.* **41**, 731–736 (1993).

592. T. Kuriyama, M. Kobiki, T. Tanaka, and Y. Imasato, Whisker growth from sodium valproate–synthetic aluminum silicate systems [in Japanese], *Yakugaku Zasshi* **107**, 627–633 (1987).

593. H. Ando, S. Watanabe, T. Ohwaki, and Y. Miyake, Crystallization of excipients in tablets, *J. Pharm. Sci.* **74**, 128–131 (1985).

594. G. P. Matthews, N. Lowther, and M. J. Shott, Crystallisation of carbamazepine in tablets stored at elevated temperatures, *Int. J. Pharm.* **50**, 111–115 (1989).

595. D. Banes, Deterioration of nitroglycerin tablets, *J. Pharm. Sci.* **57**, 893–894 (1968).

596. S. A. Fusari, Nitroglycerin sublingual tablets I: Stability of conventional tablets, *J. Pharm. Sci.* **62**, 122–129 (1973).

597. S. A. Fusari, Nitroglycerin sublingual tablets II: Preparation and stability of a new, stabilized, sublingual, molded nitroglycerin tablet, *J. Pharm. Sci.* **62**, 2012–2021 (1973).

598. A. Mitrevej and R. G. Hollenbeck, Influence of hydrophilic excipients on the interaction of aspirin and water, *Int. J. Pharm.* **14**, 243–250 (1983).

599. L. Van Campen, G. L. Amidon, and G. Zografi, Moisture sorption kinetics for water-soluble substances I: Theoretical considerations of heat transport control, *J. Pharm. Sci.* **72**, 1381–1388 (1983).

600. L. Van Campen, G. L. Amidon, and G. Zografi, Moisture sorption kinetics for water-soluble substances II: Experimental verification of heat transport control, *J. Pharm. Sci.* **72**, 1388–1393 (1983).

601. L. Van Campen, G. L. Amidon, and G. Zografi, Moisture sorption kinetics for water-soluble substances III: Theoretical and experimental studies in air, *J. Pharm. Sci.* **72**, 1394–1398 (1983).

602. M. J. Kontny and G. Zografi, Moisture sorption kinetics for water-soluble substances IV: Studies with mixtures of solids, *J. Pharm. Sci.* **74**, 124–127 (1985).

603. V. Andronis, M. Yoshioka, and G. Zografi, Effect of sorbed water on the crystallization of indomethacin from the amorphous state, *J. Pharm. Sci.* **86**, 346–351 (1997).

604. Y. Aso, S. Yoshioka, T. Totsuka, and S. Kojima, The physical stability of amorphous nifedipine determined by isothermal microcalorimetry, *Chem. Pharm. Bull.* **43**, 300–303 (1995).
605. E. Schmitt, C. W. Davis, and S. T. Long, Moisture-dependent crystallization of amorphous lamotrigine mesylate, *J. Pharm. Sci.* **85**, 1215–1219 (1996).
606. B. C. Hancock and G. Zografi, The relationship between the glass transition temperature and the water content of amorphous pharmaceutical solids, *Pharm. Res.* **11**, 471–477 (1994).
607. S. Yoshioka, Y. Aso, and S. Kojima, The effect of excipients on the molecular mobility of lyophilized formulations, as measured by glass transition temperature and NMR relaxation-based critical mobility temperature, *Pharm. Res.* **16**, 135–140 (1999).
608. B. C. Hancock, S. L. Shamblin, and G. Zografi, Molecular mobility of amorphous pharmaceutical solids below their glass transition temperatures, *Pharm. Res.* **12**, 799–806 (1995).
609. V. Andronis and G. Zografi, Molecular mobility of supercooled amorphous indomethacin, determined by dynamic mechanical analysis, *Pharm. Res.* **14**, 410–414 (1997).
610. J. S. Hancock and J. H. Sharp, Method of comparing solid-state kinetic data and its application to the decomposition of kaolinite, brucite, and $BaCO_3$, *J. Am. Ceram. Soc.* **55**, 74–77 (1972).
611. T. Umeda, N. Ohnishi, T. Yokoyama, K. Kuroda, T. Kuroda, E. Tatsumi, and Y. Matsuda, Kinetics of the thermal transition of carbamazepine polymorphic forms in the solid state [in Japanese], *Yakugaku Zasshi* **104**, 786–792 (1984).
612. Y. Matsuda, E. Tatsumi, E. Chiba, and Y. Miwa, Kinetic study of the polymorphic transformations of phenylbutazone, *J. Pharm. Sci.* **73**, 1453–1460 (1984).
613. S. Kitamura, T. Tada, Y. Okamoto, and T. Yasuda, Kinetic study of the transformation from tetrahydrate to monohydrate of a new antiallergic, sodium 5-(4-oxo-phenoxy-4*H*-quinolizine-3-carboxamide)-tetrazolate, *Pharm. Res.* **9**, 138–142 (1992).
614. C. O. Agbada and P. York, Dehydration of theophylline monohydrate powder—effects of particle size and sample weight, *Int. J. Pharm.* **106**, 33–40 (1994).
615. M. Totsuka, M. Onoe, and Y. Matsuda, Hygroscopic stability and dissolution properties of spray-dried solid dispersions of furosemide with Eudragit, *J. Pharm. Sci.* **82**, 32–38 (1993).
616. T. Tsai, C. Shen, and C. Chen, Determination and UV spectral identification of 18-α-glycyrrhetinic acid and 18-β-glycyrrhetinic acid for stability studies, *Int. J. Pharm.* **84**, 279–281 (1992).
617. A. T. Serajuddin, P. Sheen, A. Mufson, D. F. Bernstein, and M. A. Augustine, Preformulation study of a poorly water-soluble drug, α-pentyl-3-(2-quinolinylmethoxy)benzenemethanol: Selection of the base for dosage form design, *J. Pharm. Sci.* **75**, 492–496 (1986).
618. D. J. Ager, K. S. Alexander, A. S. Bhatti, J. S. Blackburn, D. Dollimorem, T. S. Koogan, K. A. Mooseman, G. M. Muhvic, B. Sims, and V. J. Webb, Stability of aspirin in solid mixtures, *J. Pharm. Sci.* **75**, 97–101 (1986).
619. T. T. Kararli, T. E. Needham, C. J. Seul, and P. M. Finnegan, Solid-state interaction of magnesium oxide and ibuprofen to form a salt, *Pharm. Res.* **6**, 804–808 (1989).
620. M. L. Cotton, D. W. Wu, and E. B. Vadas, Drug–excipient interaction study of enalapril maleate using thermal analysis and scanning electron microscopy, *Int. J. Pharm.* **40**, 129–142 (1987).
621. R. J. Prankerd and S. M. Ahmed, Physicochemical interactions of praziquantel, oxaminiquine and tablet excipients, *J. Pharm. Pharmacol.* **44**, 259–261 (1992).
622. T. Durig and A. R. Fassihi, Identification of stabilization effects of excipient–drug interactions in solid dosage form design, *Int. J. Pharm.* **97**, 161–170 (1993).
623. R. H. Pryce-Jones, G. M. Eccleston, and B. B. Abu-Bakar, Aminophylline suppository decomposition: An investigation using differential scanning calorimetry, *Int. J. Pharm.* **86**, 231–237 (1992).
624. M. J. Pikal and K. M. Dellerman, Stability testing of pharmaceuticals by high-sensitivity isothermal calorimetry at 25°C: Cephalosporins in the solid and aqueous solution states, *Int. J. Pharm.* **50**, 233–252 (1989).
625. M. Angberg and C. Nystrom, Evaluation of heat-conduction microcalorimetry in pharmaceutical stability studies. I. Precision and accuracy for static experiments in glass vials, *Acta Pharm. Suec.* **25**, 307–320 (1988).
626. M. Angberg, C. Nystrom, and S. Castensson, Evaluation of heat-conduction microcalorimetry in pharmaceutical stability studies. II. Methods to evaluate the microcalorimetric response, *Int. J. Pharm.* **61**, 67–77 (1990).
627. R. Oliyai and S. Lindenbaum, Stability testing of pharmaceuticals by isothermal heat conduction calorimetry: Ampicillin in aqueous solution, *Int. J. Pharm.* **73**, 33–36 (1991).

628. M. Angberg, C. Nystrom, and S. Castensson, Evaluation of heat-conduction microcalorimetry in pharmaceutical stability studies. VII. Oxidation of ascorbic acid in aqueous solution, *Int. J. Pharm.* **90**, 19–23 (1993).
629. M. J. Koenigbauer, S. H. Brooks, G. Rullo, and R.A. Couch, Solid-state stability testing of drugs by isothermal calorimetry, *Pharm. Res.* **9**, 939–944 (1992).
630. L. D. Hansen, E. A. Lewis, D. J. Eatough, R. G. Bergstrom, and D. DeGraft-Johnson, Kinetics of drug decomposition by heat conduction calorimetry, *Pharm. Res.* **6**, 20–27 (1989).
631. X. Tan, N. Meltzer, and S. Lindenbaum, Solid-state stability studies of 13-*cis*-retinoic acid and all-*trans*-retinoic acid using microcalorimetry and HPLC analysis, *Pharm. Res.* **9**, 1203–1208 (1992).
632. T. Otsuka, S. Yoshioka, Y. Aso, and T. Terao, Application of microcalorimetry to stability testing of meclofenoxate hydrochloride and *dl*-α-tocopherol, *Chem. Pharm. Bull.* **42**, 130–132 (1994).
633. M. Angberg, C. Nystrom, and S. Castensson, Evaluation of heat-conduction microcalorimetry in pharmaceutical stability studies. III. Crystallographic changes due to water vapor uptake in anhydrous lactose powder, *Int. J. Pharm.* **73**, 209–220 (1991).
634. G. Kortum and G. Schreyer, Uber die Gewinnung "typischer Farbkurven" von Pulvern aus Reflexionsmessungen, *Angew. Chem.* **67**, 694–698 (1955).
635. G. Kortum and W. Braun, Reflexionsmessungen an adsorbierten Molekulverbindungen, *Z. Phys. Chem. N. F.* **18**, 242–254 (1958).
636. J. L. Lach and M. Bornstein, Diffuse reflectance studies of solid–solid interactions. Interactions of oxytetracycline, phenothiazine, anthracene and salicylic acid with various adjuvants, *J. Pharm. Sci.* **54**, 1730–1736 (1965).
637. M. Bornstein and J. L. Lach, Diffuse reflectance studies of solid–solid interactions II. Interaction of metallic and nonmetallic adjuvants with anthracene, prednisone, and hydrochlorothiazide, *J. Pharm. Sci.* **55**, 1033–1039 (1966).
638. J. L. Lach and M. Bornstein, Diffuse reflectance studies of solid–solid interactions III. Interaction studies of oxytetracycline with metallic and nonmetallic adjuvants, *J. Pharm. Sci.* **55**, 1040–1045 (1966).
639. W. Wu, T. Chin, and J. L. Lach, Interaction of ascorbic acid with silicic acid, *J. Pharm. Sci.* **59**, 1122–1125 (1970).
640. W. Wu, T. Chin, and J. L. Lach, Interaction of isoniazid with magnesium oxide and lactose, *J. Pharm. Sci.* **59**, 1234–1242 (1970).
641. J. L. Lach and L. D. Bighley, Diffuse reflectance studies of solid–solid interactions, *J. Pharm. Sci.* **59**, 1261–1264 (1970).
642. J. D. McCallister, T. Chin, and J. L. Lach, Diffuse reflectance studies of solid–solid interactions IV. Interaction of bishydroxycoumarin, furosemide, and other medicinal agents with various adjuvants, *J. Pharm. Sci.* **59**, 1286–1289 (1970).
643. S. M. Blaug and W. Huang, Interaction of dextroamphetamine sulfate with spray-dried lactose, *J. Pharm. Sci.* **61**, 1770–1775 (1972).
644. D. G. Pope and J. L. Lach, Diffuse reflectance: On its application to the evaluation and control of quality of solid dosage forms, *Can. J. Pharm. Sci.* **10**, 109–111 (1975).
645. D. G. Pope and J. L. Lach, Diffuse reflectance: On its quantitative application to *N*-acetyl-*p*-aminophenol excipient-induced degradation, *Can. J. Pharm. Sci.* **10**, 114–121 (1975).
646. S. Yoshioka, K. Kimura, H. Ogata, T. Shibazaki, and A. Ejima, Application of diffuse reflectance spectroscopy to stability test of solid dosage forms [in Japanese], *Yakuzaigaku* **39**, 75–82 (1979).
647. A. J. Ferdous, N. A. Dickinson, R. D. Waigh, H NMR as an analytical tool for the investigation of hydrolysis rates: A method for the rapid evaluation of the shelf-life of aqueous solutions of drugs with ester groups, *J. Pharm. Pharmacol.* **43**, 860–862 (1991).
648. J. K. Drennen and R. A. Lodder, Nondestructive near-infrared analysis of intact tablets for determination of degradation products, *J. Pharm. Sci.* **79**, 622–627 (1990).
649. L. Chafetz and K. P. Shah, Stability of diltiazem in acid solution, *J. Pharm. Sci.* **80**, 171–172 (1991).
650. V. D. Labhasetwar, S. V. Joshi, and A. K. Dorle, A study on the dielectric constant of microcapsules during ageing, *J. Microencapsul.* **5**, 339–341 (1988).
651. V. D. Labhasetwar and A. K. Dorle, A study on the dielectric constant of salicylic acid and aspirin compacts during aging, *Int. J. Pharm.* **47**, 261–262 (1988).
652. K. Mizuno, K. Kumaki, S. Hata, Y. Nishii, and S. Tomioka, Estimation of drug stability based on extra-weak chemiluminescence. I. Construction of apparatus for measuring of extra-weak chemiluminescence using photo-electron counting method [in Japanese], *Yakugaku Zasshi* **98**, 513–519 (1978).

653. S. S. Kornblum and M. A. Zoglio, Pharmaceutical heterogeneous systems I. Hydrolysis of aspirin in combination with tablet lubricant in an aqueous suspension, *J. Pharm. Sci.* **56**, 1569–1575 (1967).
654. H. V. Maulding, M. A. Zoglio, F. E. Pigois, and M. Wagner, Pharmaceutical heterogeneous systems IV. A kinetic approach to the stability screening of solid dosage forms containing aspirin, *J. Pharm. Sci.* **58**, 1359–1362 (1969).
655. J. Tingstad, J. Dudzinski, L. Lachman, and E. Shami, Simplified method for determining chemical stability of drug substances in pharmaceutical suspensions, *J. Pharm. Sci.* **62**, 1361–1363 (1973).
656. C. Ahlneck and P. Lundgren, Methods for the evaluation of solid state stability and compatibility between drug and excipient, *Acta Pharm. Suec.* **22**, 305–314 (1985).
657. S. Bolton, Factorial designs in pharmaceutical stability studies, *J. Pharm. Sci.* **72**, 362–366 (1983).
658. R. L. Plackett and J. P. Burman, The design of optimum multifactorial experiments, *Biometrika* **33**, 305–325 (1946).
659. J. A. Plaizier-Vercammen and R. E. De Neve, Interaction of povidon with aromatic compounds I: Evaluation of complex formation by factorial analysis, *J. Pharm. Sci.* **69**, 1403–1408 (1980).
660. H. M. El-Banna, A. A. Ismail, and M. A. F. Gadalla, Factorial design of experiment for stability studies in the development of a tablet formulation, *Pharmazie* **39**, 163–165 (1984).
661. J. Waltersson, Factorial designs in pharmaceutical preformulation studies, I. Evaluation of the application of factorial designs to a stability study of drugs in suspension form, *Acta Pharm. Suec.* **23**, 129–138 (1986).
662. C. Ahlneck and J. Waltersson, Factorial designs in pharmaceutical preformulation studies, *Acta Pharm. Suec.* **23**, 139–150 (1986).
663. C. T. Hung, F. C. Lam, D. G. Perrier, and A. Souter, A stability study of amphotericin B in aqueous media using factorial design, *Int. J. Pharm.* **44**, 117–123 (1988).
664. E. Sanchez, C. M. Evora, and M. Llabres, Effect of humidity and packaging on the long-term aging of commercial sustained-release theophylline tablets, *Int. J. Pharm.* **83**, 59–63 (1992).
665. J. E. Rees and E. Shotton, Some observations on the ageing of sodium chloride compacts, *J. Pharm. Pharmacol.* **22**, 17S–23S (1970).
666. A. M. Molokhia, H. I. Al-Shora, and A. A. Hammad, Aging of tablets prepared by direct compression of bases with different moisture content, *Drug Dev. Ind. Pharm.* **13**, 1933–1946 (1987).
667. J. Akbuga and A. Gursoy, The effect of moisture sorption and desorption on furosemide tablet properties, *Drug Dev. Ind. Pharm.* **13**, 1827–1845 (1987).
668. G. L. Amidon and K. R. Middleton, Accelerated physical stability testing and long-term predictions of changes in the crushing strength of tablets stored in blister packages, *Int. J. Pharm.* **45**, 79–89 (1988).
669. Y. Matsunaga, R. Ohta, N. Bando, H. Yamada, H. Yuasa, and Y. Kanaya, Effects of water content on physical and chemical stability of tablets containing an anticancer drug TAT-59, *Chem. Pharm. Bull.* **41**, 720–724 (1993).
670. J. T. Wang, G. K. Shiu, T. Ong-Chen, C. T. Viswanathan, and J. P. Skelly, Effects of humidity and temperature on in vitro dissolution of carbamazepine tablets, *J. Pharm. Sci.* **83**, 1002–1005 (1993).
671. B. F. Johnson, P. V. Mcauley, P. M. Smith, and J. A. G. French, The effects of storage upon in vitro and in vivo characteristics of soft gelatin capsules containing digoxin, *J. Pharm. Pharmacol.* **29**, 576–578 (1977).
672. M. Dey, R. Enever, M. Kraml, D. G. Prue, D. Smith, and R. Weierstall, The dissolution and bioavailability of etodolac from capsules exposed to conditions of high relative humidity and temperatures, *Pharm. Res.* **10**, 1295–1300 (1993).
673. H. W. Gouda, M. A. Moustafa, and H. I. Al-Shora, Effect of storage on nitrofurantoin solid dosage forms, *Int. J. Pharm.* **18**, 213–215 (1984).
674. J. T. Jacob and E. M. Plein, Factors affecting dissolution rate of medicaments from tablets II. Effect of binder concentration, tablet hardness, and storage conditions on the dissolution rate of phenobarbital from tablets, *J. Pharm. Sci.* **57**, 802–805 (1968).
675. A. S. Alam and E. L. Parrott, Effect of aging on some physical properties of hydrochlorothiazide tablets, *J. Pharm. Sci.* **60**, 263–266 (1971).
676. D. S. Desai, B. A. Rubitski, S. A. Varia, and M. H. Huang, Effect of formaldehyde formation on dissolution stability of hydrochlorothiazide bead formulations, *Int. J. Pharm.* **107**, 141–147 (1994).
677. J. L. Vila-Jato, A. Concheriro, and B. Seijo, Effect of aging on the bioavailability of nitrofurantoin tablets containing carbopol 934, *Drug Dev. Ind. Pharm.* **13**, 1315–1327 (1987).
678. P. V. Marshall, D. G. Pope, and J. T. Carstensen, Methods for the assessment of the stability of tablet disintegrants, *J. Pharm. Sci.* **80**, 899–903 (1991).

679. J. T. Rubino, L. M. Halterlein, and J. Blanchard, The effects of aging on the dissolution of phenytoin sodium capsule formulations, *Int. J. Pharm.* **26**, 165–174 (1985).

680. J. Herman, N. Visavarungroj, and J. P. Remon, Instability of drug release from anhydrous theophylline–microcrystalline cellulose formulations, *Int. J. Pharm.* **55**, 143–146 (1989).

681. M. Landin, R. Martinez-Pacheco, J. L. Gomez-Amoza, C. Souto, A. Concheiro, and R. C. Rowe, Influence of microcrystalline cellulose source and batch variation on the tabletting behaviour and stability of prednisone formulations, *Int. J. Pharm.* **91**, 143–149 (1993).

682. S. Kadir, N. Yata, M. Kawata, and S. Goto, Effect of humidity aging on disintegration, dissolution and cumulative urinary excretion of calcium *p*-aminosalicylate formations, *Chem. Pharm. Bull.* **34**, 5102–5109 (1986).

683. S. T. Horhota, J. Burgio, L. Lonski, and C. T. Rhodes, Effect of storage at specified temperature and humidity on properties of three directly compressible tablet formulations, *J. Pharm. Sci.* **65**, 1746–1749 (1976).

684. C. J. Taborsky-Urdinola, V. A. Gray, and L. T. Grady, Effects of packaging and storage on the dissolution of model prednisone tablets, *Am. J. Hosp. Pharm.* **38**, 1322–1327 (1981).

685. M. W. Gouda, M. A. Moustafa, and A. M. Molokhia, Effect of storage conditions on erythromycin tablets marketed in Saudi Arabia, *Int. J. Pharm.* **5**, 345–347 (1980).

686. C. Remunan, M. J. Bretal, A. Nunez, and J. L. V. Jato, Accelerated stability study of sustained-release nifedipine tablets prepared with Gelucire, *Int. J. Pharm.* **80**, 151–159 (1992).

687. E. S. Saers, C. Nystrom, and M. Alden, Physicochemical aspects of drug release. XVI. The effect of storage on drug dissolution from solid dispersions and the influence of cooling rate and incorporation of surfactant, *Int. J. Pharm.* **90**, 105–118 (1993).

688. I. Ghebre-Sellassie, U. Iyer, D. Kubert, and M. B. Fawzi, Characterization of a new water-based coating for modified-release preparations, *Pharm. Tech.* 96–106 (1988).

689. Y. Insemi and S. Mori, Dissolution tests of commercial tablets containing potassium: Influences of humidity on release of potassium salts from coated tablets [in Japanese], *Yakuzaigaku* **41**, 177–183 (1981).

690. K. Saika, J. Mizutani, K. Nakajima, and T. Ida, Studies on sugar coating. IV. Prolongation of disintegration time of sugar coated tablets [in Japanese], *Yakuzaigaku* **28**, 149–153 (1968).

691. D. Barrett and J. T. Fell, Effect of aging on physical properties of phenylbutazone tablets, *J. Pharm. Sci.* **64**, 335–337 (1975).

692. J. R. Hoblitzell, K. D. Thakker, and C. T. Rhodes, Effects of moderate stress storage conditions on the dissolution profile of enteric-coated aspirin tablets, *Pharm. Acta Helv.* **60**, 28–32 (1985).

693. C. O. Ondari, V. K. Prasad, V. P. Shah, and C. T. Rhodes, Effects of short-term moderated storage stress on the disintegration and dissolution of four types of compressed tablets, *Pharm. Acta Helv.* **59**, 149–153 (1984).

694. O. M. Al-Gohary, S. S. Al-Gamal, A. Hammad, and A. M. Molokhia, Effect of storage on tabletted microencapsulated aspirin granules, *Int. J. Pharm.* **55**, 47–52 (1989).

695. S. A. Khalil, L. M. Ali, and M. M. Abdel-Khalek, Effects of ageing and relative humidity on drug release, *Pharmazie* **29**, 36–37 (1974).

696. M. Georgarakis, P. Htzipantou, and J. E. Kountourelis, Effect of particle size, content in lubricant, mixing time and storage relative humidity on drug release from hard gelatin ampicillin capsules, *Drug Dev. Ind. Pharm.* **14**, 915–923 (1988).

697. L. Chafetz, W. Hong, D. C. Tsilifonis, A. K. Taylor, and J. Hilip, Decrease in the rate of capsule dissolution due to formaldehyde from polysorbate 80 autoxidation, *J. Pharm. Sci.* **73**, 1186–1187 (1984).

698. K. S. Murthy, R. G. Reisch, and M. B. Fawzi, Dissolution stability of hard-shell capsule products, Part I: The effect of exaggerated storage conditions, *Pharm. Technol.* **13**(3), 72–86 (1989).

699. J. W. Cooper, H. C. Ansel, and D. E. Cadwallader, Liquid and solid solution interactions of primary certified colorants with pharmaceutical gelatins, *J. Pharm. Sci.* **62**, 1156–1164 (1973).

700. K. S. Murthy, N. A. Enders, and M. B. Fawzi, Dissolution stability of hard-shell capsule products, Part II: The effect of dissolution test conditions on in vitro drug release, *Pharm. Technol.* **13**, 53–58 (1989).

701. T. C. Dahl, I. T. Sue, and A. Yum, The effect of pancreatin on the dissolution performance of gelatin-coated tablets exposed to high-humidity conditions, *Pharm. Res.* **8**, 412–414 (1991).

702. K. Tanabe, H. Yamazaki, S. Yoshida, K. Yamamoto, E. Hiraoka, S. Itoh, M. Yamazaki, and M. Sawanoi, Comparative study of dosage forms on release of commercially available ketoprofen suppositories [in Japanese], *Yakuzaigaku* **47**, 23–30 (1987).

703. K. Nakabayashi, T. Shimamoto, H. Mima, and J. Okada, Stability of packaged solid dosage forms. V. Prediction of the effect of aging on the disintegration of packaged tablets influenced by moisture and heat, *Chem. Pharm. Bull.* **29**, 2051–2056 (1981).
704. K. Nakabayashi, S. Hanatani, and T. Shimamoto, Stability of packaged solid dosage forms. VI. Shelf-life prediction of packaged prednisolone tablets in relation to dissolution properties, *Chem. Pharm. Bull.* **29**, 2057–2061 (1981).
705. C. M. Sinko, A. F. Yee, and G. L. Amidon, Prediction of physical aging in controlled-release coatings: The application of the relaxation coupling model to glassy cellulose acetate, *Pharm. Res.* **8**, 698–705 (1991).
706. J. Guo, R. E. Robertson, and G. L. Amidon, Influence of physical aging on mechanical properties of polymer free films: The prediction of long-term aging effects on the water permeability and dissolution rate of polymer film-coated tablets, *Pharm. Res.* **8**, 1500–1504 (1991).
707. K. Thoma and P. Serno, Temperaturabhangigkeit der Schmelzzeitverlangerung von Hartfettsuppositorien 5.Mitt. uber pharmazeutische Probleme bei Suppositorien, *Pharm. Ind.* **44**, 1074–1080 (1982).
708. K. Thoma and P. Serno, Beziehungen zwischen der Schmelzzeitzunahme von Suppositorien und den Eigensschaften der Hartfettgrundmasse 6.Mitt. uber pharmazeutische Probleme bei Suppositorien, *Pharm. Ind.* **45**, 192–196 (1982).
709. K. Thoma and P. Serno, Thermoanalytischer Nachweis der Polymorphie der Suppositoriengrundlage Hartfett 8.Mitt. uber pharmazeutische Probleme bei Suppositorien, *Pharm. Ind.* **45**, 990–994 (1983).
710. M. Iijima, T. Hakata, S. Kimura, T. Okabe, Y. Uchino, H. Sato, K. Sugibayashi, and Y. Morimoto, Stability evaluation of effervescent suppositories [in Japanese], *Yakuzaigaku* **53**, 102–108 (1993).
711. R. Jalil and J. R. Nixon, Microencapsulation using poly(DL-lactic acid) IV: Effect of storage on the microcapsule characteristics, *J. Microencapsul.* **7**, 375–383 (1990).
712. V. Rosilio, J. P. Benoit, M. Deyme, C. Thies, and G. Madelmont, A physicochemical study of the morphology of progesterone-loaded microspheres fabricated from poly(D,L-lactide-*co*-glycolide), *J. Biomed. Mater. Res.* **25**, 667–682 (1991).
713. Y. Aso, S. Yoshioka, and T. Terao, Effects of storage on the physicochemical properties and release characteristics of progesterone-loaded poly(*l*-lactide) microspheres, *Int. J. Pharm.* **93**, 153–159 (1993).
714. S. Yoshioka, A. Kishida, S. Izumikawa, Y. Aso, and Y. Takeda, Base-induced polymer hydrolysis in poly(β-hydroxybutyrate/β-hydroxyvalerate) matrices, *J. Controll. Release* **16**, 341–348 (1991).
715. S. Izumikawa, S. Yoshioka, Y. Aso, and Y. Takeda, Preparation of poly(*l*-lactide) microspheres of different crystalline morphology and effect of crystalline morphology on drug release rate, *J. Controll. Release* **15**, 133–140 (1991).
716. M. Grit and D. J. A. Crommelin, The effect of aging on the physical stability of liposome dispersions, *Chem. Phys. Lipids* **62**, 113–122 (1992).
717. M. Fresta, A. Villari, G. Puglisi, and G. Gavallaro, 5-Fluorouracil: Various kinds of loaded liposomes: Encapsulation efficiency, storage stability and fusogenic properties, *Int. J. Pharm.* **99**, 145–156 (1993).
718. G. V. Betageri and L. S. Burrell, Stability of antibody-bearing liposomes containing dideoxyinosine triphosphate, *Int. J. Pharm.* **98**, 149–155 (1993).
719. T. Hernandez-Caselles, J. Villalain, and J. C. Gomez-Fernandez, Stability of liposomes on long term storage, *J. Pharm. Pharmacol.* **42**, 397–400 (1990).
720. M. Pajean, A. Huc, and D. Herbage, Stabilization of liposomes with collagen, *Int. J. Pharm.* **77**, 31–40 (1991).
721. K. Nakamori, T. Nakajima, M. Odawara, I. Koyama, M. Nemoto, T. Yoshida, H. Ohshima, and K. Inoue, Stable positively charged liposome during long-term storage, *Chem. Pharm. Bull.* **41**, 1279–1283 (1993).
722. O. L. Johnson, C. Washington, and S. S. Davis, Long-term stability studies of fluorocarbon oxygen transport emulsions, *Int. J. Pharm.* **63**, 65–72 (1990).
723. C. Washington and T. Sizer, Stability of TPN mixtures compounded from Lipofundin S and Aminoplex amino-acid solutions: Comparison of laser diffraction and Coulter counter droplet size analysis, *Int. J. Pharm.* **83**, 227–231 (1992).
724. C. Washington, J. A. Ferguson, and S. E. Irwin, Computational prediction of the stability of lipid emulsions in total nutrient admixtures, *J. Pharm. Sci.* **82**, 808–812 (1993).
725. N. S. S. Magalhaes, G. Cave, M. Seiller, and S. Benita, The stability and in vitro release kinetics of a clofibride emulsion, *Int. J. Pharm.* **76**, 225–237 (1991).
726. S. J. Holland and J. K. Warrack, Low-temperature scanning electron microscopy of the phase inversion process in a cream formulation, *Int. J. Pharm.* **65**, 225–234 (1990).

727. J. H. Wood and G. Catacalos, Prediction of the rheologic aging of cosmetic lotions, *J. Soc. Cosmet. Chem.* **1962**, 147–156.
728. P. York, Analysis of moisture sorption hysteresis in hard gelatin capsules, maize starch, and maize–drug powder mixtures, *J. Pharm. Pharmacol.* **33**, 269–273 (1981).
729. T. Takeda, Mechanism of water vapor permeation through sugar coating layer [in Japanese], *Yakuzaigaku* **36**, 216–221 (1976).
730. H. Seki, T. Kagami, T. Hayashi, and N. Okusa, Stability of drugs in aqueous solutions. II. Application of Weibull probability paper to predictions of coloration of parenteral solutions, *Chem. Pharm. Bull.* **29**, 3680–3687 (1981).
731. M. L. Shively and S. Myers, Solid-state emulsions: The effects of process and storage conditions, *Pharm. Res.* **10**, 1071–1075 (1993).
732. L. Lachman, Physical and chemical stability testing of tablet dosage forms, *J. Pharm. Sci.* **54**, 1519–1526 (1965).
733. I. Matsuura and M. Kawamata, Studies on the prediction of shelf life. III. Moisture sorption of pharmaceutical preparation under the shelf condition [in Japanese], *Yakugaku Zasshi* **98**, 986–996 (1978).
734. H. Miura, O. Shinozaki, Y. Watanabe, and N. Hayashi, Stability estimation by water vapor pressure rising method, *Housou Kenkyu* **4**, 15–25 (1983).
735. K. Nakabayashi, T. Tuchida, and H. Mima, Stability of packaged solid dosage forms. I. Shelf-life prediction of packaged tablets liable to moisture damage, *Chem. Pharm. Bull.* **28**, 1090–1098 (1980).
736. K. Nakabayashi, T. Shimamoto, and H. Mima, Stability of packaged solid dosage forms. II. Shelf-life prediction for packaged sugar-coated tablets liable to moisture and heat damage, *Chem. Pharm. Bull.* **28**, 1099–1106 (1980).
737. K. Nakabayashi, T. Shimamoto, and H. Mima, Stability of packaged solid dosage forms. III. Kinetic studies by differential analysis on the deterioration of sugar-coated tablets under the influence of moisture and heat, *Chem. Pharm. Bull.* **28**, 1107–1111 (1980).
738. K. Nakabayashi, T. Tuchida, and H. Mima, Stability of packaged solid dosage forms. IV. Shelf-life prediction of packaged aluminum tablets under the influence of moisture and heat, *Chem. Pharm. Bull.* **29**, 2027–2034 (1981).
739. M. J. Kontny, S. Koppenol, and E. T. Graham, Use of the sorption–desorption moisture transfer model to assess the utility of a desiccant in a solid product, *Int. J. Pharm.* **84**, 261–271 (1992).
740. W. A. Parker and M. E. MacCara, Compatibility of diazepam with intravenous fluid containers and administration sets, *Am. J. Hosp. Pharm.* **37**, 496–500 (1980).
741. T. Mizutani, K. Wagi, and Y. Terai, Estimation of diazepam adsorbed on glass surfaces and silicone-coated surfaces as models of surfaces of containers, *Chem. Pharm. Bull.* **29**, 1182–1183 (1981).
742. M. J. Pikal, D. A. Bibler, and B. Rutherford, Polymer sorption of nitroglycerin and stability of molded nitroglycerin tablets in unit-dose packaging, *J. Pharm. Sci.* **66**, 1293–1297 (1977).
743. P. Yuen, S. L. Denman, T. D. Sokoloski, and A. M. Burkman, Loss of nitroglycerin from aqueous solution into plastic intravenous delivery systems, *J. Pharm. Sci.* **68**, 1163–1166 (1979).
744. A. W. Malick, A. H. Amann, D. M. Baaske, and R. G. Stoll, Loss of nitroglycerin from solution to intravenous plastic containers: A theoretical treatment, *J. Pharm. Sci.* **70**, 798–800 (1981).
745. T. D. Sokoloski, C. Wu, and A. M. Burkman, Rapid adsorptive loss of nitroglycerin from aqueous solution to plastic, *Int. J. Pharm.* **6**, 63–76 (1980).
746. E. A. Kowaluk, M. S. Roberts, H. D. Blackburn, and A. E. Polack, Interactions between drugs and polyvinyl chloride infusion bags, *Am. J. Hosp. Pharm.* **38**, 1308–1314 (1981).
747. L. Illum and H. Bundgaard, Sorption of drugs by plastic infusion bags, *Int. J. Pharm.* **10**, 339–351 (1982).
748. L. Illum, H. Bundgaard, and S. S. Davis, A constant partition model for examining the sorption of drugs by plastic infusion bags, *Int. J. Pharm.* **17**, 183–192 (1983).
749. H. C. Atkinson and S. B. Duffull, Prediction of drug loss from PVC infusion bags, *J. Pharm. Pharmacol.* **43**, 374–376 (1990).
750. M. S. Roberts, E. A. Kowaluk, and A. E. Polack, Prediction of solute sorption by polyvinyl chloride plastic infusion bags, *J. Pharm. Sci.* **80**, 449–455 (1991).
751. D. R. Jenke, Modeling of solute sorption by polyvinyl chloride plastic infusion bags, *J. Pharm. Sci.* **82**, 1134–1139 (1993).
752. N. E. Richardson and B. J. Meakin, The sorption of benzocaine from aqueous solution by nylon 6 powder, *J. Pharm. Pharmacol.* **26**, 166–174 (1974).

753. A. M. Yahya, J. C. McElnay, and P. F. D'Arcy, Binding of chloroquine to glass, *Int. J. Pharm.* **25**, 217–223 (1985).
754. W. T. Wing, An examination of rubber used as a closure for containers of injectable solutions. Part I. Factors affecting the absorption of phenol, *J. Pharm. Pharmacol.* **7**, 648–660 (1955).
755. W. T. Wing, An examination of rubber used as a closure for containers of injectable solutions. Part II. The absorption of chlorocresol, *J. Pharm. Pharmacol.* **8**, 734–744 (1956).
756. W. T. Wing, An examination of rubber used as a closure for containers of injectable solutions, *J. Pharm. Pharmacol.* **8**, 738–744 (1956).
757. H. Vromans and J. A. H. van Laarhoven, A study on water permeation through rubber closures of injection vials, *Int. J. Pharm.* **79**, 301–308 (1992).
758. M. J. Pikal and J. E. Lang, Rubber closures as a source of haze in freeze dried parenterals: Test methodology for closure evaluation, *J. Parenter. Drug Assoc.* **32**, 162–173 (1978).
759. R. W. O. Jaehnke, J. Kreuter, and G. Ross, Interaction of rubber closures with powders for parenteral administration, *J. Parenter. Sci. Tech.* **44**, 282–288 (1990).
760. R. W. O. Jaehnke, J. Kreuter, and G. Ross, Content/container interactions: The phenomenon of haze formation on reconstitution of solids for parenteral use, *Int. J. Pharm.* **77**, 47–55 (1991).
761. P. Moorhatch and W. L. Chiou, Interactions between drugs and plastic intravenous fluid bags. II: Leaching of chemicals from bags containing various solvent media, *Am. J. Hosp. Pharm.* **31**, 149–152 (1974).
762. R. Venkataramanan, G. J. Burckart, R. J. Ptachcinski, R. Blaha, L. W. Logue, A. Bahnson, C. Giam, and J. E. Brady, Leaching of diethylhexyl phthalate from polyvinyl chloride bags into intravenous cyclosporine solution, *Am. J. Hosp. Pharm.* **43**, 2800–2802 (1986).
763. L. A. Cruz, M. P. Jenke, R. A. Kenley, M. J. Chen, and D. R. Jenke, Influence of solute degradation on the accumulation of solutes migrating into solution from polymeric parenteral containers, *Pharm. Res.* **7**, 967–972 (1990).
764. A. J. Woolfe and H. E. C. Worthington, The determination of product expiry dates from short term storage at room temperature, *Drug Dev. Commun.* **1**, 185–210 (1974–1975).
765. J. T. Carstensen and E. Nelson, Terminology regarding labeled and contained amounts in dosage forms, *J. Pharm. Sci.* **65**, 311–312 (1976).
766. J. P. R. Tootill, A slope-ratio design for accelerated storage tests, *J. Pharm. Pharmacol.* **13**, 75T–86T (1961).
767. L. Kennon, Use of models in determining chemical pharmaceutical stability, *J. Pharm. Sci.* **53**, 815–818 (1964).
768. N. G. Lordi and M. W. Scott, Stability charts, design and application to accelerated stability testing of pharmaceuticals, *J. Pharm. Sci.* **54**, 531–537 (1965).
769. A. K. Amirjahed, Simplified method to study stability of pharmaceutical preparations, *J. Pharm. Sci.* **66**, 785–789 (1977).
770. S. Bakar and S. Niazi, Simplified method to study stability of pharmaceutical systems, *J. Pharm. Sci.* **67**, 141 (1978).
771. I. G. Tucker, Simplified method to study stability of pharmaceutical systems, *J. Pharm. Sci.* **71**, 599–600 (1982).
772. S. Niazi, Simplified method to study stability of pharmaceutical systems: A response, *J. Pharm. Sci.* **71**, 600 (1982).
773. J. T. Carstensen, J. B. Johnson, D. C. Spera, and M. J. Frank, Equilibrium phenomena in solid dosage forms, *J. Pharm. Sci.* **57**, 23–27 (1968).
774. K. Kowalski, M. Beno, C. Bergstrom, and H. Gaud, The application of multiresponse estimation to drug stability studies, *Drug Dev. Ind. Pharm.* **13**, 2823–2838 (1987).
775. S. Yoshioka, Y. Aso, and Y. Takeda, Statistical evaluation of accelerated stability data obtained at a single temperature. II. Estimation of shelf-life from remaining drug content, *Chem. Pharm. Bull.* **38**, 1760–1762 (1990).
776. X. Y. Su, A. Liwanpo, and S. Yoshioka, Predicting shelf-lives of pharmaceutical products: Monte Carlo simulation using the simulation package SIMAN, *J. Therm. Anal.* **41**, 713–724 (1994).
777. S. Yoshioka, Y. Aso, and Y. Takeda, Statistical evaluation of accelerated stability data obtained at a single temperature. I. Effect of experimental errors in evaluation of stability data obtained, *Chem. Pharm. Bull.* **38**, 1757–1759 (1990).
778. I. Matsuura and M. Kawamata, Studies on the prediction of shelf life. I. Prediction with an analog computer [in Japanese], *Yakugaku Zasshi* **95**, 1255–1265 (1975).

779. I. Matsuura and M. Kawamata, Studies on the prediction of shelf life. II. Prediction with a digital computer [in Japanese], *Yakugaku Zasshi* **95**, 1369–1373 (1975).

780. J. T. Carstensen and C. T. Rhodes, Cyclic testing in stability programs, *Drug Dev. Ind. Pharm.* **12**, 1219–1225 (1986).

781. J. D. Haynes, Worldwide virtual temperatures for product stability testing, *J. Pharm. Sci.* **60**, 927–929 (1971).

782. M. Scher, Kinetic ratio as a parameter for product stability calculations, *J. Pharm. Sci.* **69**, 325–327 (1980).

783. Y. Ogawa, E. Yasutaka, A. Horiuchi, and M. Terao, Artificial climate laboratory for stability studies [in Japanese], *Yakugaku Zasshi* **95**, 1424–1430 (1975).

784. N. Okusa, Prediction of stability of drugs. IV. Prediction of stability by multilevel nonisothermal method, *Chem. Pharm. Bull.* **23**, 803–809 (1975).

785. M. Terao, K. Aoki, and Y. Ueki, A proposed method for the prediction of stability based on actual field temperature, *Chem. Pharm. Bull.* **30**, 2971–2979 (1982).

786. S. Lapanje, *Physicochemical Aspects of Protein Denaturation*, Wiley-Interscience, New York, 1978.

787. P. L. Privalov, Stability of protein. Small globular protein, *Adv. Protein Chem.* **33**, 167–241 (1979).

788. T. J. Ahern and M. C. Manning, *Stability of Protein Pharmaceuticals. Part A. Chemical and Physical Pathways of Protein Degradation*, Plenum Press, New York, 1992.

789. N. P. Bhatt, K. Patel, and R. T. Borchardt, Chemical pathways of peptide degradation. I. Deamidation of adrenocorticotropic hormone, *Pharm. Res.* **7**, 593–599 (1990).

790. K. Patel and R. T. Borchardt, Chemical pathways of peptide degradation. II. Kinetics of deamidation of an asparaginyl residue in a model hexapeptide, *Pharm. Res.* **7**, 703–711 (1990).

791. K. Patel and R. T. Borchardt, Chemical pathways of peptide degradation. III. Effect of primary sequence on the pathways of deamidation of asparaginyl residues in hexapeptides, *Pharm. Res.* **7**, 787–793 (1990).

792. J. Brange, L. Langkjaer, S. Havelund, and A. Volund, Chemical stability of insulin. 1. Hydrolytic degradation during storage of pharmaceutical preparations, *Pharm. Res.* **9**, 715–726 (1992).

793. R. T. Darrington and B. D. Anderson, The role of intramolecular nucleophilic catalysis and the effects of self-association on the deamidation of human insulin at low pH, *Pharm. Res.* **11**, 784–793 (1994).

794. T. Tsuda, M. Uchiyama, T. Sato, H. Yoshino, Y. Tsuchiya, S. Ishikawa, M. Ohmae, S. Watanabe, and Y. Miyake, Degradation peptides of secretin after storage in acid and neutral aqueous solutions, *J. Pharm. Sci.* **79**, 53–56 (1990).

795. T. Tsuda, M. Uchiyama, T. Sato, H. Yoshino, Y. Tsuchiya, S. Ishikawa, M. Ohmae, S. Watanabe, and Y. Miyake, Mechanism and kinetics of secretin degradation in aqueous solutions, *J. Pharm. Sci.* **79**, 223–227 (1990).

796. C. Oliyai and R. T. Borchardt, Chemical pathways of peptide degradation. IV. Pathways, kinetics, and mechanism of degradation of an aspartyl residue in a model hexapeptide, *Pharm. Res.* **10**, 95–102 (1993).

797. C. Oliyai and R. T. Borchardt, Chemical pathways of peptide degradation. VI. Effect of the primary sequence on the pathways of degradation of aspartyl residues in model hexapeptides, *Pharm. Res.* **11**, 751–758 (1994).

798. Z. Shahrokh, G. Eberlein, D. Buckley, M. V. Paranandi, D. W. Aswad, P. Stratton, R. Mischak, and Y. J. Wang, Major degradation products of basic fibroblast growth factor: Detection of succinimide and *iso*-aspartate in place of aspartate, *Pharm. Res.* **11**, 936–944 (1994).

799. M. Friedman and P. M. Masters, Kinetics of racemization of amino acid residues in casein, *J. Food Sci.* **47**, 760–764 (1982).

800. A. R. Oyler, R. E. Naldi, J. R. Lloyd, D. A. Graden, C. J. Shaw, and M. L. Cotter, Characterization of the solution degradation products of histrelin, a gonadotropin releasing hormone (LH/RH) agonist, *J. Pharm. Sci.* **80**, 271–275 (1991).

801. M. G. Motto, P. F. Hamburg, D. A. Graden, C. J. Shaw, and M. L. Cotter, Characterization of the degradation products of luteinizing hormone releasing hormone, *J. Pharm. Sci.* **80**, 419–423 (1991).

802. R. G. Strickley, M. Brandl, K. W. Chan, K. Straub, and L. Gu, High-performance liquid chromatographic and HPLC–mass spectrometric analysis of the degradation of the luteinizing hormone-releasing hormone antagonist RS-26303 in aqueous solution, *Pharm. Res.* **7**, 530–536 (1990).

803. J. A. Schrier, R. A. Kenley, R. Williams, R. J. Corcoran, Y. Kim, R. P. Northey, D. D'Augusta, and M. Huberty, Degradation pathways for recombinant human macrophage colony-stimulating factor in aqueous solution, *Pharm. Res.* **10**, 933–944 (1993).

804. R. A. Kenley and N. W. Warne, Acid-catalyzed peptide bond hydrolysis of recombinant human interleukin 11, *Pharm. Res.* **11**, 72–76 (1994).

805. T. J. Ahern and A. M. Klibanov, The mechanism of irreversible enzyme inactivation at 100°C, *Science* **228**, 1280–1284 (1985).
806. D. B. Volkins and A. M. Klibanov, Thermal destruction processes in proteins involving cystine residues, *J. Biol. Chem.* **262**, 2945–2950 (1987).
807. W. R. Liu, R. Langer, and A. M. Klibanov, Moisture-induced aggregation of lyophilized proteins in the solid state, *Biotechnol. Bioeng.* **37**, 177–184 (1991).
808. T. Moreira, L. Cabrera, and A. Gutierrez, Role of temperature and moisture, on monomer content of freeze-dried human albumin, *Acta Pharm. Nord.* **4**, 59–60 (1992).
809. G. M. Jordan, S. Yoshioka, and T. Terao, The aggregation of bovine serum albumin in solution and in the solid state, *J. Pharm. Pharmacol.* **46**, 1–4 (1993).
810. H. R. Costantino, R. Langer, and A. M. Klibanov, Moisture-induced aggregation of lyophilized insulin, *Pharm. Res.* **11**, 21–29 (1994).
811. S. Yoshioka, Y. Aso, K. Izutsu, and T. Terao, Aggregates formed during storage of β-galactosidase in solution and in the freeze-dried state, *Pharm. Res.* **10**, 687–691 (1993).
812. M. W. Townsend and P. P. DeLuca, Use of lyoprotectants in the freeze-drying of a model protein, ribonuclease A, *J. Parenter. Sci. Technol.* **42**, 190–199 (1988).
813. M. W. Townsend and P. P. DeLuca, Nature of aggregates formed during storage of freeze-dried ribonuclease A, *J. Pharm. Sci.* **80**, 63–66 (1991).
814. M. S. Hora, R. K. Rana, and F. W. Smith, Lyophilized formulations of recombinant tumor necrosis factor, *Pharm. Res.* **9**, 33–36 (1992).
815. J. Brange, S. Havelund, and P. Hougaard, Chemical stability of insulin. 2. Formation of higher molecular weight transformation products during storage of pharmaceutical preparations, *Pharm. Res.* **9**, 727–734 (1992).
816. R. T. Darrington and B. D. Anderson, Evidence for a common intermediate in insulin deamidation and covalent dimer formation: Effects of pH and aniline trapping in dilute acidic solutions, *J. Pharm. Sci.* **84**, 275–282 (1995).
817. Z. Shahrokh, P. Stratton, G. A. Eberlein, and Y. J. Wang, Approaches to analysis of aggregates and demonstrating mass balance in pharmaceutical protein (basic fibroblast growth factor) formulations, *J. Pharm. Sci.* **83**, 1645–1650 (1994).
818. S. J. Tomazie and A. M. Klibanov, Mechanisms of irreversible thermal inactivation of bacillus α-amylases, *J. Biol. Chem.* **263**, 3086–3091 (1988).
819. M. Coltrera, M. Rosenblatt, and J. T. Potts, Jr., Analogues of parathyroid hormone containing D-amino acids: Evaluation of biological activity and stability, *Biochemistry*, **19**, 4380–4385 (1980).
820. T. H. Nguyen, J. Burnier, and W. Meng, The kinetics of relaxin oxidation by hydrogen peroxide, *Pharm. Res.* **10**, 1563–1571 (1993).
821. M. W. Townsend, P. R. Byron, and P. P. DeLuca, The effects of formulation additives on the degradation of freeze-dried ribonuclease A, *Pharm. Res.* **7**, 1086–1091 (1990).
822. S. Li, C. Schoneich, G. S. Wilson, and R. T. Borchardt, Chemical pathways of peptide degradation. V. Ascorbic acid promotes rather than inhibits the oxidation of methionine to methionine sulfoxide in small model peptides, *Pharm. Res.* **10**, 1572–1579 (1993).
823. B. S. Chang, C. S. Randall, and Y. S. Lee, Stabilization of lyophilized porcine pancreatic elastase, *Pharm. Res.* **10**, 1478–1483 (1993).
824. D. C. Dubost, M. J. Kaufman, J. A. Zimmerman, M. J. Bogusky, A. B. Coddington, and S. M. Pitzenberger, Characterization of a solid state reaction product from a lyophilized formulation of a cyclic heptapeptide. A novel example of an excipient-induced oxidation, *Pharm. Res.* **13**, 1811–1814 (1996).
825. M. J. Pikal, K. M. Dellerman, M. L. Roy, and R. M. Riggin, The effects of formulation variables on the stability of freeze-dried human growth hormone, *Pharm. Res.* **8**, 427–436 (1991).
826. A. Riggin, D. Clodfelter, A. Maloney, E. Rickard, and E. Massey, Solution stability of the monoclonal antibody–vinca alkaloid conjugate, KS1/4-DAVLB, *Pharm. Res.* **8**, 1264–1269 (1991).
827. V. Sluzky, J. A. Tamada, A. M. Klibanov, and R. Langer, Kinetics of insulin aggregation in aqueous solutions upon agitation in the presence of hydrophobic surfaces, *Proc. Natl. Acad. Sci. U.S.A* **88**, 9377–9381 (1991).
828. W. Jiskoot, E. C. Beuvery, A. A. M. de Koning, J. N. Herron, and D. J. A. Crommelin, Analytical approaches to the study of monoclonal antibody stability, *Pharm. Res.* **7**, 1234–1241 (1990).
829. P. J. M. van den Oetelaar, P. S. L. Jansen, P. A. T. A. Melgers, G. N. Wagenaars, and P. B. W. ten Kortenaar, Stability assessment of peptide and protein drugs, *J. Controll. Release* **21**, 11–22 (1992).

830. D. J. Kroon, A. Baldwin-Ferro, and P. Lalan, Identification of sites of degradation in a therapeutic monoclonal antibody by peptide mapping, *Pharm. Res.* **9**, 1386–1393 (1992).

831. T. Arakawa, L. Hung, V. Pan, T. P. Horan, C. G. Kolvenbach, and L. O. Narhi, Analysis of the heat-induced denaturation of proteins using temperature gradient gel electrophoresis, *Anal. Biochem.* **208**, 255–259 (1993).

832. T. J. Racey, P. Rochon, D. V. C. Awang, and G. A. Neville, Aggregation of commercial heparin samples in storage, *J. Pharm. Sci.* **76**, 314–318 (1987).

833. S. S. Weber, W. A. Wood, and E. A. Jackson, Availability of insulin from parenteral nutrient solutions, *Am. J. Hosp. Pharm.* **34**, 353–357 (1977).

834. T. Mizutani, Estimation of protein and drug adsorption onto silicone-coated glass surfaces, *J. Pharm. Sci.* **70**, 493–496 (1981).

835. C. J. Burke, B. L. Steadman, D. B. Volkin, P. Tsai, M. W. Bruner, and C. R. Middaugh, The adsorption of proteins to pharmaceutical container surfaces, *Int. J. Pharm.* **86**, 89–93 (1992).

836. P. Wang, G. O. Udeani, and T. P. Johnston, Inhibition of granulocyte colony stimulating factor (G-CSF) adsorption to polyvinyl chloride using a nonionic surfactant, *Int. J. Pharm.* **114**, 177–184 (1995).

837. M. R. Duncan, J. M. Lee, and M. P. Warchol, Influence of surfactants upon protein/peptide adsorption to glass and polypropylene, *Int. J. Pharm.* **120**, 179–188 (1995).

838. J. P. Patel, Urokinase: Stability studies in solution and lyophilized formulations, *Drug Dev. Ind. Pharm.* **16**, 2613–2626 (1990).

839. V. J. Helm and B. W. Muller, Stability of the synthetic pentapeptide thymopentin in aqueous solution: Effect of pH and buffer on degradation, *Int. J. Pharm.* **70**, 29–34 (1991).

840. H. Takenaka, H. Ito, A. Kojima, S. Watanabe, Y. Taguchi, and M. Sugiura, Pharmaceutical studies on enzymes. II. Proteases produced by microorganisms [in Japanese], *Yakuzaigaku* **29**, 221–224 (1969).

841. K. Asahara, H. Yamada, S. Yoshida, and S. Hirose, Stability of urokinase in solutions containing sodium bisulfite, *Chem. Pharm. Bull.* **38**, 1675–1679 (1990).

842. K. Asahara, H. Yamada, and S. Yoshida, Stability of human insulin in solutions containing sodium bisulfite, *Chem. Pharm. Bull.* **39**, 2662–2666 (1991).

843. R. E. Johnson, L. A. Lanaski, V. Gupta, M. J. Griffin, H. T. Gaud, T. E. Needham, and H. Zia, Stability of atriopeptin III in poly(D,L-lactide-*co*-glycolide) microspheres, *J. Controll. Release* **17**, 61–68 (1991).

844. C. Oliyai, J. P. Patel, L. Carr, and R. T. Borchardt, Chemical pathways of peptide degradation. VII. Solid state chemical instability of an aspartyl residue in a model hexapeptide, *Pharm. Res.* **11**, 901–908 (1994).

845. M. Paborji, N. L. Pochopin, W. P. Coppola, and J. B. Bogardus, Chemical and physical stability of chimeric L6, a mouse-human monoclonal antibody, *Pharm. Res.* **11**, 764–771 (1994).

846. S. Cohen, T. Yoshioka, M. Lucarelli, L. H. Hwang, and R. Langer, Controlled delivery systems for proteins based on poly(lactic/glycolic acid) microspheres, *Pharm. Res.* **8**, 713–720 (1991).

847. M. W. Townsend and P. P. DeLuca, Stability of ribonuclease A in solution and the freeze-dried state, *J. Pharm. Sci.* **79**, 1083–1086 (1990).

848. M. J. Pikal, K. Dellerman, and M. L. Roy, Formulation and stability of freeze-dried proteins: Effects of moisture and oxygen on the stability of freeze-dried formulations of human growth hormone, *Dev. Biol. Stand.* **74**, 21–38 (1991).

849. M. L. Roy, M. J. Pikal, E. C. Rickaard, and A. M. Maloney, The effects of formulation and moisture on the stability of a freeze-dried monoclonal antibody–vinca conjugate: A test of the WLF glass transition theory, *Dev. Biol. Stand.* **74**, 323–340 (1991).

850. C. C. Hsu, C. A. Ward, R. Pearlman, H. M. Nguyen, D. A. Yeung, and J. G. Curley, Determining the optimum residual moisture in lyophilized protein pharmaceuticals, *Dev. Biol. Stand.* **74**, 255–271 (1991).

851. Y. Nakai, S. Yoshioka, Y. Aso, and S. Kojima, Solid-state rehydration-induced recovery of bilirubin oxidase activity in lyophilized formulations reduced during freeze-drying, *Chem. Pharm. Bull.* **46**, 1031–1033 (1998).

852. H. R. Costantino, R. Langer, and A. M. Klibanov, Aggregation of a lyophilized pharmaceutical protein, recombinant human albumin: Effect of moisture and stabilization by excipients, *Bio/Technology* **13**, 493–496 (1995).

853. S. Yoshioka, Y. Aso, K. Izutsu, and T. Terao, Stability of β-galactosidase, a model protein drug, is related to water mobility as measured by ^{17}O nuclear magnetic resonance (NMR), *Pharm. Res.* **10**, 103–108 (1993).

854. S. Yoshioka, Y. Aso, and S. Kojima, Determination of molecular mobility of lyophilized bovine serum albumin and γ-globulin by solid-state ^{1}H NMR and relation to aggregation-susceptibility, *Pharm. Res.* **13**, 926–930 (1996).
855. S. Yoshioka, Y. Aso, and S. Kojima, Softening temperature of lyophilized bovine serum albumin and γ-globulin as measured by spin–spin relaxation time of protein protons, *J. Pharm. Sci.* **86**, 470–474 (1997).
856. S. Yoshioka, Y. Aso, and S. Kojima, Dependence of the molecular mobility and protein stability of freeze-dried γ-globulin formulations on the molecular weight of dextran, *Pharm. Res.* **14**, 736–741 (1997).
857. S. Yoshioka, Y. Aso, Y. Nakai, and S. Kojima, Effect of high molecular mobility of poly(vinyl alcohol) on protein stability of lyophilized γ-globulin formulations, *J. Pharm. Sci.* **87**, 147–151 (1998).
858. F. Franks, Freeze drying: From empiricism to predictability, *Cryo-Letters* **11**, 93–110 (1990).
859. J. F. Carpenter and J. H. Crowe, An infrared spectroscopic study of the interactions of carbohydrates with dried proteins, *Biochemistry* **28**, 3916–3922 (1989).
860. A. W. Ford and P. J. Dawson, The effect of carbohydrate additives in the freeze-drying of alkaline phosphatase, *J. Pharm. Pharmacol.* **45**, 86–93 (1993).
861. M. E. Ressing, W. Jiskoot, H. Talsma, C. W. van Ingen, E. C. Beuvery, and D. J. A. Crommelin, The influence of sucrose, dextran, and hydroxypropyl-β-cyclodextrin as lyoprotectants for a freeze-dried mouse IgG2a monoclonal antibody (MN12), *Pharm. Res.* **9**, 266–270 (1992).
862. B. S. Chang, G. Reeder, and J. F. Carpenter, Development of a stable freeze-dried formulation of recombinant human interleukin-1 receptor antagonist, *Pharm. Res.* **13**, 243–249 (1996).
863. K. Izutsu, S. Yoshioka, and Y. Takeda, The effects of additives on the stability of freeze-dried β-galactosidase stored at elevated temperature, *Int. J. Pharm.* **71**, 137–146 (1991).
864. K. Izutsu, S. Yoshioka, and S. Kojima, Physical stability and protein stability of freeze-dried cakes during storage at elevated temperatures, *Pharm. Res.* **11**, 995–999 (1994).
865. L. N. Bell, M. J. Hageman, and L. M. Muraoka, Thermally induced denaturation of lyophilized bovine somatotropin and lysozyme as impacted by moisture and excipients, *J. Pharm. Sci.* **84**, 707–712 (1995).
866. S. J. Prestrelski, K. A. Pikal, and T. Arakawa, Optimization of lyophilization conditions for recombinant human interleukin-2 by dried-state conformational analysis using Fourier-transform infrared spectroscopy, *Pharm. Res.* **12**, 1250–1259 (1995).
867. H. Uedaira and H. Uedaira, The effect of sugars on the thermal denaturation of lysozyme, *Bull. Chem. Soc. Jpn.* **53**, 2451–2455 (1980).
868. J. C. Lee and S. N. Timasheff, The stabilization of proteins by sucrose, *J. Biol. Chem.* **256**, 7193–7201 (1981).
869. S. A. Charman, K. L. Mason, and W. N. Charman, Techniques for assessing the effects of pharmaceutical excipients on the aggregation of porcine growth hormone, *Pharm. Res.* **10**, 954–962 (1993).
870. P. Wang and T. P. Johnston, Thermal-induced denaturation of two model proteins: Effect of poloxamer 407 on solution stability, *Int. J. Pharm.* **96**, 41–49 (1993).
871. P. K. Tsai, D. B. Vokin, J. M. Dabora, K. C. Thompson, M. W. Bruner, J. O. Gress, B. Matuszewska, M. Keogan, J. V. Bondi, and R. Middaugh, Formulation design of acidic fibroblast growth factor, *Pharm. Res.* **10**, 649–659 (1993).
872. M. Vrkljan, T. M. Foster, M. E Powers, J. Henkin, W. R. Porter, H. Staack, J. F. Carpenter, and M. C. Manning, Thermal stability of low molecular weight urokinase during heat treatment. II. Effect of polymeric additives, *Pharm. Res.* **11**, 1004–1008 (1994).
873. W. R. Gombotz, S. C. Pankey, D. Phan, R. Drager, K. Donaldson, K. P. Antonsen, A. S. Hoffman, and H. V. Raff, The stabilization of a human IgM monoclonal antibody with poly(vinylpyrrolidone), *Pharm. Res.* **11**, 624–632 (1994).
874. B. Chen, T. Arakawa, C. F. Morris, W. C. Kenney, C. M. Wells, and C. G. Pitt, Aggregation pathway of recombinant human keratinocyte growth factor and its stabilization, *Pharm. Res.* **11**, 1581–1587(1994).
875. B. Chen and T. Arakawa, Stabilization of recombinant human keratinocyte growth factor by osmolytes and salts, *J. Pharm. Sci.* **85**, 419–422 (1996).
876. K. C. Lee, Y. J. Lee, H. M. Song, C. J. Chun, and P. P. DeLuca, Degradation of synthetic salmon calcitonin in aqueous solution, *Pharm. Res.* **9**, 1521–1523 (1992).
877. M. A. Hoitink, J. H. Beijnen, A. Bult, O. A. G. J. van der Houwen, J. Nijholt, and W. J. M. Underberg, Degradation kinetics of gonadorelin in aqueous solution, *J. Pharm. Sci.* **85**, 1053–1059 (1996).
878. I. S. Ha, B. H. Woo, J. T. Lee, T. S. Kang, K. K. Tak, C. K. Oh, and K. C. Lee, Degradation kinetics of growth hormone-releasing hexapeptide (GHRP-6) in aqueous solution, *Int. J. Pharm.* **144**, 91–97 (1996).

879. S. Yoshioka, K. Izutsu, Y. Aso, and Y. Takeda, Inactivation kinetics of enzyme pharmaceuticals in aqueous solution, *Pharm. Res.* **8**, 480–484 (1991).

880. M. Sugiura, M. Kurobe, S. Tamura, and S. Ikeda, Kinetics of digestive enzyme stability in the solid state. II. Quantitative prediction of enzyme inactivation, *Chem. Pharm. Bull.* **29**, 2096–2100 (1981).

881. M. Sugiura, M. Kurobe, S. Tamura, and S. Ikeda, Kinetics of digestive enzyme stability in solid state I: Application of Weibull distribution function to solid-state enzyme inactivation, *J. Pharm. Sci.* **68**, 1381–1383 (1979).

882. B. R. Cole and L. Leadbeater, Estimation of the stability of dry horse serum cholinesterase by means of an accelerated storage test, *J. Pharm. Pharmacol.* **20**, 48–53 (1968).

883. V. J. Helm and B. W. Muller, Stability of gonadorelin and triptorelin in aqueous solution, *Pharm. Res.* **7**, 1253–1256 (1990).

884. A. S. Kearney, S. C. Mehta, and G. W. Radebaugh, Aqueous stability and solubility of CI-988, a novel "dipeptoid" cholecystokinin-B receptor antagonist, *Pharm. Res.* **9**, 1092–1095 (1992).

885. L. C. Gu, E. A. Erdos, H. Chiang, T. Calderwood, K. Tsai, G. C. Visor, J. Duffy, W. C. Hsu, and L. C. Foster, Stability of interleukin 1 beta (IL-1 beta) in aqueous solution: Analytical methods, kinetics, products, and solution formulation implications, *Pharm. Res.* **8**, 485–490 (1991).

886. P. Jameson, D. Creiff, and S. E. Crossberg, Thermal stability of freeze-dried mammalian interferons. Analysis of freeze-drying conditions and accelerated storage tests for murine interferon, *Cryobiology* **16**, 301–314 (1979).

887. J. Geigert, D. L. Ziegler, B. M. Panschar, A. A. Creasey, and C. R. Vitt, Potency stability of recombinant (serine-17) human interferon-β, *J. Interferon Res.* **7**, 203–211 (1987).

888. S. Yoshioka, Y. Aso, K. Izutsu, and T. Terao, Application of accelerated testing to shelf-life prediction of commercial protein preparations, *J. Pharm. Sci.* **83**, 454–456 (1994).

889. T. Le Bihan and C. Gicquaud, Kinetic study of the thermal denaturation of G actin using differential scanning calorimetry and intrinsic fluorescence spectroscopy, *Biochem. Biophys. Res. Commun.* **194**, 1065–1073 (1993).

890. S. Yoshioka, Y. Aso, K. Izutsu, and S. Kojima, Is stability prediction possible for protein drugs? Denaturation kinetics of β-galactosidase in solution, *Pharm. Res.* **11**, 1721–1725 (1994).

891. K. Thoma and R. Kerker, Photoinstabilitat von Arzneimitteln, 1. Mittelung uber das verhalten von nur im UV-Bereich absorbierenden Substanzen bei der Tageslichtsimulation, *Pharm. Ind.* **54**, 169–177 (1992).

892. S. Yoshioka, Y. Ishihara, T. Terazono, N. Tsunakawa, M. Murai, T. Yasuda, S. Kitamura, Y. Kunihiro, K. Sakai, Y. Hirose, K. Tonooka, K. Takayama, F. Imai, M. Godo, M. Matsuo, K. Nakamura, Y. Aso, S. Kojima, Y. Takeda, and T. Terao, Quinine actinometry as a method for calibrating ultraviolet radiation intensity in light-stability testing of pharmaceuticals, *Drug Dev. Ind. Pharm.* **20**, 2049–2062 (1994).

893. H. D. Drew, J. F. Brower, W. E. Juhl, and L. K. Thornton, Quinine photochemistry: A proposed chemical actinometer system to monitor UV-A exposure in photostability studies of pharmaceutical drug substances and drug products, *Pharmacop. Forum (U.S.A.)* **24**, 6334–6346 (1998).

894. S. W. Baertschi, Commentary on the quinine actinometry system described in the ICH draft guideline on photostability testing of new drug substances and products, *Drug Stability* **1**, 193–195 (1997).

895. P. Helboe, New designs for stability testing programs: Matrix or factorial designs. Authorities' viewpoint on the predictive value of such studies, *Drug Inform. J.* **26**, 629–634 (1992).

896. S. Yoshioka, Y. Aso, and S. Kojima, Statistical evaluation of shelf-life of pharmaceutical products estimated by matrixing, *Drug Stability* **1**, 147–151 (1996).

897. S. Yoshioka, Y. Aso, S. Kojima, and A. Li Wan Po, Power of analysis of variance for assessing batch-variation of stability data of pharmaceuticals, *Chem. Pharm. Bull.* **44**, 1948–1950 (1996).

898. S. Yoshioka, Y. Aso, and S. Kojima, Assessment of shelf-life equivalence of pharmaceutical products, *Chem. Pharm. Bull.* **45**, 1482–1484 (1997).

Index

GPSR Compliance
The European Union's (EU) General Product Safety Regulation (GPSR) is a set of rules that requires consumer products to be safe and our obligations to ensure this.

If you have any concerns about our products, you can contact us on

ProductSafety@springernature.com

In case Publisher is established outside the EU, the EU authorized representative is:

Springer Nature Customer Service Center GmbH
Europaplatz 3
69115 Heidelberg, Germany

www.ingramcontent.com/pod-product-compliance
Ingram Content Group UK Ltd.
Pitfield, Milton Keynes, MK11 3LW, UK
UKHW051130260726
13967UKWH00010B/2965

* 9 7 8 1 4 7 5 7 8 6 7 1 2 *